Stars, Nebulae and the Interstellar Medium

Observational physics and astrophysics

C R Kitchin

Senior Lecturer in Astronomy
Hatfield Polytechnic Observatory

Adam Hilger, Bristol and Boston

British Library Cataloguing in Publication Data

Kitchin, Christopher R.
Stars, nebulae and the interstellar medium: observational physics and astrophysics.
1. Galaxies
I. Title
523.1′12 QB857

ISBN 0-85274-580-X
ISBN 0-85274-581-8 Pbk

Consultant Editor: **Professor A J Meadows**

Published under the Adam Hilger imprint by IOP Publishing Ltd
Techno House, Redcliffe Way, Bristol BS1 6NX, England
PO Box 230, Accord, MA 02018, USA

Typeset by Mathematical Composition Setters Ltd, Salisbury
Printed in Great Britain by J W Arrowsmith Ltd, Bristol

LIVERPOOL JMU LIBRARY
3 1111 00945 3794

A hypothesis or theory is clear, decisive and positive, but it is believed by no one but the man who created it.

Experimental findings, on the other hand, are messy, inexact things which are believed by everyone except the man who did the work.

Harlow Shapley

This book is dedicated to the galaxy of messy observers and experimentalists whose contributions form its basis, and whose results are now believed even by themselves (at least for today!).

Contents

Preface

This book was originally envisaged as a companion to the author's *Astrophysical Techniques* (Adam Hilger 1984), summarising the results of the techniques and methods discussed therein. It has, however, developed a life of its own whilst being written, and is now a much more extensive and comprehensive account of our current understanding of constituents of the Universe ranging between planets and galaxies in their masses. Much background in astrophysics, physics, and mathematics has been added to explain how and where the quoted results originate. Nonetheless, the bias of the treatment remains towards that of the observer. While it is now intended to be a book that stands on its own, perusal of the previously mentioned text will add considerably to the reader's appreciation of the problems and errors behind the slickly presented astronomical results and data to be found in the literature.

The organisation of the book is intended to proceed from the most fundamental properties governing the behaviour of stars towards their more detailed aspects, finally linking them to the wider features of the Galaxy and Universe, and showing how the often disparate stellar and nebular types join together in the grand picture of stellar evolution (see note 1, Appendix II; hereinafter references to this appendix will be indicated by an asterisk, e.g. *1). In spite of the intended observational approach to the topics, more theoretically biased astrophysics continually needs to be involved; sometimes explicitly as in the inferred properties of stellar interiors, more often as a guide to which observational properties should be dealt with in depth, and which mentioned only briefly or not at all. Without such a structure the book would degenerate into a 'stamp album' of lists of data. Some prior knowledge of astronomy is assumed but not a great deal. In the absence of such knowledge, a quick skim through an up-to-date general astronomy book (see below) together with recourse, when required, to a dictionary of astronomy, should suffice to enable the otherwise scientifically literate reader to cope with the material in this book.

The level of treatment is generally that appropriate to the latter years of most physical science, astronomy, mathematics, engineering or similar undergraduate courses. However, the sections containing large portions of physical and mathematical background material have been gathered together and separated from the more astronomical sections. The 'Physics' chapters are indicated by a P, the 'Astronomy' chapters by an A, the physics chapters preceding the astronomical chapters to which they have the most relevance. Thus readers with adequate prior knowledge of physics may proceed directly to the astronomy chapters. Other readers may well prefer to do likewise, only working through the background material after an appreciation of the astronomy has been acquired. Even non-mathematically inclined readers should find that the A-type chapters are sufficiently readable to be useful and interesting, if they accept the derivations etc as presented. The author hopes therefore that a wide range of people can glean the odd interesting grain of knowledge from the book; certainly the research for the material herein was a fascinating task.

Many thousands of observational and theoretical astronomers (the difference is not as wide as it sometimes may appear), over many generations, have contributed to the material summarised in this book. In lieu of trying, almost certainly without success, to give adequate and just credits to them, this book is dedicated to them and to their successors who will, it is to be hoped, soon render most of it out of date!

C R Kitchin
March 1986

Suitable introductory books

The Universe I Nicolson and P Moore (1985)
London: Collins

Universe W J Kaufmann III (1985)
San Francisco: Freeman

Contemporary Astronomy J M Pasachoff (1985)
Saunders College Publishing (3rd edn)

The Physical Universe F H Shu (1982)
Mill Valley, California: University Science Books

1A Fundamentals

SUMMARY

Methods of determining stellar masses and their results. The compositions of the main stellar types at their surfaces and throughout their interiors; the effects of aging.

See also degeneracy (§2P.1), mass limits (§2A.1), mass–luminosity relation (§2A.4), HR diagram (§3A.4), abundances (§3A.6), and stellar formation and evolution (§§8A.1, 8A.2).

The organisation of this book is primarily around the observed properties of stars, starting with the most fundamental ones, and proceeding towards the finer details. The parameter of a star of a given age, which determines its nature in most respects, is its mass. Almost as fundamental is its composition. In this chapter we therefore begin by looking at the masses of stars, and the ways in which they may be found, and then continue on to look at the compositions of the principal star types.

1A.1 MASS

The sophisticated modern astronomer using huge telescopes and vast amounts of computer time in the attempt to probe the depths of the Universe rightly regards the maunderings of present-day astrologers with contempt. It has not always been so, however. The only direct and unequivocal method of measuring the most fundamental property of stars and other objects has its roots in the work of a man who was half mystic and astrologer and half scientist, and who lived three and a half centuries ago.

Johannes Kepler (1571–1630) produced his three famous laws of planetary motion during his many years of work on the prediction of the position of Mars in the sky. In modern phraseology they are:

I The orbit of a planet is an ellipse with the Sun at one of the foci.
II The Sun–planet line sweeps out area at a constant rate.
III The square of the planet's orbital period is proportional to the cube of the semi-major axis of the orbital ellipse.

Newton (1643–1727) synthesised these laws into his single law of gravitation:

$$|F| = G\mathcal{M}_1\mathcal{M}_2/R^2 \tag{1A.1.1}$$

where F is the force of mutual gravitational attraction between the two bodies, $\mathcal{M}_1$ and $\mathcal{M}_2$ are their masses, R is the separation of their centres of gravity and G is a constant whose value in SI units is

$$G = 6.670 \times 10^{-11}\ \mathrm{N\,m^2\,kg^{-2}}.$$

Newton's synthesis may of course be reversed and Kepler's three laws derived from it. When this is done it is found that the third law (*1) is only an approximation, and that its correct form is:

$$a^3 = (G/4\pi^2)T^2(\mathcal{M}_1 + \mathcal{M}_2) \tag{1A.1.2}$$

where a is the semi-major axis of the orbital ellipse and T is the orbital period.

For the planets, the additional term $(\mathcal{M}_1 + \mathcal{M}_2)$ is almost constant at a value equal to the mass of the Sun. Thus the third law is a very good approximation in this case. Equation (1A.1.2), however, is applicable to any pair of mutually orbiting bodies, and when these comprise objects other than the Sun and a planet, the extra factor becomes very important and can enable the objects' individual masses to be found.

The mass of a star can thus potentially be found if it happens to be a member of a pair of stars which are orbiting each other. Such stars are commonly termed *binary stars* or *binary systems*(*2). The characteristics of the binary then need to be such that the parameters of its orbit may be found. Since a majority of the stars in the Galactic plane are members of binary systems, or of low-number multiple systems (*3), it might not seem too difficult to find suitable candidates for mass determinations. Unfortunately this is not the case, and of all the many millions of stars in the Sun's region of the Galaxy, only a few hundred have had their masses determined directly and without ambiguity through the use of Kepler's third law. The major problem arises in finding the inclination of the plane of the orbit to the line of sight. Only in two circumstances may this be found with any precision, and these two cases are called *visual* binary systems and *eclipsing* binary systems.

A visual binary is a nearby system, with the two stars sufficiently widely separated to be resolved. They may thus be seen directly from Earth as two images. Most binaries are too close together and/or too far away from the Earth to be resolved by present-day telescopes, and they therefore appear as a single image when observed directly (their binary nature is revealed by spectroscopic or photometric observations). The orbital periods of visual binaries are typically very long; from a few decades to many centuries, because of the large physical separations of the stars. By observing the relative positions of the binary's components over many years, the apparent orbit may be determined. The apparent orbit is the projection of the true orbit on the celestial sphere, and since a linear projection of a conic section is simply another conic section, it is usually observed to be another ellipse (figure 1A.1). The foci of the true ellipse do not, however, project on to the foci of the apparent ellipse. The true orbit may therefore be found by reversing the projection until the primary star is at a focus of the ellipse. In practice, difficulties and uncertainties abound in this procedure. In particular, the measuring errors of the data producing the apparent orbit are often large, and permit a range of estimates of the true orbit to be fitted to the apparent orbit. In this situation, the data may be constrained to fit Kepler's second law in order to lessen the errors, since ratios of areas are unchanged under linear projection. Even so, considerable uncertainties will often still remain, and hence the inclination of the orbital plane is also uncertain. Given a well-determined true orbit, however, only the distance of the system (Appendix I) is needed in order to obtain the required values of a and T for insertion into equation (1A.1.2.) The sum of the masses of the two stars then follows directly. To find the actual individual masses, we simply have to remember that both components of the system are in fact

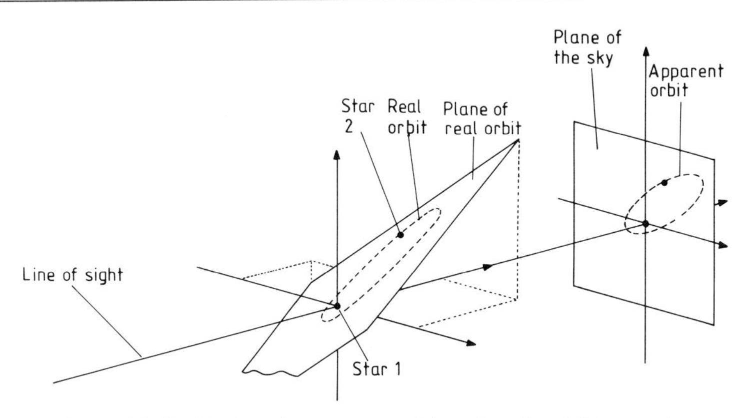

Figure 1A.1 Real and apparent orbits of a visual binary star.

orbiting their common centre of mass, and that their distances from it are inversely proportional to their masses (figure 1A.2).

$$\mathcal{M}_1/\mathcal{M}_2 = R_2/R_1. \qquad (1A.1.3)$$

Thus with the sum and ratio of the star's masses both known, their individual masses may be calculated.

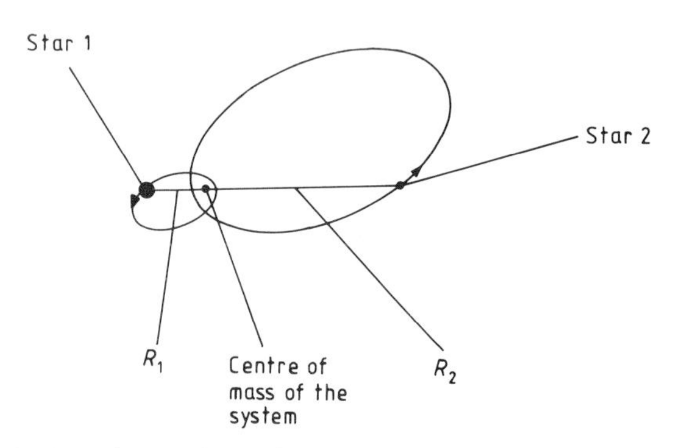

Figure 1A.2 Members of a binary system co-orbiting their common centre of mass.

In principle, therefore, we have a simple and straightforward method for obtaining measurements of stars' masses. In practice only a few stellar masses are found in this way. Numerous problems intervene; the orbital period may be too long, the measurement errors may be too large to provide a usable true orbit, the distance may not be known, etc, etc; and these prevent many of the known visual binary stars from having their masses determined.

The majority of stellar mass determinations are thus obtained from eclipsing binary systems. These belong to a much larger class of stars, which are known as the *photometric* binaries. Such systems appear to be single stars when viewed directly through even the largest telescopes, but their binary nature is revealed by the form of the regular changes in their brightnesses with time. (See figure 1A.3.) The inclination of the orbital plane to the plane of the sky must be near 90° for an eclipsing system, since otherwise it would not be possible for the eclipses to be visible from the Earth. The period and inclination thus being known, only the semi-major axis remains to be found in order to provide the data required to determine the masses. Since the binary appears as a single point image when viewed directly, its semi-major axis cannot be found from the photometric observations alone, and we must turn to spectroscopy in order to make any further progress.

If two stars in a close binary system are within about one stellar

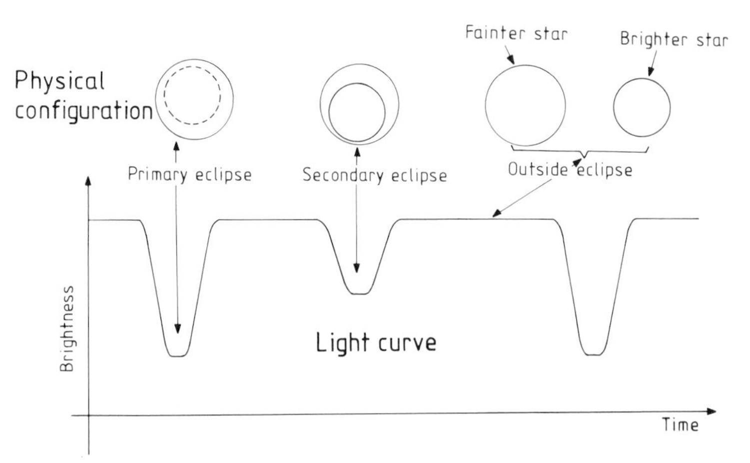

Figure 1A.3. Schematic example of an eclipsing binary system and its light curve.

magnitude (a factor of two to three; see §3A.1) of each other in brightness, then the spectrum will be composite, containing observable contributions from both stars. Each star's motion along the line of sight will change the observed wavelengths of its spectrum lines via the Doppler effect

$$\Delta\lambda = (v/c)\lambda \tag{1A.1.4}$$

where $\Delta\lambda$ is the change in the line's wavelength from its laboratory value, λ, c is the velocity of light (2.9979×10^8 m s^{-1}) and v is the line-of-sight velocity (see Chapter 3P for a fuller discussion of spectroscopy). Except during eclipses, there will be a component of one star's motion towards the observer, whilst the other star will be moving away. Thus the composite spectrum will generally exhibit pairs of lines whose separations change as the stars orbit each other (figure 1A.4). With near-circular orbits (which is generally the case for close binaries, due to their tidal interactions), the size of the relative orbit is easily retrievable from the maximum apparent relative velocities (*4):

$$a = \Delta v_{\max} T/2\pi \tag{1A.1.5}$$

where $\Delta v_{\max}$ is the maximum apparent relative velocity of the two stars (figure 1A.5).

Thus equation (1A.1.2) may again be used to obtain the sum of the masses, while the ratio of the masses now comes from the ratio of the stars' individual maximum radial velocities:

$$\mathcal{M}_1/\mathcal{M}_2 = |(\Delta v_{\max})_2|/|(\Delta v_{\max})_1| \tag{1A.1.6}$$

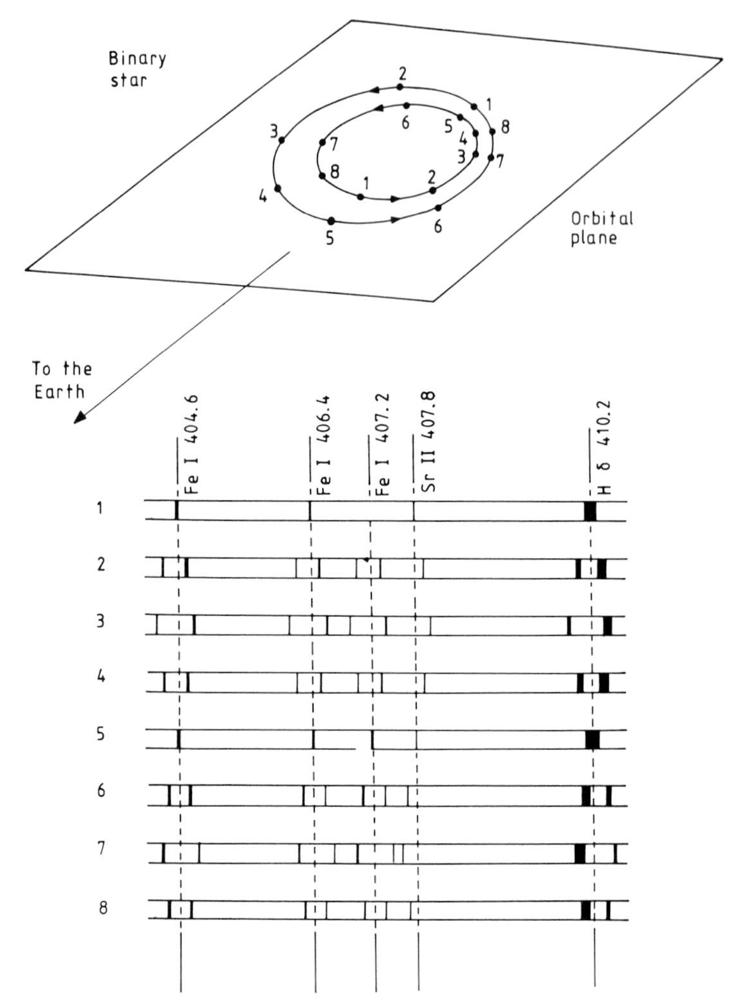

Figure 1A.4 Spectroscopic binary star. The lower part of the figure shows portions of the stellar spectra as they might appear at the eight points of the orbit indicated in the upper part of the figure.

(see figure 1A.5 for the definitions of the quantities), and so the individual masses may be found as before.

The above analysis is for the ideal case; normally the actual situation is far from ideal, and various errors and uncertainties are thereby introduced into the mass estimates. The three major problem areas arising for the determination of masses from eclipsing binary systems are discussed individually below.

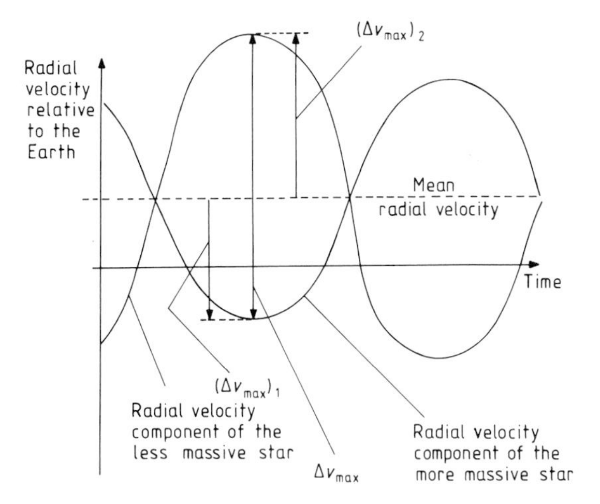

Figure 1A.5 Schematic velocity curve of a spectroscopic binary curve.

A The line of sight not in the plane of the orbit. By selecting eclipsing binaries, we have guaranteed that the orbital plane is close to the line of sight, since otherwise no eclipse would be visible from the Earth. It is possible, nonetheless, that a small inclination may exist even though there are still observable eclipses occurring, especially amongst the close binaries (figure 1A.6). The effect of such an inclination is to introduce an extra factor into equation (1A.1.5):

$$a = \Delta v_{\max} T/2\pi \sin i \tag{1A.1.7}$$

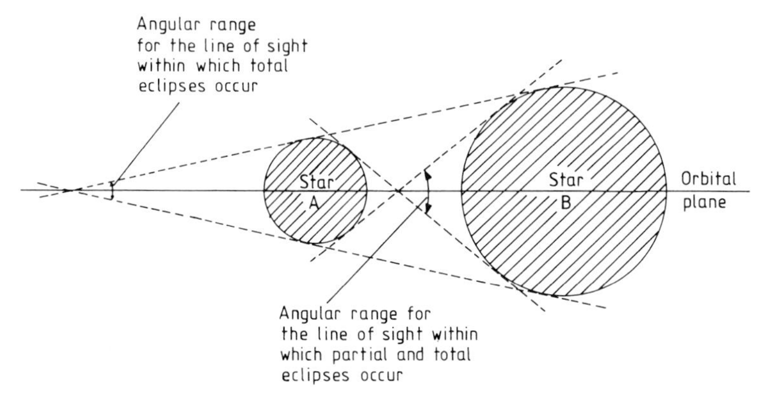

Figure 1A.6 Possible deviation of the line of sight from the orbital plane of a binary star, within which it may remain an eclipsing binary.

since the observed value of Δv_{max} is reduced in comparison with the true value by a factor of sin i. (N.B. By convention, the inclination of the orbital plane, i, is measured from the plane of the celestial sphere, and *not* with respect to the line of sight.) The mass equation (1A.1.2) thus becomes:

$$\mathcal{M}_1 + \mathcal{M}_2 = \Delta v_{max}{}^3 T/2\pi G \sin^3 i \tag{1A.1.8}$$

and using the velocity ratios to obtain the individual masses,

$$\mathcal{M}_1 \sin^3 i = (\Delta v_{max})_2 [(\Delta v_{max})_1 + (\Delta v_{max})_2]^2 T/2\pi G \tag{1A.1.9}$$

$$= \mathcal{M}_1(90) \tag{1A.1.10}$$

$$\mathcal{M}_2 \sin^3 i = (\Delta v_{max})_1 [(\Delta v_{max})_1 + (\Delta v_{max})_2]^2 T/2\pi G \tag{1A.1.11}$$

$$= \mathcal{M}_2(90) \tag{1A.1.12}$$

where $\mathcal{M}_1$ (90) and $\mathcal{M}_2$ (90) are the masses which would be obtained if the inclination were 90° (line of sight in the plane of the orbit). Thus if i is unknown, we have

$$\mathcal{M} \geqslant \mathcal{M}(90) \tag{1A.1.13}$$

and if i is assumed to be 90°, then the resulting estimates of the stars' masses will be too low by a factor of $\sin^3 i$ (figure 1A.7). Thus for an inclination of 80° (line of sight at an angle of 10° to the orbital plane), the underestimate will be by only 5%, but for an inclination of 45°, the masses will be too low by a factor of 2.8. Hence totally eclipsing binary stars will give good estimates of the masses, but partially eclipsing binaries (*5) or

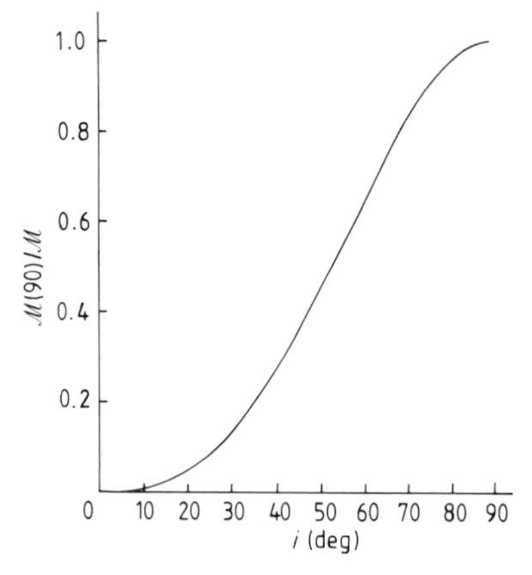

Figure 1A.7 Error introduced into estimates of the masses of stars in spectroscopic binaries, when the line of sight is not coincident with the orbital plane.

non-eclipsing spectroscopic binaries will lead only to lower limits for the masses (unless an estimate for the actual inclination may be obtained in some other way).

B Non-circular orbits. The wider binaries may have orbits with significant eccentricities. This will lead to asymmetric velocity curves (e.g. figure 1A.8). Generally, however, these may still be interpreted in such a way that the semi-major axis and mass ratio may be found. The masses of the components are therefore still individually determinable.

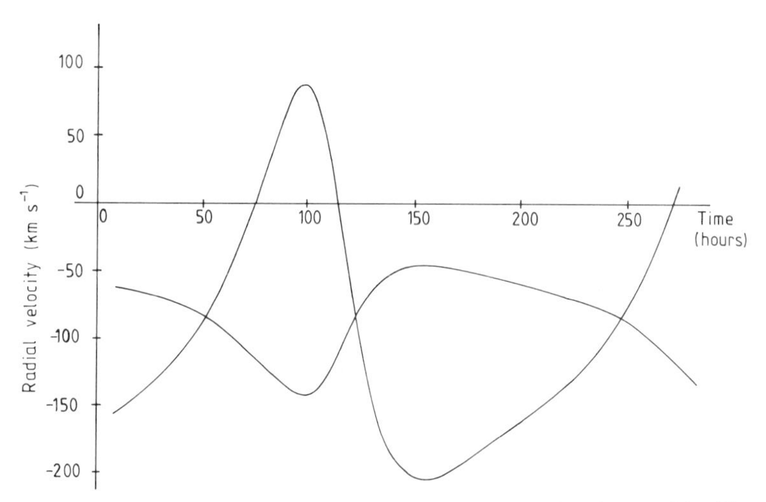

Figure 1A.8 Radial velocity curves for a spectroscopic binary with elliptical orbits (example).

C Only one component detectable. As already noted, if the two stars differ in brightness by more than about one stellar magnitude, then only the more luminous star's spectrum will be seen. The velocity variations of the brighter star will still affect its spectrum, so that one of the velocity curves may be found. In this situation, if there are no eclipses, then only the mass function, $f(\mathcal{M})$ may be found:

$$f(\mathcal{M}) = (\mathcal{M}_2 \sin i)^3/(\mathcal{M}_1 + \mathcal{M}_2)^2 \qquad (1A.1.14)$$

$$= (\Delta v_{\max})_1^3\, T/2\pi G \qquad (1A.1.15)$$

where $(\Delta v_{\max})_1$ is defined from a base line which divides the velocity curve into two equal and opposite areas (figure 1A.9).

If the binary contains normal main sequence components (§3A.4), which have not undergone any mass exchange, then the brighter component will

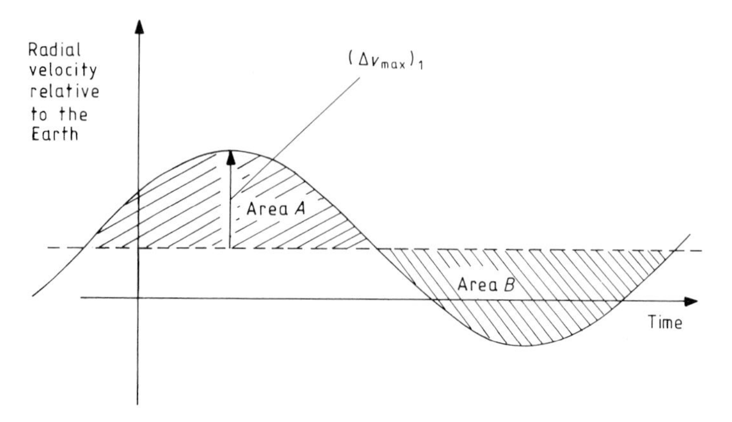

Figure 1A.9 Schematic velocity curve of a spectroscopic binary star, where only the spectrum of one component is visible. The mean velocity is defined by setting area A = area B.

also be the more massive (§8A.2), and the empirical mass–luminosity law (equation 2A.4.2) may be used to estimate the stars' masses (*6).

$$\mathcal{M} \propto L^{0.39} \tag{1A.1.16}$$

or

$$\mathcal{M}_2/\mathcal{M}_1 = 10^{0.156(m_1 - m_2)} \tag{1A.1.17}$$

where L is the star's luminosity and m is the star's magnitude (Chapter 3A). Since the observed star is at least one stellar magnitude brighter than the other, we have

$$m_1 - m_2 \leqslant -1 \tag{1A.1.18}$$

so that

$$\mathcal{M}_2 \leqslant 0.70\mathcal{M}_1 \tag{1A.1.19}$$

and

$$\mathcal{M}_1 \geqslant 3.4(\Delta v_{max})_1^3 \, T/\pi G \sin^3 i. \tag{1A.1.20}$$

Frequently, however, complex mass interchanges and unusual evolutionary sequences occur for the stars in close binary systems, and there can then be no certainty that the brighter star is the more massive.

Pulsation masses

Pulsating stars such as the Cepheids (§3A.5) may have their masses determined independently of Kepler's law from linear pulsation theory. A relatively composition-independent relationship between the fundamental mode of pulsation (period P), and the star's luminosity (L) and surface temperature (T) is then found.

$$\mathcal{M}_{\mathrm{p}} \simeq 3 \times 10^{17} (L^5/P^6 T^{21})^{1/4} \qquad (1A.1.21)$$

where $\mathcal{M}$ is in solar masses, L is in solar luminosities and P is in days.

There is no overlap with the masses determined via Kepler's law to provide any means of cross-checking the validity of the two methods. However, the pulsation masses are found to be significantly lower than those to be expected for the stars on evolutionary grounds (§§3A.5, 8A.2). The reason for the discrepancy is uncertain, but it may lie in an underestimation of the stars' luminosities by some 20%, thereby leading to an underestimate of the pulsation mass (equation 1A.1.21).

Results

Conventional visual and eclipsing binaries provide the main sources of estimates of the masses of stars. Other possibilities include X-ray binary pulsars (§3A.5) which have a neutron star (§§2P.1, 2A.1, 3A.5) observable in the x-ray region, in orbit around a more normal star which is detectable in the optical region. Thus both stars' orbits may be determined, and so their masses found. There are large errors in this procedure, however. The best estimates when averaged suggest masses in the region of 1.2–1.6 $\mathcal{M}_{\odot}$ for the neutron stars in such systems. In a few cases, theory can provide some help. Thus the mass of a white dwarf star (Chapters 2A, 3A) must be less than about 1.4 $\mathcal{M}_{\odot}$ if it is not to collapse down to an even more condensed object. Another example occurs for the members of a star cluster, where average masses may be found via the virial theorem (§2P.5). Cepheids (above and §3A.5) and other pulsating stars can have their masses inferred from the hydromagnetic theory of their pulsations. These additional hints are, however, of only marginal use in comparison with the direct method of finding the masses.

The observed masses lie in the range 0.01 to 100 times the mass of the Sun (about 10^{28}–10^{32} kg), with most stars having masses a small fraction of that of the Sun (figure 1A.10). The distribution is proportional to $\mathcal{M}^{-2.35}$ for stars of over about half a solar mass, perhaps reducing to $\mathcal{M}^{-1.35}$ for the lower mass stars. The observed mean stellar mass is about $0.35\mathcal{M}_{\odot}$ (7×10^{29} kg). At first sight, such a low average mass may seem a little surprising, since the sky appears to be full of massive, bright stars, with far fewer of the fainter, low mass stars. The true stellar population, however, can only be found by looking at all the stars out to a given distance from the Earth, not by taking all the stars brighter than some limiting magnitude. The apparent preponderance of the more massive stars results from their being visible out to much greater distances than the less massive stars, through their much greater brightnesses.

There is one further *caveat* which must be added to this discussion of stars' masses, and that is that as we have seen, almost all the estimates are obtained from binary stars, with many of the determinations coming from very close binaries. Since many processes occur in close binaries (e.g. mass interchange etc) which do not occur in single stars or in widely separated binaries, great caution is needed before the results are accepted as applicable to *single* stars without further thought. Fortunately, at least at the moment, there is no evidence for any systematic variation between the two groups of stars, so this possible complication is usually ignored.

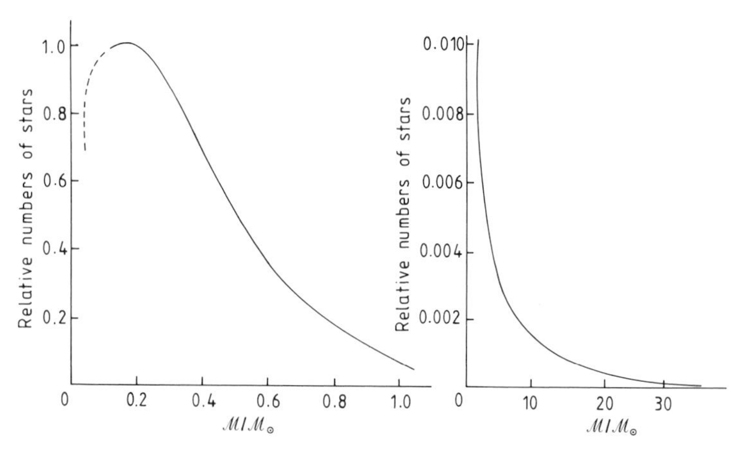

Figure 1A.10 Relative numbers of stars, in terms of mass, within our Galaxy (from measurements corrected for observed fractions of total).

1A.2 COMPOSITION

The mass of a star is the most important single factor influencing its past, present and future properties and behaviour. The effects of it are modified, however, by the composition of the star, and that, in turn, depends upon both the star's initial composition and its age.

The determination of the composition of a star (usually expressed as the abundances of elements within the star) is a twofold process. First the surface composition may be found by direct observation of the star's spectrum. Then the bulk composition must be inferred from this and from theoretical models of the conditions inside the star. The details of both processes are extremely complex, and an adequate coverage is beyond the scope of this book. Some aspects of the problem are covered elsewhere in this work; here just the main results are outlined.

The majority of stars are fairly similar to the Sun in their compositions (Chapter 4A), with most of the remainder being the white dwarfs (Chapters 2A, 3A). The few stars not included in these two categories are very rare indeed, and include such objects as Wolf–Rayet stars, pulsars, etc. Although these latter systems form only a fraction of the total stellar population, they often are amongst the most interesting objects, and many of them are discussed elsewhere in this book (Chapters 3A, 8A etc).

'Solar'-type stars

In this context, 'solar type' includes all stars of essentially gaseous composition: main sequence stars, giants, supergiants, subdwarfs, etc (Chapter 3A). The surface composition is primarily hydrogen, with helium as a minor constituent. The other elements in total only amount to between 0.1% and 2% of the star's mass. The differences between the types of stars in this group are largely a consequence of their differing internal compositions; primarily the helium/hydrogen ratio at their centres. It is common practice to term all the elements other than hydrogen and helium as metals, irrespective of whether they are actually metallic or not. In an astrophysical situation this rarely causes confusion. The custom arose from the fact that the vast majority of lines in spectra of stars of cool and intermediate temperatures are due to metallic elements, particularly iron.

The composition of the surface layers of the Sun is shown in figure 1A.11 (see also Chapter 4A). The relative abundances, by mass, for the more significant constituents are

hydrogen	73.4%	
helium	24.9%	
carbon	0.29%	1.7% (carbon to rest)
nitrogen	0.10%	
oxygen	0.77%	
neon	0.12%	
iron	0.16%	
rest	0.29%	

The solar surface composition is typical of that of most stars. The major departures from it occur amongst the metals and are of two types: global increases or decreases in the total metal abundance, and changes in the relative abundances of only one or two elements.

The global changes correlate well with the age of the star. The interstellar medium (Chapter 7A) is continually being enriched in the heavier elements through the ejection by supernovae, long-period variables, planetary nebulae, etc of material which has undergone nucleosynthetic reactions inside stars (Chapter 2A). Thus stars forming in the interstellar gas clouds (§8A.1) do so from material which becomes richer in the heavy elements, the later in time that this happens. Stars being 'born' today contain about 4% metals by mass. The Sun, now some 4.5 thousand million years old, contains about 1.7% metals by mass. Even older stars have even lower proportions. Figure 1A.12 shows (very approximately) the relationship between the metal abundance and age (*7).

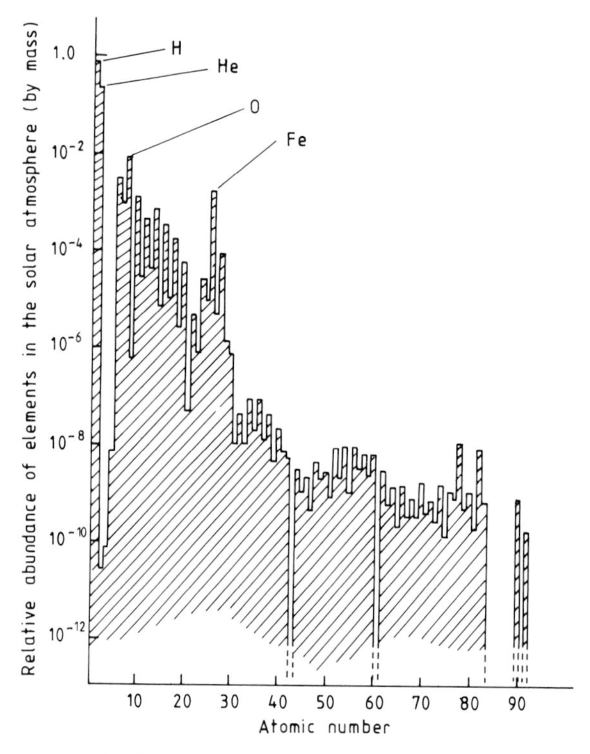

Figure 1A.11 Composition of the solar atmosphere.

A few stars appear to be unusual, and to have noticeable variations from the main type of composition. Thus the Ap stars are hot stars which have abnormally low abundances of iron and some other elements. This appears to be due to element separation by diffusion processes in the outer layers of the stars, or less certainly to an unusual pattern of nucleosynthesis, perhaps caused by the very strong magnetic fields often associated with these stars. The Wolf–Rayet stars are even hotter and appear to lack hydrogen almost completely. Here, the cause seems to be that the outer layers of the star have been stripped away in a violent stellar wind (Chapter 6A), revealing the highly processed material which originally formed the core of the star. Amongst the cooler stars, the subdwarfs appear to be metal deficient. However, this may not be a real underabundance, but due to reduced shortwave line blanketing (Chapter 3P) changing the energy balance within the spectrum. If true, then the stars are probably relatively normal low mass Population I stars (Chapter 3A).

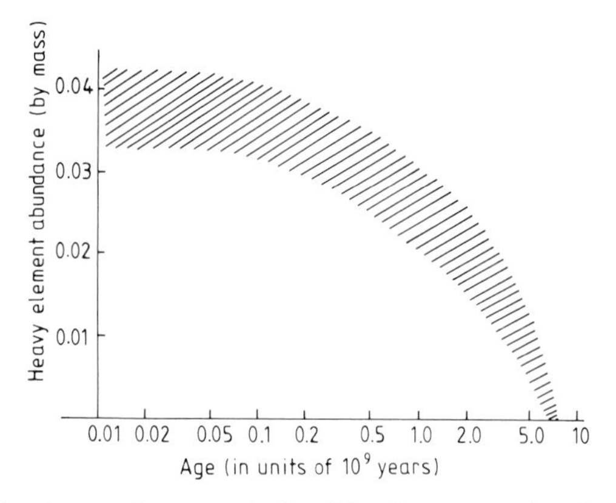

Figure 1A.12 Approximate relationship between the heavy-element ('metal') abundance and the ages of objects.

Changes in the abundances of individual elements can occur for many reasons, for example, unusual nucleosynthesis processes, diffusion, turbulence, magnetic fields, etc causing unusual mixtures of material, or accretion of interstellar material. Ap stars have already been mentioned as possible cases of the first two processes. T Tauri stars have an overabundance of lithium by up to a factor of one hundred. This appears to be attributable to their recent formation from the interstellar medium so that there has been insufficient time for the lithium to have been destroyed by the nuclear reactions at the centre of the star.

The nuclear reactions, which have been mentioned several times already, occur only in the central regions of the star; for example, in the Sun, the central 0.5% of the mass generates 90% of its energy. In the absence of complete mixing of the material inside the star, which is the normal situation, the composition must therefore alter between its surface and centre. There is currently no observing technique capable of determining the composition of the material inside stars; even neutrino telescopes (*Astrophysical*

Techniques, Chapter 1.5) can only give details of certain nuclear reactions, not the abundances. Thus we must rely upon stellar modelling to suggest the manner of this compositional variation. The postulated composition of the Sun is shown in figure 1A.13; the hydrogen mass fraction (X) decreases, and the helium mass fraction (Y) increases due to the conversion of hydrogen into helium (§2A.2). The heavy element abundance hardly changes at all. More evolved stars than the Sun take this process further until the core may become almost solely composed of helium. Further nuclear reactions (§2A.2) may then successively produce carbon, oxygen, magnesium, aluminium, silicon, and finally iron and its neighbouring elements. In the final stages of the life of a large star, the composition may grade from hydrogen and helium at the surface through the above list of elements to a core of almost pure iron-group elements. Only a very small proportion of the stars progress along this process further than the helium core; indeed, many do not even produce $^4_2$He, but stop at $^3_2$He (Chapters 2A and 8A). The star needs a mass greater than about 0.35$\mathcal{M}_\odot$ if the helium is to undergo nuclear reactions. Solar mass stars can probably produce elements up to oxygen, but for the complete sequence to occur an initial mass of over 20 $\mathcal{M}_\odot$ is probably needed.

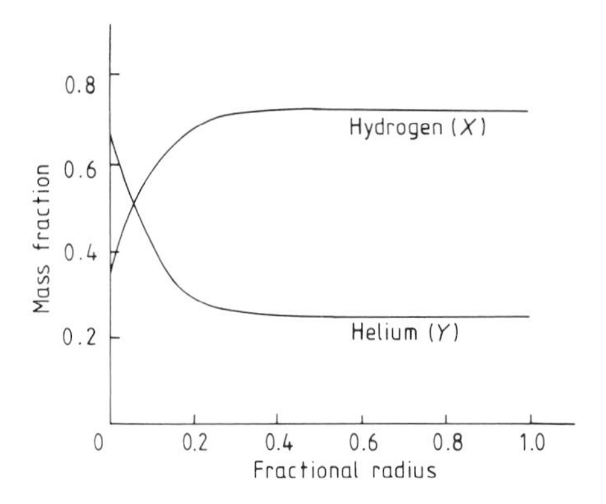

Figure 1A.13 Fraction of the solar composition by mass, in the form of hydrogen and helium, as a function of the solar radius.

An important physical change may also occur within the star as it develops one of these single-element cores; the material at the centre may become electron-degenerate (Chapter 2P), and this produces major changes in the star's structure and behaviour. The actual elemental composition is not altered by this change unless the material becomes baryon-degenerate (see neutron stars, Chapters 2P and 2A).

White dwarfs

The material in white dwarfs, which form the other major class of stars in terms of numbers, is largely in an electron-degenerate form (§2P.1). These are physically small stars; typically having a radius of 1% or less of that of the Sun, but still with a mass comparable with that of the Sun (Chapters 2A, 3A). Their mean densities are therefore very large: 10^5 to 10^7 kg m^{-3} (100 to 10 000 times the density of water). The central densities can be as high as 10^9 kg m^{-3}, or a million times that of water. The surface layers, however, remain gaseous and some estimate of their composition may be made, although the very high pressure of the gas leads to severe problems in the analysis and interpretation of the spectra. There is a wide variation; from almost solar-type compositions, through helium-rich stars (i.e. depleted in hydrogen) to calcium-rich stars (depleted in hydrogen and helium). Some white dwarf spectra are totally devoid of lines, so that then no clue to their composition at all is available. It has been suggested that their inner structure ranges from almost pure helium to almost pure iron, but very great uncertainties still remain in the determination of the structure of any given individual white dwarf.

PROBLEMS

1A.1 The true orbits of the members of a visual binary system are shown in figure 1A.14. From the orbits and the data below determine the masses of the two components of the binary.

Dates of the observations

Observation	Date	Observation	Date
1	5/4/1823	7	20/4/1927
2	2/7/1832	8	10/8/1954
3	28/9/1847	9	25/11/1956
4	6/1/1852	10	3/1/1964
5	10/7/1864	11	22/10/1981
6	3/1/1900	12	30/3/1987

Distance of binary = 11 pc.

1A.2 Using the data shown in figure 1A.4, and assuming the spectra to be spaced at equal time intervals, plot the velocity curve of the binary. If its orbital period is 8.1 hours, what are the masses of the two components in terms of the solar mass?

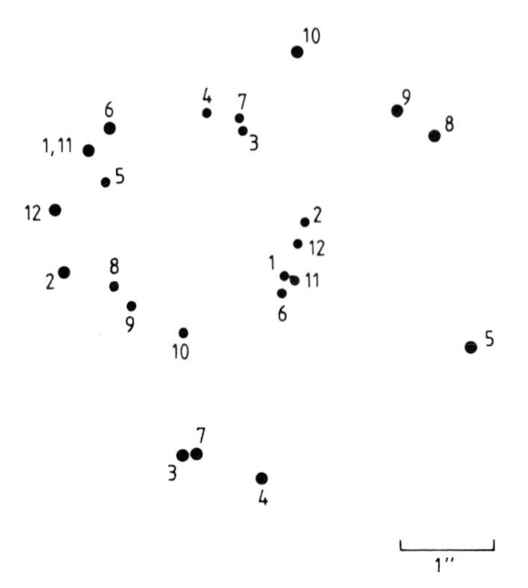

Figure 1A.14 Positions of two members of a visual binary system in the sky, the plane of the orbit being perpendicular to the line of sight.

1A.3 Show that the mass function, $f(\mathcal{M})$ (equation 1A.14) is indeed given by equation (1A.15), starting from equation (1A.2).

1A.4 Assuming that the binary whose velocity curve is shown in figure 1A.8 is viewed along the plane of its orbit, use the data in that figure to determine the parameters of the relative orbit. Hence estimate the masses of the components.

2P Physical Background

SUMMARY

Physics of gases, equation of state, degenerate material, nuclear reactions of importance in nucleosynthesis, black body radiation laws, convection, the virial theorem.

Provides background material for Chapter 2A in particular.

See also stellar masses (§1A.1), mass limits (§2A.1), energy sources (§2A.2), energy transfer (§2A.3), collapsed objects (§§3A.5, 3A.9), coronal stellar winds (§§4A.5, 5A.1, 5A.2), radiation-driven stellar winds (§5A.2), Strömgren spheres (§6A.1), post-main sequence evolution (§8A.2).

These chapters are intended only as a background, and not as a coherent discourse on physics. The data and equations are therefore presented with only sufficient explanation to enable them to be used in the related astrophysics sections. The reader is referred to specialist physics books (see Appendix III) should additional material in this area be required.

2P.1 PHYSICS OF GASES

It is an extremely fortunate circumstance that the material inside most stars behaves like a gas, and that this behaviour can generally be described adequately by the equations developed from experiments in terrestrial laboratories. Were this not the case, then we should still be standing on the shore of the sea of astronomical knowledge, instead of having at least got one toe wet. In situations where terrestrial experience provides no help (such as radiative interactions in planetary nebulae, for example) our quantitative descriptions of the processes which may be occurring remain very inadequate. Since the conditions inside stars are so different from the terrestrial environment (temperature 1.5×10^7 K, density 1.6×10^5 kg m^{-3}, pressure 3.4×10^{16} N m^{-2} or 3.4×10^{11} atm at the centre of the Sun, for example) this wide applicability of terrestrial experience may seem surprising. It arises, however, from the fact that the vast majority of atoms are totally ionised inside stars. The individual nuclei and electrons are far smaller than the original atoms (e.g. the Bohr 'radius' of a hydrogen atom is 5×10^{-11} m compared with the classical electron radius of 3×10^{-15} m). Thus, relative to their size, the particles at the centre of the Sun have as much 'room' as the atoms in a room-temperature gas whose pressure is only 0.024 N m^{-2} (2.4×10^{-7} atm, 1.8×10^{-4} mmHg), and thus obey similar equations.

Particulate weights

The mean particulate weight (or molecular weight) in a completely ionised gas is easily obtained. Writing X, Y and Z for the mass fractions of hydrogen, helium and the heavier elements respectively, we have the following numbers of particles per kilogram of the material when the elements are ionised:

hydrogen $(X/m_{\mathrm{H}}) \times (1 \text{ nucleus} + 1 \text{ electron})$
helium $(Y/4m_{\mathrm{H}}) \times (1 \text{ nucleus} + 2 \text{ electrons})$
'metals' $(Z/\varphi m_{\mathrm{H}}) \times (1 \text{ nucleus} + \sim\varphi/2 \text{ electrons})$

where φ is the mean atomic mass of the heavy elements when not ionised and m_{H} is the mass of the hydrogen atom, 1.6×10^{-27} kg. The form of the contribution from the metals arises because there are roughly half as many electrons in these atoms as particles (protons and neutrons) in the nucleus. Hence the number N of particles in 1 kg of completely ionised gas is given by:

$$N \simeq \frac{2X}{m_{\mathrm{H}}} + \frac{3Y}{4m_{\mathrm{H}}} + \frac{\varphi Z}{2\varphi m_{\mathrm{H}}} \qquad (2P.1.1)$$

and so the mean particulate weight (*1) is

$$\bar{\mu} \simeq \frac{m_{\mathrm{H}}}{(2X + \frac{3}{4}Y + \frac{1}{2}Z)} \quad (\mathrm{kg}) \qquad (2P.1.2)$$

$$\simeq (2X + \tfrac{3}{4}Y + \tfrac{1}{2}Z)^{-1} \quad (\mathrm{amu}) \qquad (2P.1.3)$$

or, since by definition

$$X + Y + Z = 1 \qquad (2P.1.4)$$

we have

$$\bar{\mu} \simeq (\tfrac{3}{4}X + \tfrac{1}{4}Y + \tfrac{1}{2})^{-1} \quad \text{(amu)}. \qquad (2P.1.5)$$

Thus for the Sun (see Chapters 1A and 4A), we have

$$\text{Surface:} \quad \bar{\mu} = 0.601 \quad \text{(amu)} \qquad (2P.1.6)$$

$$\text{Centre:} \quad \bar{\mu} = 0.837 \quad \text{(amu)}. \qquad (2P.1.7)$$

At the visible surface of the Sun, most elements are un-ionised, and the mean particulate weight is about 1.25 amu, but equation (2P.1.6) becomes applicable at a depth of only 7000 km ($0.01R_{\odot}$) or less. Similar results obtain for most other stars.

Equation of state

Since the material inside a star is usually behaving like a rarefied gas, its properties may be described by the perfect gas law

$$PVT^{-1} = \text{constant}. \qquad (2P.1.8)$$

If the quantity of material being studied is one mole (*2), then the constant in equation (2P.1.8) is called the gas constant, $\mathscr{R}$, and has the value

$$\mathscr{R} = 8.314 \ (\text{J K}^{-1}\,\text{mol}^{-1}) \qquad (2P.1.9)$$

or

$$R = 8.314/\mu \ (\text{J K}^{-1}\,\text{kg}^{-1}) \qquad (2P.1.10)$$

where R is the gas constant per kilogram.

The law is more commonly used in a slightly different form by astrophysicists, and we may derive that form in a straightforward manner as follows. The volume of one kilogram of material is

$$1/\rho \ (\text{m}^3) \qquad (2P.1.11)$$

so that

$$P = R\rho T \qquad (2P.1.12)$$

$$= R\bar{\mu}m_{\text{H}}nT \qquad (2P.1.13)$$

where n is the number of particles per cubic metre. Hence

$$P = 1000\mathscr{R}m_{\text{H}}nT \qquad (2P.1.14)$$

or

$$P = nkT \qquad (2P.1.15)$$

where $k = 1000\mathscr{R}m_{\text{H}} = 1.3806 \times 10^{-23}$ J K^{-1} and is known as Boltzmann's constant or the gas constant per particle. Equation (2P.1.15) is the most generally useful form of the perfect gas law for astrophysics, and with the addition of the effects of radiation, turbulence, magnetic fields, etc when necessary, forms the *equation of state* of the material.

Change of state

When material moves within a star, its physical conditions must generally change in some manner. Two limiting cases are found to be useful: adiabatic and isothermal changes. In the former, there is no interchange of energy between the moving element and its surroundings, while in the latter case, energy interchange occurs at a sufficient rate to maintain a constant temperature within the moving element.

In an adiabatic change, there is no alteration of the internal energy of the system under consideration, or in terms of the first law of thermodynamics

$$\mathrm{d}Q = \mathrm{d}U + \mathrm{d}W = 0 \tag{2P.1.16}$$

where $\mathrm{d}Q$ is the energy transferred to or from the system, $\mathrm{d}U$ is the internal energy of the system and $\mathrm{d}W$ is the work done by the system. If we imagine the element of material moving so that it experiences a changing pressure, then we will have

$$\mathrm{d}W = PA\mathrm{d}x \tag{2P.1.17}$$

where A is the surface area of the element and $\mathrm{d}x$ is the linear movement of the element's boundaries, or

$$\mathrm{d}W = P\mathrm{d}V \tag{2P.1.18}$$

giving

$$\mathrm{d}U + P\mathrm{d}V = 0 \tag{2P.1.19}$$

for adiabatic changes.

Now, the specific heat of a gas, C, is defined as the heat energy added for a unit temperature rise

$$C = \mathrm{d}Q/\mathrm{d}T \tag{2P.1.20}$$

and is customarily defined either under conditions of constant volume (C_V), or constant pressure (C_P). From equations (2P.1.16) and (2P.1.18), we have

$$\frac{\mathrm{d}Q}{\mathrm{d}T} = \frac{\mathrm{d}U}{\mathrm{d}T} + P\frac{\mathrm{d}V}{\mathrm{d}T} \tag{2P.1.21}$$

and at constant volume

$$\frac{\mathrm{d}V}{\mathrm{d}T} = 0 \tag{2P.1.22}$$

so that

$$\left.\frac{\mathrm{d}Q}{\mathrm{d}T}\right|_V = C_V = \frac{\mathrm{d}U}{\mathrm{d}T}. \tag{2P.1.23}$$

Similarly

$$\left.\frac{\mathrm{d}Q}{\mathrm{d}T}\right|_P = C_P = \left.\frac{\mathrm{d}U}{\mathrm{d}T}\right|_P + P\left.\frac{\mathrm{d}V}{\mathrm{d}T}\right|_P. \tag{2P.1.24}$$

From equation (2P.1.8), we have

$$PV/T = \Omega\,\mathscr{R} \tag{2P.1.25}$$

where Ω is the number of moles of the element involved. Thus

$$P\frac{\mathrm{d}V}{\mathrm{d}T} + V\frac{\mathrm{d}P}{\mathrm{d}T} = \Omega\,\mathscr{R} \tag{2P.1.26}$$

and so

$$P\left.\frac{\mathrm{d}V}{\mathrm{d}T}\right|_P = \Omega\,\mathscr{R}. \tag{2P.1.27}$$

For gas formed from a single type of particle (monoparticulate gas), with no change in the ionisation or excitation states, if any, the changes in the internal energy must arise from changes in the particles' thermal velocities alone. The mean thermal energy per particle in an ideal gas is

$$\tfrac{3}{2}kT \tag{2P.1.28}$$

so that

$$U = \text{constant} + \tfrac{3}{2}\,\Omega\mathscr{R}T \tag{2P.1.29}$$

and

$$\frac{\mathrm{d}U}{\mathrm{d}T} = \tfrac{3}{2}\,\Omega\,\mathscr{R}. \tag{2P.1.30}$$

Thus equations (2P.1.23), (2P.1.24), (2P.1.27) and (2P.1.30) give

$$C_V = \tfrac{3}{2}\,\Omega\,\mathscr{R} \tag{2P.1.31}$$

and

$$C_P = \tfrac{5}{2}\,\Omega\,\mathscr{R} \tag{2P.1.32}$$

for a gas containing a single particle type. Mixtures of various types of single particle gases give the same result. If the particles are composite (i.e. molecules) or significant quantities of energy are diverted into ionisation etc, then the specific heats will increase from these values since not all the energy contributed to the system will be taken up by the increase in the particles' thermal motions.

Returning to the adiabatic equation (2P.1.19), we have

$$\frac{dU}{dT} + P\frac{dV}{dT} = 0 \tag{2P.1.33}$$

so that with equations (2P.1.23) and (2P.1.26) we get

$$C_V + \Omega\mathscr{R} - V\frac{dP}{dT} = 0 \tag{2P.1.34}$$

and equation (2P.1.31) then gives

$$V\frac{dP}{dT} = \tfrac{5}{2}\Omega\mathscr{R}. \tag{2P.1.35}$$

Using equation (2P.1.25),

$$\Omega\mathscr{R}\frac{T}{P}\frac{dP}{dT} = \tfrac{5}{2}\Omega\mathscr{R} \tag{2P.1.36}$$

and so

$$\frac{dP}{P} = \frac{5}{2}\frac{dT}{T} \tag{2P.1.37}$$

which upon integration gives

$$\log_e P = \text{constant} + \tfrac{5}{2}\log_e T \tag{2P.1.38}$$

or

$$P = \text{constant} \times T^{5/2} \tag{2P.1.39}$$

for adiabatic changes. Looking back through this analysis, we see that the index in equation (2P.1.39) arises from the two specific heats. Writing

$$\gamma = C_P/C_V \tag{2P.1.40}$$

a more general form of equation (2P.1.39) is

$$P = \text{constant} \times T^{\gamma/(\gamma-1)} \tag{2P.1.41}$$

in which $\gamma = \frac{5}{3}$ for fully ionised gases (and other monoparticulate mixtures).

The relationship between pressure and density for adiabatic changes is also useful. Equation (2P.1.12) gives

$$T \propto \frac{P}{\rho} \tag{2P.1.42}$$

and so

$$P = \text{constant} \times \rho^{\gamma}. \tag{2P.1.43}$$

A pressure law of this type is called polytropic, and polytropes formed the basis of much stellar modelling work before the advent of large, fast computers. They may still provide useful guides to the behaviour of the material in some regions of some stars, the main examples being regions wherein the main form of energy transport is by convection, and inside white dwarfs.

During an isothermal change, the temperature of the material remains constant. From equation (2P.1.12), we then have simply

$$P = K\rho \qquad (2P.1.44)$$

where K is a constant which depends upon the actual value of the temperature. Such a situation might be physically realised inside the non-energy generating core of an evolved star, for example.

Degeneracy

Subatomic particles, including composites such as nuclei, fall into two groups depending upon whether their spins are integer multiples or half-integer (i.e. half-odd integer) multiples of $h/2\pi$ (*3). The former particles are termed *bosons*, and include photons, pions, and nuclei of even atomic mass number. The others are called *fermions*, and include electrons, protons, neutrons, and nuclei with odd atomic mass numbers. The behaviour of the two groups is quite different and is described by Bose–Einstein statistics and Fermi–Dirac statistics respectively. The difference which is of relevance here is that the fermions obey the Pauli exclusion principle (*4), while the bosons do not.

Free electrons inside a star have therefore to obey the exclusion principle, and in their case it may be stated in the form that only two electrons (with opposite spins) can have the same position and momentum, within the limits given by the uncertainty principle. We may thus envisage an electron as 'occupying' a six-dimensional volume in position/momentum space. If we denote position by (x_1, x_2, x_3), and momentum by (p_1, p_2, p_3) in three independent spatial directions, 1, 2, and 3, then the 'volume' occupied by the electron is about

$$\delta x_1 \delta x_2 \delta x_3 \delta p_1 \delta p_2 \delta p_3 \simeq h^3. \qquad (2P.1.45)$$

At low densities the physical separations of the electrons are large, so that their momentum differences can be very small without violating equation (2P.1.45), and there are many more available momentum states than electrons to occupy them. Under these conditions the momentum distribution then approximates the Maxwell–Boltzmann form:

$$\frac{N(p)\,\mathrm{d}p}{N} = \left(\frac{2}{\pi m^3 k^3}\right)^{1/2} T^{-3/2} p^2 \exp(-p^2/2mkT)\,\mathrm{d}p \qquad (2P.1.46)$$

where $N(p)$ is the number density of particles with momenta in the range p to $p + dp$ and N is the total number density of particles.

As the density starts to increase, however, the physical separations must decrease, and so the momentum separations must increase to compensate (2P.1.45). The number of available momentum values within a given range therefore decreases. Eventually a stage will be reached at which the Maxwellian distribution is no longer an adequate approximation. The more general form of the distribution of momenta as given by the Fermi–Dirac formula is then required:

$$N(p)\,\mathrm{d}p/N = (4\pi gm/h^3\rho)p^2\,\mathrm{d}p \times \left(\frac{gm(2\pi mkT)^{3/2}}{\rho h^3}\exp(p^2/2mkT)+1\right)^{-1} \tag{2P.1.47}$$

where g is the statistical weight, and takes a value of 2 for an electron gas, and ρ is the density of the fermions.

At sufficiently low densities we have

$$[gm(2\pi mkT)^{3/2}/\rho h^3]\exp(p^2/2mkT) \gg 1 \tag{2P.1.48}$$

so that equation (2P.1.47) becomes

$$\frac{N(p)\mathrm{d}p}{N} = \frac{4\pi gm}{h^3\rho}\,\frac{\rho h^3}{gm(2\pi mkT)^{3/2}}\exp(-p^2/2mkT)p^2\,\mathrm{d}p \tag{2P.1.49}$$

$$= \left(\frac{2}{\pi m^3k^3}\right)^{1/2} T^{-3/2}p^2\exp(-p^2/2mkT)\,\mathrm{d}p \tag{2P.1.50}$$

and is then identical to Maxwell's equation (2P.1.46), as indeed ought to be the case.

When the low-density approximation does not hold, the matter is often described as *degenerate*. The strict use of this term within physics, however, is to describe multiple states with the same energy. In its most widespread use in astronomy, the material forming white dwarfs is called electron-degenerate. Since these stars are thought to have very intense magnetic fields (§3A.8), in reality the degeneracy is broken by a very large factor. In this book, the looser astronomical meaning of the term degenerate matter will be retained and this should not lead to confusion.

With this *caveat*, the velocities of the particles at higher densities are given by

$$N(v)\,\mathrm{d}v/N = (4\pi gm^4/h^3\rho)v^2\,\mathrm{d}v \times \left(\frac{gm(2\pi mkT)^{3/2}}{\rho h^3}\exp(mv^2/2kT)+1\right)^{-1} \tag{2P.1.51}$$

and the manner of the resulting variation with density is shown in figure 2P.1. Degeneracy effects can clearly be seen at electron densities of $10\ \mathrm{kg\,m^{-3}}$ and above; the lower-velocity ends of the curves are asymptotic

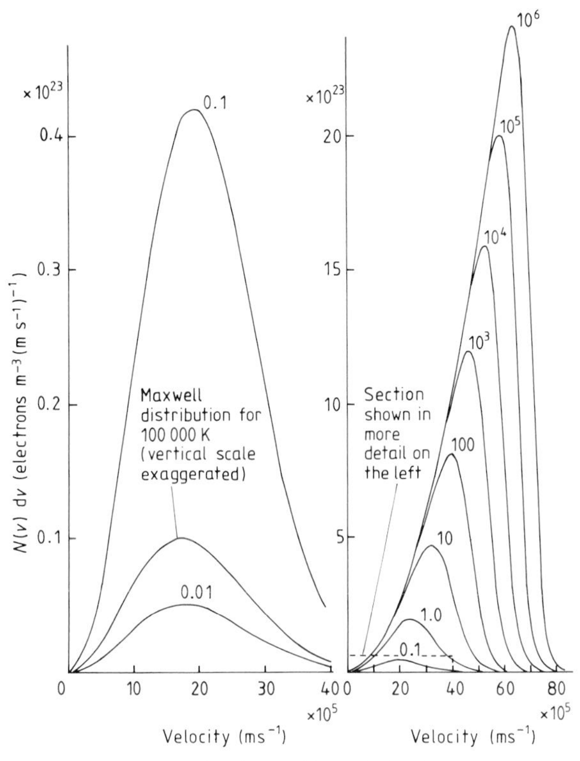

Figure 2P.1 Particle distribution with velocity for an electron gas at a temperature of 100 000 K at various densities. The density (in units of kg m^{-3}) is indicated by the number near the peak of each curve. The left-hand frame shows in greater detail the section of the right-hand frame enclosed within the broken lines.

to the parabola

$$N(v)\,\mathrm{d}v = \left(\frac{4\pi g m^3}{h^3}\right) v^2\,\mathrm{d}v \qquad (2P.1.52)$$

since all the available states are filled. Although less marked, significant deviations from the Maxwellian distribution continue to lower densities; the distribution is skewed to the right, and the maximum moved towards the higher velocities. Thus, in the example of figure 2P.1, the peak of the Maxwell distribution occurs at 1.74×10^6 m s^{-1}, but even for electron densities as low as 0.1 kg m^{-3} (such as might result in a hydrogen plasma with a density of 200 kg m^{-3}), it is shifted to 1.9×10^6 m s^{-1}.

Since protons and other nuclei have masses 2000 or more times that of an electron, degeneracy does not begin to have significant effects upon their distribution until much higher densities. The onset of significant degeneracy is signalled by the behaviour of the second denominator on the right-hand side of equations (2P.1.47) or (2P.1.51). To a good approximation, degeneracy effects may become important when

$$\frac{gm(2\pi mkT)^{3/2}}{\rho h^3}\exp(mv^2/2kT) < 1 \tag{2P.1.53}$$

or

$$\rho > \frac{gm(2\pi mkT)^{3/2}}{h^3} \tag{2P.1.54}$$

(where a velocity equal to the Maxwellian mean velocity,

$$v = (2kT/m)^{1/2} \tag{2P.1.55}$$

has been used). Thus for a hydrogen plasma at 100 000 K, the onset of proton degeneracy requires a density some 10^5 times greater than that needed for the onset of electron degeneracy.

The pressure inside a material is given by the sum of the contributions from each of its constituent particles. If some particles are degenerate, while others are not, then we simply sum the partial pressures:

$$P_{\text{Total}} = P_{\text{Non-degenerate component}} + P_{\text{Degenerate component}}. \tag{2P.1.56}$$

Radiation and other pressures are also added in if significant. The expression for the pressure in a degenerate material may be obtained from equation (2P.1.47), since pressure is just rate of change of momentum per unit area. Thus

$$P_{\text{D}} = \frac{4\pi g}{3m}\int_0^\infty p^4\left(\frac{gm(2\pi mkT)^{3/2}}{\rho h^3}\exp(p^2/2mkT)+1\right)^{-1}\text{d}p. \tag{2P.1.57}$$

By convention, we write

$$U = p^2/2mkT \tag{2P.1.58}$$

$$\psi = \log_{\text{e}}[\rho h^3/gm(2\pi mkT)^{3/2}] \tag{2P.1.59}$$

so that

$$P_{\text{D}} = \frac{2\pi g(2mkT)^{5/2}}{3mh^3}\int_0^\infty \frac{u^{3/2}\text{d}u}{\text{e}^{u-\psi}+1} \tag{2P.1.60}$$

$$= \frac{2\pi g(2mkT)^{5/2}}{3mh^3}F_{3/2}(\psi) \tag{2P.1.61}$$

where $F_{3/2}(\psi)$ is called the Fermi–Dirac integral.

We may also obtain an expression for the total density from the total number of particles per unit volume, which in turn is obtained by integrating equation (2P.1.47):

$$\rho = N\bar{\mu}_e m_H \tag{2P.1.62}$$

$$= \frac{4\pi g\bar{\mu}_e m_H}{h^3}\int_0^\infty p^2\left(\frac{gm(2\pi mkT)^{3/2}}{\rho h^3}\exp(p^2/2mkT)+1\right)^{-1}\mathrm{d}p \tag{2P.1.63}$$

$$= \frac{2\pi g\bar{\mu}_e m_H(2mkT)^{3/2}}{h^3}\int_0^\infty \frac{u^{1/2}\,\mathrm{d}u}{\mathrm{e}^{u-\psi}+1} \tag{2P.1.64}$$

$$= \frac{2\pi g\bar{\mu}_e m_H(2mkT)^{3/2}}{h^3}F_{1/2}(\psi) \tag{2P.1.65}$$

where $\bar{\mu}_e$ is the mean molecular weight per free electron. In completely ionised gases its value is 1 for hydrogen, and about 2 for all other elements. Thus equations (2P.1.61) and (2P.1.64) give

$$P_D = \frac{2kT}{3\bar{\mu}_e m}\frac{F_{3/2}(\psi)}{F_{1/2}(\psi)}\rho\,. \tag{2P.1.66}$$

The variation of $F_{3/2}(\psi)/F_{1/2}(\psi)$ with ψ is shown in figure 2P.2. At low values of ψ (low densities), the ratio is asymptotic to 1.5, giving

$$P_D = \frac{\rho}{\bar{\mu}_e m_H}kT \tag{2P.1.67}$$

$$= nkT \tag{2P.1.68}$$

and this is the perfect gas law (equation 2P.1.15), as would be expected. At large values of ψ

$$F_{3/2}(\psi) \to \tfrac{2}{5}\psi^{5/2} \tag{2P.1.69}$$

and

$$F_{1/2}(\psi) \to \tfrac{2}{3}\psi^{3/2} \tag{2P.1.70}$$

giving

$$P_D = \left(\frac{4\pi g}{15mh^3}\right)(2mkT)^{5/2}\psi^{5/2} \tag{2P.1.71}$$

and

$$\rho = \left(\frac{4\pi g\bar{\mu}_e m_H}{3h^3}\right)(2mkT)^{3/2}\psi^{3/2} \tag{2P.1.72}$$

so that finally

$$P_D = \left(\frac{4\pi g}{15mh^3}\right)\left(\frac{\rho}{4\pi g\bar{\mu}_e m_H/3h^3}\right)^{5/3} \tag{2P.1.73}$$

$$= \left(\frac{9h^6}{2000\pi^2 g^2 m^3 m_H^5 \bar{\mu}_e^5}\right)^{1/3}\rho^{5/3} \tag{2P.1.74}$$

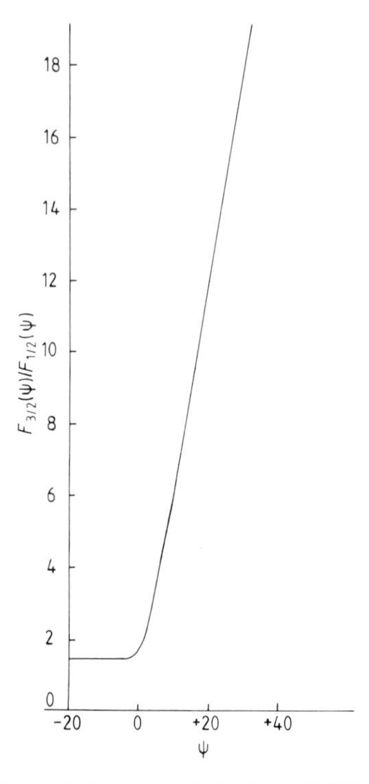

Figure 2P.2 Variation of the ratio of the Fermi–Dirac integrals with ψ. (After R Härm and M Schwarzschild (1955), *Astrophys. J. Suppl.* **1** 319.)

and we see that the pressure exerted by a completely degenerate gas depends only on the density; it is independent of temperature. Equation (2P.1.74) gives the pressure for particles with non-relativistic velocities. Above velocities of about 10% of the speed of light, relativistic effects become significant. Using the relativistic expression for momentum,

$$p = \frac{mv}{(1 - v^2/c^2)^{1/2}} \tag{2P.1.75}$$

we then obtain

$$P_{\mathrm{D}} = \left(\frac{3h^3c^3}{256\pi g m_{\mathrm{H}}^4 \bar{\mu}_{\mathrm{e}}^4}\right)^{1/3} \rho^{4/3}. \tag{2P.1.76}$$

Equations (2P.1.74) and (2P.1.76) are both of polytropic form (equation 2P.1.44), and so white dwarfs (Chapters 2A and 3A) whose internal pressure is largely due to a relativistic degenerate electron gas may be modelled by polytropes with an index of 3 (*5).

At very high densities ($\sim 10^{16}$ kg m^{-3}), the protons and electrons mostly combine with each other by inverse beta decay to produce neutrons, leaving only a few per cent of the mass in the form of individual protons and electrons. Since neutrons are also fermions, they may become degenerate. They belong to the group of particles called *baryons*, and so matter in which the neutrons are degenerate is usually known as baryon-degenerate material. From equation (2P.1.74), we then have for a non-relativistic degenerate neutron gas

$$P_{\mathrm{D}} = \left(\frac{9h^6}{200\pi^2 g^2 m_{\mathrm{n}}^8}\right)^{1/3} \rho^{5/3} \qquad \text{(2P.1.77)}$$

where m_{n} is the neutron mass (1.675×10^{-27} kg), g has a value of 2 for neutrons and ρ is the neutron density. In the case of neutron stars (Chapters 2A and 3A), other sources of pressure such as electrons and nuclei may be significant, so that it is only towards the centre of such objects that baryon degeneracy pressure predominates.

2P.2 PARTICLE PHYSICS

Of all the very great development that has occurred in particle physics over the last few decades, we are concerned here with only a very small section: that required to follow the nuclear reactions thought to be occurring inside stars. The details of these reactions are discussed in §2A.2.

Notation

Nuclei are denoted by the chemical symbols of the corresponding atom, with their total number of protons and neutrons as a superscript (approximately the same as the atomic mass), and with the atomic number as subscript. Thus, for example, the nucleus of the common isotope of oxygen would be symbolised by

$$^{16}_{8}\mathrm{O}. \qquad \text{(2P.2.1)}$$

Particles other than nuclei are symbolised as follows:

neutron	n	
proton	p^+	(occasionally as $^1_1\mathrm{H}$)
electron	e^-	
positron	e^+	
neutrino	ν_{e}	These are both electron-
anti-neutrino	$\tilde{\nu}_{\mathrm{e}}$	associated particles
photon	γ	

Nuclear reactions are indicated by an arrow, →, in order to distinguish them from apparently similar equations describing chemical reactions (where an equals sign is used). Thus, for example, one stage of the CNO bi-cycle (§2A.2) involves the radioactive decay of fluorine:

$$^{17}_{9}\mathrm{F} \rightarrow {}^{17}_{8}\mathrm{O} + \mathrm{e}^{+} + \nu_{\mathrm{e}}. \qquad (2P.2.2)$$

Nuclear reactions

Energy may be released or absorbed (exothermic or endothermic reactions) during nuclear processes. Most of such an energy change occurs due to changes in the binding energy of the nuclei. This gives rise to the difference between the actual mass of the nucleus and the sum of the masses of its constituents: the *mass defect*. Thus protons and neutrons have masses of

$$m_{\mathrm{p}} = 1.67352 \times 10^{-27}\ \mathrm{kg} \qquad (2P.2.3)$$

$$m_{\mathrm{n}} = 1.67492 \times 10^{-27}\ \mathrm{kg} \qquad (2P.2.4)$$

which gives

$$2m_{\mathrm{p}} + 2m_{\mathrm{n}} = 6.69688 \times 10^{-27}\ \mathrm{kg}. \qquad (2P.2.5)$$

The mass of the helium nucleus, however, is

$$m_{\mathrm{He}} = 6.64446 \times 10^{-27}\ \mathrm{kg} \qquad (2P.2.6)$$

leading to a mass defect of 5.242×10^{-29} kg. The formation of a helium nucleus from its constituents therefore releases about 4.72×10^{-12} J. This energy appears in the form of other particles (including photons) and/or increased kinetic energies, and it is the energy which would need to be added to the nucleus in order to disrupt it back into individual protons and neutrons.

Units with greater convenience than joules and kilograms are generally used when discussing sub-atomic processes, and these are the electron volt (eV), and the atomic mass unit (amu). The electron volt is the energy acquired by an electron in 'falling' through a potential difference of one volt. It has a value of 1.602192×10^{-19} J. Through the application of Einstein's equation for the equivalence of mass and energy,

$$E = mc^2 \qquad (2P.2.7)$$

the electron volt may also be used as a measure of mass, and this is very useful when energetic interactions are being considered, since the effect in energy terms of the production or destruction of particles is immediately apparent. The atomic mass unit is defined to be one twelfth of the mass of the $^{12}_{6}\mathrm{C}$ atom, and has a value of 1.660531×10^{-27} kg. The values of the masses of some of the commonly encountered particles in astrophysics are listed in table 2P.1

Table 2P.1

Particle	Mass (amu)	Mass (MeV)	
electron	5.486×10^{-4}	0.511010	
positron	5.486×10^{-4}	0.511010	
proton	1.007275	938.2574	
neutron	1.008664	939.5513	
helium ($^4_2$He)	4.00140	3 727.228	nuclei only
carbon ($^{12}_6$C)	12.00000	11 177.77	nuclei only
nitrogen ($^{14}_7$N)	14.00307	13 043.59	nuclei only
oxygen ($^{16}_8$O)	15.99491	14 898.95	nuclei only

Some rough values worth remembering are

electron/positron	0.5 MeV
proton/neutron/hydrogen nucleus	1000 MeV.

Whether a nuclear reaction is exothermic or not depends upon how the binding energy changes during the reaction. The binding energy per nucleon for the quasi-stable nuclei is shown in figure 2P.3, and if the reaction is such that the products are higher on that diagram than the precursors, then there will be a net energy release. Thus energy may be generated during nuclear reactions by the combination of light elements into heavier ones (fusion), or by the disintegration of heavy elements into lighter ones (fission). In astrophysics, it is almost always the first of these which is important. Figure 2P.3 shows that there is a maximum binding energy per nucleon which is

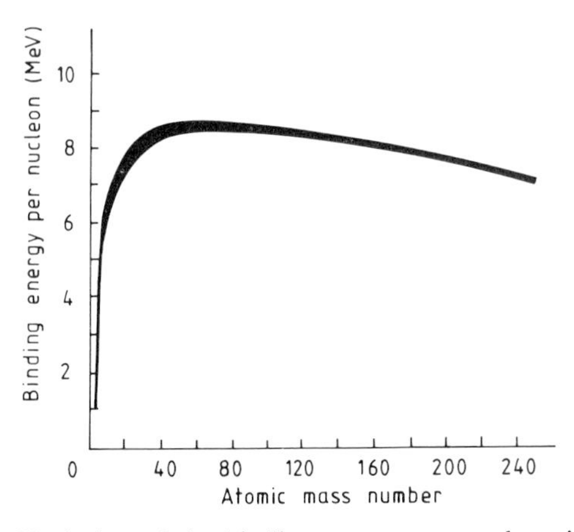

Figure 2P.3 Variation of the binding energy per nucleon in nuclei.

close to iron and the nearby elements. Thus any reaction which builds up heavier elements from iron *or* which splits iron to light elements must be endothermic, and iron-group elements must be the end product of any continuing series of exothermic reactions. This limit to the potential availability of energy generating nucleosynthetic reactions is of great significance in the evolution of massive stars (Chapters 3A and 8A).

The main reactions inside stars cause one ${}^4_2$He nucleus to be formed from four ${}^1_1$H nuclei (§2A.2). A small fraction of the released energy is carried away by neutrinos, and so does not contribute to the star's internal energy. This fraction varies, depending upon the sequence of reactions which may be involved; the energy which is available for heating the interior of the star when one helium nucleus is produced therefore ranges from 25 to 26 MeV. The conversion of 1 kg of hydrogen into helium thus releases about 6×10^{14} J, and there is a mass loss of about 0.007 kg. The same amount of energy, if it were to be produced by burning coal, for example, would require the combination of about 18 000 tons of coal with 48 000 tons of oxygen; another comparison is that it is equivalent to the energy released by the explosion of about 15 000 tons of TNT.

Whether or not a reaction will be significant in a stellar interior depends largely upon the rate of the reaction, or reaction probability. Thus, for example, one might make ${}^4_2$He by the simultaneous collision and cohesion of four ${}^1_1$H nuclei. However, even if this were to be a stable reaction, the involvement of four particles as input to the reaction makes its probability of occurrence very low indeed, and other routes for the conversion of hydrogen into helium will actually occur in practice (Chapter 2A). The probability of a reaction is governed by the mean free paths of the particles, and by their collision or interaction cross-sectional areas. The mean free path is the average distance travelled by the particle between collisions or interactions. We may find an approximate value for it from an idealisation of the situation. Assume that the particles have a certain collision radius, a, within which a collision occurs, and outside which there is no interaction. Imagine a 1 m^2 column of the material 'viewed' by a particle travelling down its length (figure 2P.4). If the number density of the particles comprising the gas is N, then a layer with a depth equal to the particle's collision diameter will contain

$$(2a/1)N \tag{2P.2.8}$$

particles. Its *unobscured* area will therefore be

$$A = 1 - 2aN\pi a^2. \tag{2P.2.9}$$

If there are q such layers superimposed, then the total unobscured area will be

$$A^q = (1 - 2\pi a^3 N)^q. \tag{2P.2.10}$$

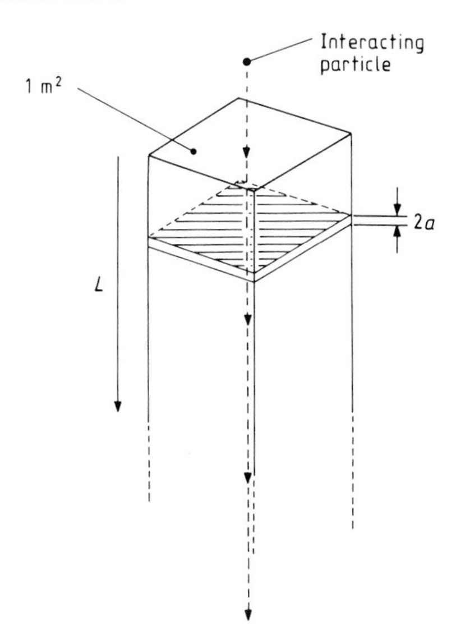

Figure 2P.4 Calculation of the collisional mean free path of a particle.

For a particle to have a 50% chance of undergoing an interaction, we therefore require

$$A^q = 0.5 \qquad (2P.2.11)$$

and so writing L for the mean free path, we have

$$0.5 = (1 - 2\pi a^3 N)^{L/2a}. \qquad (2P.2.12)$$

The expansion of this is

$$0.5 = 1 - \left(\frac{L}{2a}\right)(2\pi a^3 N) + \frac{1}{2!}\left(\frac{L}{2a}\right)\left(\frac{L}{2a} - 1\right)(2\pi a^3 N)^2 + \ldots \qquad (2P.2.13)$$

or, since

$$L/2a \gg 1 \qquad (2P.2.14)$$

in most circumstances, we then have

$$0.5 = 1 - (\pi a^2 LN) + \frac{1}{2!}(\pi a^2 LN)^2 \ldots \qquad (2P.2.15)$$

$$= e^{-\pi a^2 LN} \qquad (2P.2.16)$$

and so

$$L \simeq (\log_e 2)/\pi a^2 N. \qquad (2P.2.17)$$

A more precise calculation, taking into account the motions of all the particles, and of the collision radius of the interacting particle, gives

$$L \simeq \frac{1}{\sqrt{2}\pi a^2 N}. \tag{2P.2.18}$$

The collision cross-sectional area is thus πa^2; it may approximate the 'physical' cross section of the particle, as in neutral atoms, or it may be very much larger than that, as within charged particles. We may quantify the radius as the minimum separation of the two particles' original paths were they to be extended, and take an effective collision to be one in which the direction of the particles' motions is changed by 90° or more (figure 2P.5). Thus for collisions between neutral hydrogen atoms we have

$$a = 5.6 \times 10^{-11}\ \mathrm{m} \tag{2P.2.19}$$

compared with the Bohr radius of

$$5.3 \times 10^{-11}\ \mathrm{m}. \tag{2P.2.20}$$

Collisions between electrons, however, give

$$a \simeq 250 v^{-2}\ (\mathrm{m}) \tag{2P.2.21}$$

so that at 30 000 K ($\bar{v} = 1000\ \mathrm{km\,s^{-1}}$)

$$a = 2.5 \times 10^{-10}\ \mathrm{m} \tag{2P.2.22}$$

compared with the classical electron radius of

$$2.8 \times 10^{-15}\ \mathrm{m}. \tag{2P.2.23}$$

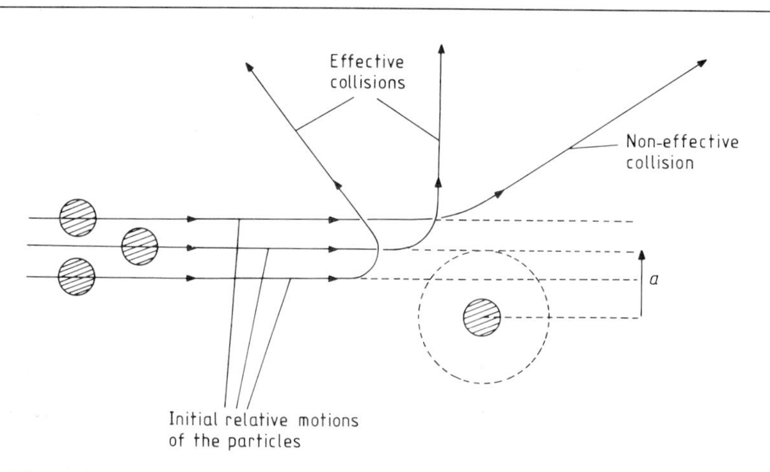

Figure 2P.5 Collisional radius of a particle in the rest frame of the target.

The cross-sectional area is often measured in units called barns, with

$$1 \text{ barn} = 10^{-28} \text{ m}^2. \tag{2P.2.24}$$

Hence neutral hydrogen has a collisional cross-sectional area of about 10^8 barns.

Interactions between particles do not occur every time that they collide. At the centre of the Sun, for example, the initial reaction in the proton–proton chain (§2A.2)

$${}^1_1\text{H} + {}^1_1\text{H} \rightarrow {}^2_1\text{H} + \text{e}^+ + \nu_\text{e} \tag{2P.2.25}$$

requires the interaction of two protons. At a temperature of 1.5×10^7 K ($\bar{v} = 5 \times 10^5$ m s^{-1}), the collisional radius is about 5.5×10^{-13} m. With a central density of 1.6×10^5 kg m^{-3}, we have a mean free path of about 7×10^{-9} m, and a collision frequency of about 10^{14} collisions per second. The interaction, however, occurs on average only about once in 5×10^9 years; thus some 10^{31} collisions occur for every interaction. The interaction cross-sectional area must thus be 10^{-31} times the collisional cross-sectional area, or about 10^{-27} barns (for solar conditions).

The reaction rate between two nuclei, a and b, is given by

$$r_{a,b} = \frac{N_a N_b \bar{\sigma} \bar{v}}{1 + \delta_{ab}} \tag{2P.2.26}$$

where $\bar{\sigma}$ is the interaction cross section and δ_{ab} is the Kronecker delta (1 if $a = b$, 0 if $a \neq b$) and N_a is the number density of particles of type a, etc. The probability of the interaction of two protons is very low; most nuclear reactions have much higher probabilities. For example, nuclei heavier than iron are thought mostly to be formed by neutron capture (Chapters 2A, 3A, and 8A) in what are called the r and s processes. The neutron-capture cross section for ${}^{56}_{26}\text{Fe}$ is then about 0.01 barns, or 10^{25} times greater than that for two protons. Neutrinos are very unreactive particles: their interaction cross section with ${}^{37}_{17}\text{Cl}$ (one of the more 'probable' reactions, and the basis of some neutrino telescopes—§4A.4) for example is only 10^{-19} barns for 10 MeV neutrinos, but this is still a million times more probable than the reaction of two protons! It is the low probability of the first stage of the proton–proton cycle which determines the star's lifetime. Were it to change (as some of the other fundamental constants of nature may do, according to some of the more *outré* cosmological theories) then the future development of stars could be very different from that currently envisaged (Chapter 8A). Thus an increase by a factor of only two or three in the values of $\bar{\sigma}$ for nucleosynthetic reactions would probably suffice to turn Jupiter and Saturn into stars, and to cause the Sun to progress eventually to a supernova.

2P.3 RADIATION LAWS

The radiation field deep inside a star is a very close approximation to that of a black body, unlike the radiation near the surface (Chapters 3P, 3A), or in interstellar space (Chapter 7A). This fortunate circumstance renders the description of the radiation field a relatively simple process.

Black body radiation is radiation emitted by a body in thermodynamic equilibrium (*6), and which is also a perfect absorber of radiation. Such an object can never be physically realised, but very close approximations may be found: the inside of a sealed opaque enclosure, or deep inside a star, etc. A small hole in an otherwise sealed enclosure will emit a very close approximation to black body radiation for wavelengths less than the size of the hole, provided that the energy which escapes is only a very small fraction of the total energy inside the enclosure.

The radiation is found to be continuous and unpolarised, and it is emitted isotropically. Its intensity at any wavelength is a function only of the temperature, and its spectral distribution peaks at a frequency whose value is again a function only of temperature (figures 2P.6 and 2P.7). The spectral distribution is described by the Planck equation,

$$B_\nu(T) = \frac{2h\nu^3\mu^2}{c^2[\exp(h\nu/kT) - 1]} \tag{2P.3.1}$$

or

$$B_\lambda(T) = \frac{2hc^2\mu^2}{\lambda^5[\exp(hc/\lambda kT) - 1]} \tag{2P.3.2}$$

where $B_\nu(T)$ is the intensity of radiation of frequency ν per unit frequency interval, for a black body at a temperature T; $B_\lambda(T)$ is similarly the intensity per unit wavelength interval; and μ is the refractive index of the medium (normally unity for most purposes). At frequencies low enough for

$$h\nu/kT \ll 1 \tag{2P.3.3}$$

then

$$e^{h\nu/kT} \simeq 1 + \frac{h\nu}{kT} \tag{2P.3.4}$$

and equation (2P.3.1) becomes

$$B_\nu(T) \simeq 2kc^{-3}\mu^2 T\nu^2 \tag{2P.3.5}$$

or, in terms of wavelength,

$$B_\lambda(T) \simeq 2kc\mu^2 T\lambda^{-4} \tag{2P.3.6}$$

and the last two equations are known as the Rayleigh–Jeans law. At

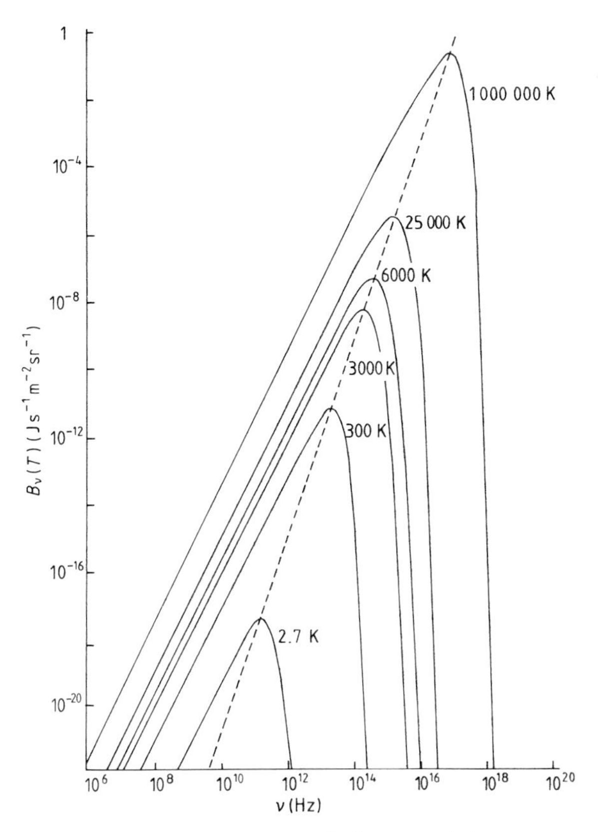

Figure 2P.6 Black body curves (per unit frequency) for temperatures of astrophysical interest (*in vacuo*).

frequencies high enough for

$$h\nu/kT \gg 1 \tag{2P.3.7}$$

then

$$e^{h\nu/kT} - 1 \simeq e^{h\nu/kT} \tag{2P.3.8}$$

and equation (2P.3.1) then becomes

$$B_\nu(T) \simeq 2hc^{-2}\mu^2 e^{-h\nu/kT}\nu^3 \tag{2P.3.9}$$

with equation (2P.3.2) similarly becoming

$$B_\lambda(T) \simeq 2hc^2\mu^2 e^{-hc/\lambda kT}\lambda^{-5}. \tag{2P.3.10}$$

These equations represent the Wien approximation. Differentiating

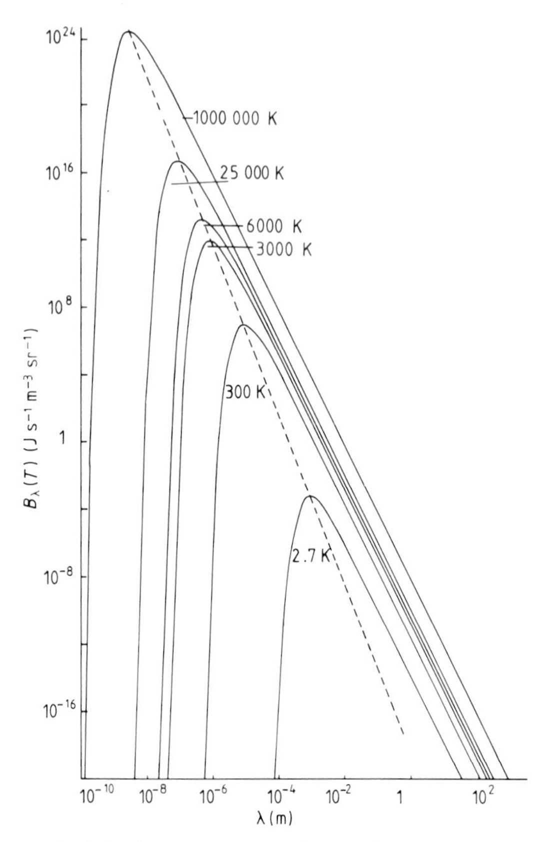

Figure 2P.7 Black body curves (per unit wavelength) for temperatures of astrophysical interest (*in vacuo*).

equation (2P.3.2), we get

$$\frac{\mathrm{d}B_\lambda(T)}{\mathrm{d}\lambda}=\frac{2hc^2\mu^2[(hc/\lambda kT-5)\mathrm{e}^{hc/\lambda kT}+5]}{\lambda^6(\mathrm{e}^{hc/\lambda kT}-1)^2} \tag{2P.3.11}$$

which is zero for $\lambda = \pm\infty$, or when

$$\left(\frac{hc}{\lambda kT}-5\right)\mathrm{e}^{hc/\lambda kT}+5=0\,. \tag{2P.3.12}$$

Using Newton–Raphson iteration, we may evaluate this equation, and we then find

$$hc/\lambda kT = 4.965\,114 \tag{2P.3.13}$$

or

$$\lambda_\mathrm{m}T = 2.8979\times10^{-3}\ (\mathrm{m\,K}) \tag{2P.3.14}$$

where λ_m is the wavelength for which $B_\lambda(T)$ is a maximum. Note that the units are metre kelvins, not millikelvins.

Equation (2P.3.14) is known as Wien's law, and it is shown by the broken line in figure 2P.7. A similar law,

$$\nu_m/T = 5.8807 \times 10^{10}\ (\mathrm{Hz\,K^{-1}}) \tag{2P.3.15}$$

where ν_m is the frequency for which $B_\nu(T)$ is a maximum, may be found from equation (2P.3.1), and is shown by the broken line in figure 2P.6.

Integrating equation (2P.3.1) or (2P.3.2) over frequency or wavelength and over a hemisphere leads to an expression for the total energy flux, $\mathscr{F}$, from a black body:

$$\mathscr{F} = \int_{\varphi=0}^{2\pi} \int_{\theta=0}^{\pi} \cos\theta \sin\theta \int_{\nu=0}^{\infty} B_\nu(T)\,\mathrm{d}\nu\,\mathrm{d}\theta\,\mathrm{d}\varphi \tag{2P.3.16}$$

$$= \pi \int_0^\infty \frac{2h\mu^2}{c^2} \frac{\nu^3}{\mathrm{e}^{h\nu/kT} - 1}\,\mathrm{d}\nu\,. \tag{2P.3.17}$$

Putting

$$x = h\nu/kT \tag{2P.3.18}$$

we get

$$\mathscr{F} = \frac{2\pi\mu^2 k^4 T^4}{h^3 c^2} \int_0^\infty \frac{x^3}{\mathrm{e}^x - 1}\,\mathrm{d}x \tag{2P.3.19}$$

and we may write this as

$$\mathscr{F} = \frac{2\pi\mu^2 k^4 T^4}{h^3 c^2} \int_0^\infty \frac{x^3 \mathrm{e}^{-x}}{(1 - \mathrm{e}^{-x})}\,\mathrm{d}x \tag{2P.3.20}$$

and then expand:

$$\mathscr{F} = \frac{2\pi\mu^2 k^4 T^4}{h^3 c^2} \int_0^\infty x^3 \mathrm{e}^{-x}(1 + \mathrm{e}^{-x} + \mathrm{e}^{-2x} + \ldots)\,\mathrm{d}x \tag{2P.3.21}$$

$$= \frac{2\pi\mu^2 k^4 T^4}{h^3 c^2} \sum_{n=1}^{\infty} \int_0^\infty x^3 \mathrm{e}^{-nx}\,\mathrm{d}x \tag{2P.3.22}$$

$$= \frac{2\pi\mu^2 k^4 T^4}{h^3 c^2} \sum_{n=1}^{\infty} \left[\mathrm{e}^{-nx}\left(\frac{-x^3}{n} - \frac{3x^2}{n^2} - \frac{6x}{n^3} - \frac{6}{n^4}\right)\right]_0^\infty \tag{2P.3.23}$$

$$= \frac{2\pi\mu^2 k^4 T^4}{h^3 c^2} \sum_{n=1}^{\infty} \frac{6}{n^4} \tag{2P.3.24}$$

$$= \frac{2\pi^5 k^4}{15 h^3 c^2} \mu^2 T^4\,. \tag{2P.3.25}$$

Equation (2P.3.25) is more commonly known as the Stefan–Boltzmann law. In SI units it takes the form

$$\mathscr{F} = 5.6696 \times 10^{-8} \mu^2 T^4 \ (\mathrm{W\,m^{-2}}). \qquad (2P.3.26)$$

The energy of a photon, E_ν, is given by

$$E_\nu = h\nu \qquad (2P.3.27)$$

and so from equation (2P.3.1) we may obtain an expression for the number of photons of frequency ν, $N_\nu(T)$, emitted per unit frequency interval by each unit area of a black body at a temperature T in unit time and into unit solid angle:

$$N_\nu(T) = 2\mu^2\nu^2/c^2(\mathrm{e}^{h\nu/kT} - 1) \qquad (2P.3.28)$$

and similarly

$$N_\lambda(T) = 2\mu^2 c/\lambda^4(\mathrm{e}^{hc/\lambda kT} - 1). \qquad (2P.3.29)$$

Radiation exerts a pressure upon the objects with which it interacts. In our normal terrestrial experience, this pressure is negligible, but in many astrophysical situations it may form a significant part of the total pressure, or even become the dominant component. Pressure is the rate of change of momentum per unit area, and we may find an expression for the momentum of a photon by an analogy; writing momentum as

$$p = mv \qquad (2P.3.30)$$

for a particle, we would then have

$$p = (E/c^2)c \qquad (2P.3.31)$$

for a photon, or

$$p = h\nu/c. \qquad (2P.3.32)$$

Now if we imagine a beam of photons striking a surface at an angle θ to its plane, the change in the momentum of a single photon after reflection is

$$\Delta p = 2(h\nu/c)\cos\theta \qquad (2P.3.33)$$

and so the radiation force is

$$F_{\mathrm{R}}(\nu) = 2(h\nu/c)\cos\theta N_\nu c \qquad (2P.3.34)$$

where N_ν is the number density of photons of frequency ν in the beam. For a light beam of cross-sectional area A and intensity I_ν the radiation pressure is thus

$$P_{\mathrm{R}}(\nu) = (2/c)I_\nu(\cos^2\theta)A^{-1}. \qquad (2P.3.35)$$

For black body radiation we have

$$I_\nu = B_\nu \qquad (2P.3.36)$$

and so the total pressure due to photons of frequency ν may be found by integrating equation (2P.3.35) over the hemisphere illuminating the surface:

$$P_R(\nu) = \int_{\text{hemisphere}} \frac{2}{c} I_\nu \cos^2\theta \, d\omega \tag{2P.3.37}$$

where $d\omega$ is the solid angle of the beam, and A has been taken to be unit area. Since black body radiation is isotropic, we have

$$P_R(\nu) = \frac{1}{c} \int_{\text{sphere}} I_\nu \cos^2\theta \, d\omega \tag{2P.3.38}$$

and so

$$P_R(\nu) = \frac{1}{c} B_\nu(T) \int_0^{2\pi} \int_0^{\pi} \cos^2\theta \sin\theta \, d\theta \, d\varphi \tag{2P.3.39}$$

$$= (4\pi/3c) B_\nu(T) . \tag{2P.3.40}$$

Finally, integrating over frequency gives the total radiation pressure (equations (2P.3.16) and (2P.3.26)):

$$P_R = (4\pi/3c) \frac{\mathscr{F}}{\pi} \tag{2P.3.41}$$

$$= (4/3c)(2\pi^5 k^4 \mu^2 / 15 h^3 c^2) T^4 \tag{2P.3.42}$$

$$= (4/3c)\sigma T^4 \tag{2P.3.43}$$

where σ is the Stefan–Boltzmann constant (equation 2P.3.26); or, using the radiation density constant (*7)

$$P_R = \tfrac{1}{3} a T^4 . \tag{2P.3.44}$$

In a non-black body radiation field, the radiation pressure must be found directly from equation (2P.3.37) for the particular function $I_\nu(\omega)$ involved.

2P.4 ENERGY TRANSPORT

From school physics we are familiar with the three possible forms of energy transport: conduction, convection, and radiation. Conduction is of little significance within an astrophysical context, except in degenerate matter. Radiation has largely been dealt with in the preceding section, and will be further considered in Chapter 3P, thus leaving only convection to be considered here.

In §2P.1 (equations 2P.1.39, 2P.1.43) we saw that adiabatic changes led to a pressure law of the form

$$P = K_1 T^{5/2} \tag{2P.4.1}$$

$$= K_2 \rho^{5/3} . \tag{2P.4.2}$$

If we imagine material moving deep inside a star sufficiently rapidly for the energy interchange between the moving element and its surroundings to be negligible, then conditions in the moving element will change adiabatically. For the motion to be unstable (i.e. for convection to occur) the density of the moving element must become less than that of its surroundings if its motion takes it upwards, or greater than that of its surroundings if moving downwards. Thus at the higher level for an upwardly moving element

$$\rho_{\text{element}} < \rho_{\text{star}} \tag{2P.4.3}$$

if convection is to occur. Since we have pressure equilibrium between the element and its surroundings, the perfect gas law (equation 2P.1.12) gives

$$T_{\text{element}} > T_{\text{star}} \tag{2P.4.4}$$

as the criterion for convection to occur. Hence the temperature gradient must be steeper inside the material of the star than in the moving element. Since the latter is changing adiabatically, we have

$$\left|\frac{\mathrm{d}T}{\mathrm{d}r}\right|_{\text{star}} > \left|\frac{\mathrm{d}T}{\mathrm{d}r}\right|_{\text{adiabatic}} \tag{2P.4.5}$$

for convection to occur. Since pressure decreases monotonically as r increases, and is equal within the element and its surroundings, this may also be written

$$\left.\frac{\mathrm{d}T}{\mathrm{d}P}\right|_{\text{star}} > \left.\frac{\mathrm{d}T}{\mathrm{d}P}\right|_{\text{adiabatic}}. \tag{2P.4.6}$$

But equation (2P.1.41) gives

$$\left.\frac{P}{T}\frac{\mathrm{d}T}{\mathrm{d}P}\right|_{\text{adiabatic}} = \frac{\gamma - 1}{\gamma} \tag{2P.4.7}$$

or

$$\left.\frac{\mathrm{d}\log T}{\mathrm{d}\log P}\right|_{\text{adiabatic}} = \frac{\gamma - 1}{\gamma}. \tag{2P.4.8}$$

So, finally, the condition for the occurrence of convection may most conveniently be written as

$$\left.\frac{\mathrm{d}\log T}{\mathrm{d}\log P}\right|_{\text{star}} > \left.\frac{\mathrm{d}\log T}{\mathrm{d}\log P}\right|_{\text{adiabatic}} \tag{2P.4.9}$$

$$> (\gamma - 1)/\gamma\,. \tag{2P.4.10}$$

Since γ has a value of $\frac{5}{3}$ for a fully ionised gas, we see that convection will occur inside a star when the temperature gradient becomes steeper than $\frac{2}{5}$, i.e.

$$\frac{\mathrm{d}\log T}{\mathrm{d}\log P} > 0.4\,. \tag{2P.4.11}$$

When convection occurs, the bulk of the energy transfer is due to its operation. Thus it is customary to speak of radiative and convective regions of stars depending upon which process predominates. In the convective regions the lapse rate is generally close to the adiabatic value (0.4). Since the radiative lapse rate is a function of the opacity (Chapter 3P), and this changes with the temperature and composition, different sections of a star may have different energy transport mechanisms, and these may change with time. Thus the Sun currently has three zones: radiative energy transport from the centre to about 0.75 $R_\odot$, convective energy transport from there to within a few hundred kilometres of the surface, followed by radiative energy transport again.

2P.5 POTENTIAL ENERGY

The only source of energy to play any significant part in the development of a star, other than nuclear energy, is the release of potential energy during gravitational contraction. This energy source is of particular importance during the formation of the star, and while the star is contracting down to white dwarf stage late in its life. The inverse of the release of gravitational potential energy is the kinetic energy required for the escape velocity. This is the minimum energy required for material to escape from the star's surface; it governs mass loss rates, etc, and is of importance throughout the star's whole life.

We may best describe the release of potential energy by considering this inverse problem of the amount of energy required to lift material against the star's gravitational field. In figure 2P.8, we have a small mass which is acted on by a force which is only infinitesimally larger than the gravitational force upon the mass. In effect

$$F = G\mathcal{M}m/r^2 \tag{2P.5.1}$$

where $\mathcal{M}$ is the mass of the star and the mass moves very slowly (quasi-stationary change) away from the star. In moving a distance Δr the work done is

$$F\Delta r \tag{2P.5.2}$$

so that in moving from a radius r to an infinite distance, the total work done by the force acting on the mass is

$$W = \int_r^\infty \frac{G\mathcal{M}m}{r^2}\,\mathrm{d}r \tag{2P.5.3}$$

$$= \frac{G\mathcal{M}m}{r}. \tag{2P.5.4}$$

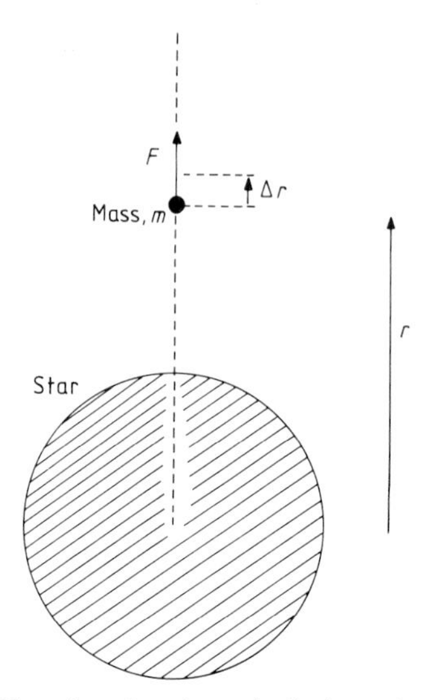

Figure 2P.8 Notation for the calculation of potential energy.

The potential energy released when a mass of one kilogram falls from infinity to a distance r from the star ($r \geqslant$ stellar radius) is therefore simply

$$\Omega = -G\mathcal{M}/r. \tag{2P.5.5}$$

Putting r equal to the star's radius, R, the escape velocity, v_e, is then obtained from

$$\tfrac{1}{2}mv_e^2 = G\mathcal{M}/R. \tag{2P.5.6}$$

$$v_e = (2G\mathcal{M}/R)^{1/2}. \tag{2P.5.7}$$

The released potential energy during a collapse must obviously reappear in other forms, and at least some as the thermal energy of the atoms of the collapsing material. If we consider a spherical layer of material, of radius r and mass $d\mathcal{M}(r)$, inside a star (figure 2P.9), then its potential energy is

$$-G\mathcal{M}(r)\, d\mathcal{M}(r)/r \tag{2P.5.8}$$

where $\mathcal{M}(r)$ is the mass of material inside the layer under consideration. The total potential energy of the star is therefore

$$\Omega = -\int_0^{\mathcal{M}} \frac{G\mathcal{M}(r)}{r}\, d\mathcal{M}(r). \tag{2P.5.9}$$

Now the thermal energy of an atom of a gas at temperature T is

$$\tfrac{3}{2}kT \tag{2P.5.10}$$

and we may relate the two energies as follows. We have inside a stable star the pressure differential across a shell multiplied by the shell's area equal to the gravitational force on that shell, so from figure 2P.9

$$-4\pi r^2 \, \mathrm{d}P = G\mathcal{M}(r) \, \mathrm{d}\mathcal{M}(r)/r^2 \qquad (2P.5.11)$$

or

$$\mathrm{d}P/\mathrm{d}r = -G\mathcal{M}(r)\rho(r)/r^2 \qquad (2P.5.12)$$

since

$$\mathrm{d}\mathcal{M}(r) = 4\pi r^2 \, \mathrm{d}r\rho(r). \qquad (2P.5.13)$$

Equation (2P.5.12) is often called the equation of hydrostatic equilibrium, and is a fundamental equation of stellar structure (Chapter 2A). The total internal kinetic energy (from 2P.5.10) is

$$\tau = \int_{\text{centre}}^{\text{surface}} \tfrac{3}{2} kT \, \mathrm{d}N \qquad (2P.5.14)$$

and since the number of particles in the shell, dN, is

$$\mathrm{d}N = 4\pi r^2 \, \mathrm{d}r\rho(r)/\bar{\mu}m_{\mathrm{H}} \qquad (2P.5.15)$$

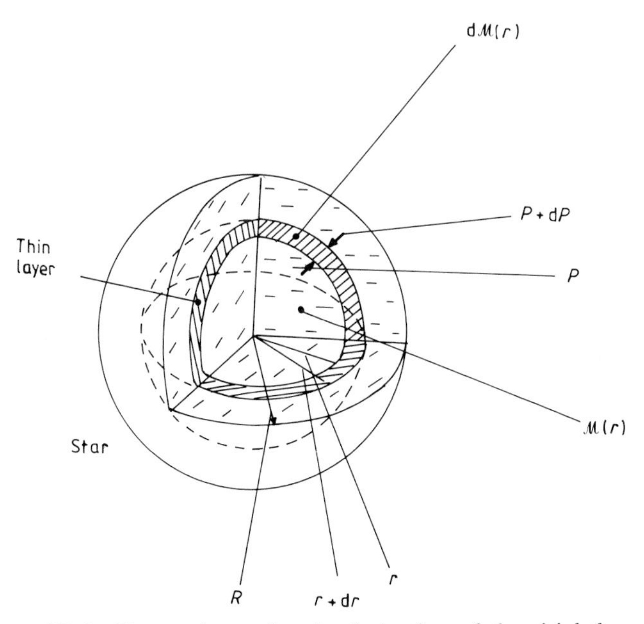

Figure 2P.9 Nomenclature for the derivation of the virial theorem.

we have

$$\tau = 6\pi \int_{\text{centre}}^{\text{surface}} \frac{\rho(r)}{\bar{\mu} m_{\text{H}}} kT(r) r^2 \, \mathrm{d}r \tag{2P.5.16}$$

$$= 6\pi \int_{\text{centre}}^{\text{surface}} P(r) r^2 \, \mathrm{d}r \tag{2P.5.17}$$

(using the perfect gas law, equation (2P.1.15)), giving

$$\tau = 6\pi \left([\tfrac{1}{3} r^3 P(r)]_{\text{centre}}^{\text{surface}} - \tfrac{1}{3} \int_{\text{centre}}^{\text{surface}} r^3 \, \mathrm{d}P \right) \tag{2P.5.18}$$

after integrating by parts. The first term of the right-hand side, however, is zero, since $r = 0$ at the centre, and $P \simeq 0$ at the surface. With the equation of hydrostatic equilibrium, we therefore have

$$\tau = 2\pi \int_{\text{centre}}^{\text{surface}} r^3 \frac{G\mathcal{M}(r)\rho(r)}{r^2} \, \mathrm{d}r \tag{2P.5.19}$$

and (2P.5.13) then gives

$$\tau = \tfrac{1}{2} \int_0^{\mathcal{M}} G\mathcal{M}(r)/r \, \mathrm{d}\mathcal{M}(r) \tag{2P.5.20}$$

and so, finally, from (2P.5.9)

$$\tau = -\tfrac{1}{2}\Omega \tag{2P.5.21}$$

or, as it is more commonly expressed,

$$2\tau + \Omega = 0 . \tag{2P.5.22}$$

In the latter form the equation is often known as the virial theorem for equilibrium configurations and it will be encountered in many astrophysical situations (Chapters 3A, 6A, 7A, 8A).

The relationship between the internal energy of a star (without other energy sources) and the energy released during collapse has two important implications. First the internal energy must rise as the collapse proceeds:

$$\bar{T} = \frac{\bar{\mu} m_{\text{H}}}{3k\mathcal{M}} \int_0^{\mathcal{M}} \frac{G\mathcal{M}(r)}{r} \, \mathrm{d}\mathcal{M}(r) \tag{2P.5.23}$$

where $\bar{T}$ is a mean temperature for the collapsing material. Secondly, only *half* the released energy goes into raising the temperature; the remainder must be disposed of in other ways. Usually this energy is radiated away, and provides a source of energy for proto-stars, white dwarf precursors, and (perhaps) Jupiter. Other loss mechanisms are possible however, for example the dissociation and ionisation of hydrogen in collapsing proto-stars.

At one time, collapse of the Sun was proposed as a source of its *current* energy release. The Sun is highly centrally condensed, so if we approximate it by a sphere of uniform density, but only one tenth of its present surface radius, then we may estimate the energy released to date by its collapse. Equation (2P.5.9) is then

$$\Omega = -\int_0^{\mathcal{M}_\odot} \frac{G\mathcal{M}(r)}{r}\,\mathrm{d}\mathcal{M}(r) \tag{2P.5.24}$$

with

$$\mathcal{M}(r) = \tfrac{4}{3}\pi r^3 \bar{\rho} \tag{2P.5.25}$$

and

$$\mathrm{d}\mathcal{M}(r) = 4\pi r^2 \bar{\rho}\,\mathrm{d}r \tag{2P.5.26}$$

so that

$$\Omega = -\int_0^{R_\odot/10} (G/r)\,\tfrac{4}{3}\pi r^3 \bar{\rho} 4\pi r^2 \bar{\rho}\,\mathrm{d}r \tag{2P.5.27}$$

$$= -\frac{16}{3} G\pi^2 \bar{\rho}^2 \int_0^{R_\odot/10} r^4\,\mathrm{d}r \tag{2P.5.28}$$

$$= -\frac{16}{15} G\pi^2 \bar{\rho}^2 \left(\frac{R_\odot}{10}\right)^5 \tag{2P.5.29}$$

and so the energy released

$$E = \frac{1}{2}\frac{16}{15} G\pi^2 \bar{\rho}^2 \left(\frac{R_\odot}{10}\right)^5 \tag{2P.5.30}$$

$$\simeq 10^{42}\ \mathrm{J}. \tag{2P.5.31}$$

Since the Sun radiates energy at a rate of 4×10^{26} W, and on geological evidence alone has done so for some 4.5×10^9 years, it has actually radiated over 5×10^{43} J. Hence gravitational energy could keep the Sun going for only 2% or so of its actually known lifetime, and nuclear energy sources (§2A.2) become the only suitable sources for stars' energies. Possible measured changes in the radius of the surface of the Sun over the last few years are discussed in §4A.1.

2A Bulk Properties of Stars

SUMMARY

Analytical and numerical methods of modelling stellar interiors, stellar mass limits in theory and practice, sources of stellar energy, nucleosynthetic reactions, energy transfer inside stars, stellar luminosities, diameters, and densities.

See also stellar masses (§1A.1), internal compositions of stars (§1A.2), physics of gases (§2P.1), equations of state (§2P.1), properties of degenerate material (§2P.1), nuclear reactions (§2P.2), energy transport (§2P.4), the virial theorem (§2P.5), stellar surface temperatures (§3A.3), variable stars (§3A.5), solar constant, luminosity and temperature (§4A.1), solar granulation (§4A.3), solar neutrino problem (§4A.4), and the formation and evolution of stars (§§8A.1, 8A.2).

INTRODUCTION

The interiors of stars, in spite of the potentialities of neutrino astronomy, remain inaccessible to direct study. The properties and behaviour of the material within stars must therefore be deduced from the laws of physics and the conditions observed at the surface of the star, together with a few global properties such as mass, size etc. The principal equations involved in this process are known as the equations of stellar structure, and in the simplest (static) case take the forms:

A The equation of hydrostatic equilibrium (equation 2P.5.12)

$$\frac{\mathrm{d}P(r)}{\mathrm{d}r} = -\frac{G\mathcal{M}(r)\rho(r)}{r^2}. \tag{2A.1}$$

B The equation of mass continuity (equation 2P.5.13)

$$\frac{\mathrm{d}\mathcal{M}(r)}{\mathrm{d}r}=4\pi r^2\rho(r). \tag{2A.2}$$

C The flux equation

$$\frac{\mathrm{d}\mathcal{F}(r)}{\mathrm{d}r}=4\pi r^2\rho(r)\varepsilon(r) \tag{2A.3}$$

where $\mathcal{F}(r)$ is the total energy flux inside the star at radius r, and $\varepsilon(r)$ is the energy generation rate at radius r.

D The temperature gradient:

$$\frac{\mathrm{d}T(r)}{\mathrm{d}r}=\frac{-3\mathcal{F}(r)\varkappa(r)\rho(r)}{64\pi\sigma r^2T(r)^3} \tag{2A.4}$$

for radiative regions and

$$\frac{\mathrm{d}T(r)}{\mathrm{d}r}=\frac{\gamma-1}{\gamma}\frac{T(r)}{P(r)}\frac{\mathrm{d}P(r)}{\mathrm{d}r} \tag{2A.5}$$

for convective regions, where $\varkappa(r)$ is the mass absorption coefficient, $T(r)$ is the temperature, σ is Stefan's constant and γ is the ratio of the specific heats.

E The pressure law

Perfect gas law (equation 2P.1.15), radiation pressure (equations 2P.3.43, 2P.3.44), degeneracy pressure (equations 2P.1.74, 2P.1.76, 2P.1.77), magnetic pressure etc as required.

F The energy generation equation

Appropriate energy generation rates for the nuclear processes expected to be occurring (§2A.2, equations 2A.2.6, 2A.2.7, 2A.2.22 etc) as required.

G The opacity law

Bound–free, free–free transitions, electron scattering etc as required (§3P.1).

These seven sets of equations must then be solved simultaneously and consistent with the known surface and centre boundary conditions, for the seven primary unknowns: $P(r)$, $\mathcal{M}(r)$, $\rho(r)$, $\mathcal{F}(r)$, $T(r)$, $\varkappa(r)$ and $\varepsilon(r)$ to produce a model for the interior of the star. Often the mass is used as the variable instead of the radius, the manner of the variation of the radius with mass then becoming one of the unknowns to be determined. This procedure has the advantage of improved relevance to normal requirements since models are usually determined for a particular mass of a star rather than for a particular radius.

Analytically, only very general upper or lower limits obtained from very rough approximations are possible. Thus the assumption of constant density for the interior of the Sun leads to a lower limit of $1.3 \times 10^{14}\ \mathrm{N\,m^{-2}}$ for its central pressure, compared with the current best estimate of $3.4 \times 10^{16}\ \mathrm{N\,m^{-2}}$. Rather more sophisticated treatment results from the assumption of polytropic behaviour for the material (equation 2P.1.43). Models may then be obtained by fairly simple numerical techniques and are useful in describing the convective and degenerate regions of stars.

Most modern work, however, requires the problem to be attacked without any such prior assumptions, and is based upon the direct numerical solution of the equations via powerful computers. The details of the computer programming which this involves are beyond the scope of this book. They are in any case generally much more concerned with optimising the available computer time, and with minimising the numerical errors, than with the physics and astrophysics of the situation. The reader interested in this aspect of the problem is referred to the bibliography, and to the research literature, for further information. The outlines of the two main approaches to the numerical modelling of stellar interiors are given below. The results are to be found throughout this book, particularly and explicitly in Chapters 2A, 4A and 8A, but also in providing much of the foundation for other chapters.

Numerical modelling of stars normally adopts a variation on one of two main methods of approach to the problem: the stepwise method or the relaxation method. The first of these is based on the first-order Taylor expansion

$$y(\Delta x) = y(0) + \frac{\mathrm{d}y(0)}{\mathrm{d}x}\,\Delta x \tag{2A.6}$$

where the boundary conditions specify the initial values, $y(0)$ and $\mathrm{d}y(0)/\mathrm{d}x$. The differential equation,

$$\mathrm{d}y/\mathrm{d}x = f(x, y) \tag{2A.7}$$

represents one or more of the equations of stellar structure (equations 2A.1 to 2A.4, etc). The algorithm provides an estimate of the value of y at a point, Δx, away from the boundary (normally the centre or surface of the star). The value of $\mathrm{d}y(\Delta x)/\mathrm{d}x$ can then be estimated, and the solution continued to a point $2\Delta x$ away from the boundary, and so on until the whole star has been covered. The procedure suffers from truncation errors due to the curtailment of the Taylor expansion. These may be reduced by reducing the size of the steps (Δx), but this then exacerbates the rounding errors involved in each stage of the calculation. Improved variants of the process, such as the Runge–Kutta method, are therefore employed in practice. The modelling is also likely to be started from both boundaries, and the results fitted together at an intermediate point between the boundaries. Normally, the intermediate values obtained from each boundary will differ from each other. The mis-match can then be used to alter the initial parameters, and the process repeated until a good fit throughout the star has been obtained. By an extension of this method, the changes in the model with time (due to evolution, or because the star is a variable) may be determined. The initial (successful) model is used to suggest time derivatives for the various parameters and the boundary conditions which will obtain a short interval of time later on. A new model is then produced for those conditions. This may then be used to predict derivatives and boundary conditions another short interval later, and so on. The method gives good results for main sequence stars, whose long lifetimes enable the initial time derivatives to be taken as zero, but is less satisfactory for more rapid changes.

The alternative and more recently developed approach to the solution of the equations of stellar structure is known as the relaxation method, since the solution 'relaxes' by successive approximations towards the correct values. The differential equations are applied to an initial assumed structure for the star. In general they will fail to be satisfied. The model is then corrected at each point by a process which is based upon the magnitude of the deviations of the equations at that point. The new model then has the differential equations applied to it, and is corrected, if required, in its turn. For a function y and variable x with the initial model digitised at preset values of x, the trial function will be

$$[y^0(x+\Delta x) - y^0(x)] - \frac{dy^0(x+\Delta x/2)}{dx}\Delta x \tag{2A.8}$$

where the zero superscript indicates a value from the initial model; x and $x+\Delta x$ are consecutive digitisation points of the model. The differential equation

$$dy/dx = f(x, y) \tag{2A.9}$$

represents one or more of the equations of stellar structure (equations 2A.1 to 2A.4 etc). For the correct solution, this should have a zero value throughout the model. To correct the initial model values, a version of Newton–Raphson iteration is employed. The process is repeated until the corrections become negligible, the solution then being taken as the final model. The method may be extended to cover periodic variables such as the Cepheids (§3A.5). A stable periodic solution is then sought by relaxing in time as well as space.

2A.1 MASS LIMITS

In a remarkable demonstration of the power of abstract thought (at least when the answer is known anyway!), Eddington in his book *The Internal*

Constitution of the Stars showed that a study of the balance between gravitational forces and radiation pressure (equations 2P.1.15, 2P.3.43) in gas spheres might lead a scientist on a completely cloud-bound planet to suspect the existence of stars. He pointed out that as such a scientist considered spheres of ever increasing mass, he would find that gas pressure dominated at low masses, and radiation pressure at high masses. Only over a restricted range of masses, 10^{30} to 10^{32} kg, would there be any comparability of the two. In this range our hypothetical scientist might expect something to happen, and to quote Eddington (*ibid.*), 'what "happens" is the stars'. We may follow his argument roughly by assuming uniform densities for the spheres. The mass inside radius r is then

$$\mathcal{M}(r) = \tfrac{4}{3}\pi r^3 \bar{\rho} \tag{2A.1.1}$$

or

$$r = (3/4\pi)^{1/3} \bar{\rho}^{-1/3} \mathcal{M}(r)^{1/3}. \tag{2A.1.2}$$

Equation (2P.5.22) then gives us an estimate of the mean temperature:

$$\bar{T} = \frac{\bar{\mu} m_{\mathrm{H}} G}{2k\mathcal{M}} \int_0^{\mathcal{M}} \mathcal{M}(r) r^{-1} \, \mathrm{d}\mathcal{M}(r) \tag{2A.1.3}$$

$$= K \bar{\rho}^{1/3} \mathcal{M}^{2/3} \tag{2A.1.4}$$

where $\mathcal{M}$ is the mass of the body, and

$$K = \left(\frac{4\pi}{3}\right)^{1/3} \frac{\bar{\mu} m_{\mathrm{H}} G}{3k}. \tag{2A.1.5}$$

The mean gas pressure, from equation (2P.1.15), is then

$$\bar{P}_{\mathrm{G}} = nk\bar{T} \tag{2A.1.6}$$

$$= (\bar{\rho}/\bar{\mu} m_{\mathrm{H}}) kK \bar{\rho}^{1/3} \mathcal{M}^{2/3} \tag{2A.1.7}$$

$$= (k/\bar{\mu} m_{\mathrm{H}}) K \bar{\rho}^{4/3} \mathcal{M}^{2/3} \tag{2A.1.8}$$

while the mean radiation pressure from equation (2P.3.43) is

$$\bar{P}_{\mathrm{R}} = \tfrac{1}{3} a \bar{T}^4 \tag{2A.1.9}$$

$$= \tfrac{1}{3} a K^4 \bar{\rho}^{4/3} \mathcal{M}^{8/3} \tag{2A.1.10}$$

and the ratio of the pressure is

$$\frac{\bar{P}_{\mathrm{R}}}{\bar{P}_{\mathrm{G}}} = \frac{aK^4}{3} \frac{\mu m_{\mathrm{H}}}{kK} \mathcal{M}^2 \tag{2A.1.11}$$

$$= 5 \times 10^{-64} \mathcal{M}^2 \tag{2A.1.12}$$

for $\mathcal{M}$ in kg. Thus the radiation and gas pressures are equal for a mass of gas of about 4×10^{31} kg (about 20 solar masses). More precise calculations give the pressure equality at about 10^{31} kg (about 5 solar masses), and the two pressures within a factor of ten of each other over the range 2×10^{29} to 2×10^{32} kg (figure 2A.1). This range is from about 0.1 to 100 solar masses, and this is precisely the observed range of stellar masses (figure 1A.10).

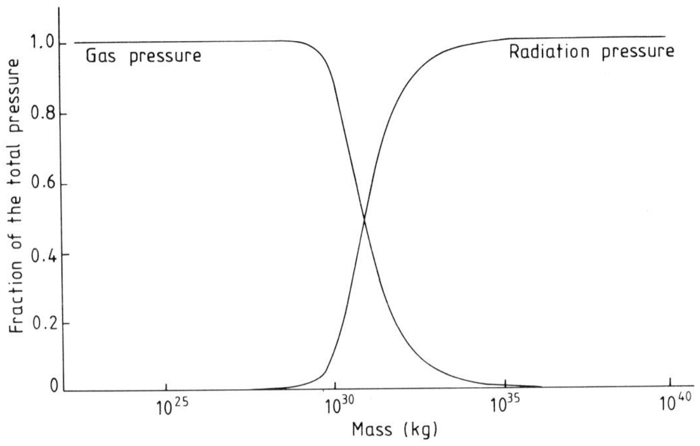

Figure 2A.1 Balance between gas and radiation pressure in gas spheres of various masses. (After A S Eddington (1930), *The Internal Constitution of the Stars*, Cambridge: Cambridge University Press.)

Although Eddington's comment obviously has significant implications for the understanding of the formation of stars, it does not of itself explain why stars should be limited to this mass range. However, Eddington also suggested that a limit to the luminosity that was possible for a star might exist. This is now known as Eddington's limit, and is the point at which radiation pressure and gravity balance each other at the star's surface. Both gravity and radiation pressure follow inverse square laws, so that if the latter exceeds the former at the star's surface, then the material will experience a net repulsion at all distances from the star. Luminosities higher than the limit would therefore lead to rapid loss of material from the star. This limit to the star's luminosity is

$$L_* = 10^{31}(\mathcal{M}_*/\mathcal{M}_\odot) \qquad \text{(W)} \tag{2A.1.13}$$

where $\mathcal{M}_*$ is the star's mass and $\mathcal{M}_\odot$ is the Sun's mass.

However, there is also an approximate empirical relationship between stellar masses and luminosities (equation 2A.4.4) of the form for high-mass stars

$$L_* \propto \mathcal{M}^{2.7}. \tag{2A.1.14}$$

Thus, using

$$L_\odot = 4 \times 10^{26}\ \mathrm{W} \tag{2A.1.15}$$

we have

$$\frac{L_*}{L_\odot} = 2.5 \times 10^4 \left(\frac{L_*}{L_\odot}\right)^{1/2.7} \tag{2A.1.16}$$

and so

$$L_* \simeq 10^7 L_\odot \tag{2A.1.17}$$

is the greatest possible luminosity for a star. From equation (2A.1.13) we therefore have

$$\mathcal{M}_* \simeq 400 \mathcal{M}_\odot \tag{2A.1.18}$$

as the greatest mass possible for a star before it may be disrupted by radiation pressure. In practice, other factors such as turbulence, surface activity, rotation, etc will cause instability to set in before the Eddington limit. Main sequence stars, for example, of over about 60 $\mathcal{M}_\odot$ are expected to be unstable to radial pulsations driven by the energy generating reactions in their cores, though this is not likely to disrupt the star in less than its main sequence lifetime. The maximum possible stellar mass is thus reduced to 150 to 200 solar masses. This is in good agreement with the observations (Chapter 1A), since the most massive stars found to date are of about 120 $\mathcal{M}_\odot$ (for example, HD 93250: 120 $\mathcal{M}_\odot$, η Car: 115 $\mathcal{M}_\odot$, Plaskett's star: about 100 $\mathcal{M}_\odot$). A possible exception may occur for extremely high masses. Relativistic effects then may permit the existence of stars of 1000 $\mathcal{M}_\odot$ or more, and these are sometimes postulated as existing in the centres of galaxies, quasars etc, but have yet to be detected with certainty.

The lower limit to the masses of stars (*1) arises from the central temperature being insufficient to initiate nuclear reactions. The proton–proton chain (equation 2A.2.6) generates energy at a rate which varies with temperature roughly according to

$$E_{\mathrm{pp}} \propto \exp(-3381/T^{1/3}). \tag{2A.1.19}$$

This is a very high sensitivity to temperature: for a central temperature of 10^6 K, the reaction occurs 10^9 times more slowly than in the centre of the Sun (central temperature 15.5×10^6 K), and has effectively therefore ceased to supply the 'star' with energy (*1). The central temperature varies with mass as shown in figure 2A.2, based upon theoretical stellar models. A central temperature of 10^6 K occurs for stars of about 0.1 to 0.2 solar masses, and this gives the lower limit to the mass possible for a star. More precise work leads to the estimate of 0.085 $\mathcal{M}_\odot$ for the lower limit for stars with solar compositions. Such very low mass stars are probably quite rare, however, since the proportion of observed stars decreases sharply as the luminosity becomes fainter than an absolute magnitude (§3A.1) of about + 14, and this corresponds to a mass of about a tenth of a solar mass. At the bottom end of this range only the first two stages of the proton–proton cycle are thought to occur, and so the star will accumulate $^3_2$He in its interior, rather than the normal isotope of helium.

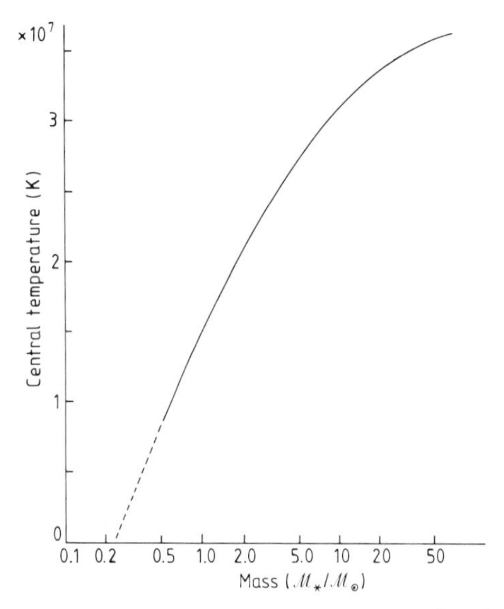

Figure 2A.2 Variation of the central temperatures of main sequence stars with mass.

The actual mass range of stars is thus from about 0.1 to 100 solar masses (2×10^{29} to 2×10^{32} kg), and covers almost precisely the region over which the total pressure inside a body contains significant contributions from both gas and radiation pressures.

Stars in which degeneracy (§2P.1) is important have much more stringent limits to their masses. The virial theorem in the form derived earlier (equation 2P.5.21) for a spherical object of radius R and uniform density and temperature is

$$3PV - \frac{3}{5}\frac{G\mathcal{M}^2}{R} = 0 \qquad (2A.1.20)$$

and the pressure in a relativistic electron-degenerate material is given by equation (2P.1.76):

$$P_{\rm D} = \left(\frac{3h^3c^3}{256\pi g m_{\rm H}^4 \mu_{\rm e}^4}\right)^{1/3} \rho^{4/3} \tag{2A.1.21}$$

so that writing

$$\rho = \mathcal{M}/\tfrac{4}{3}\pi R^3 \tag{2A.1.22}$$

we get

$$3\left(\frac{3h^3c^3}{256\pi g m_{\rm H}^4 \mu_{\rm e}^4}\right)^{1/3}\left(\frac{\mathcal{M}}{\frac{4}{3}\pi R^3}\right)^{4/3} \tfrac{4}{3}\pi R^3 = \tfrac{3}{5}\frac{G\mathcal{M}^2}{R} \tag{2A.1.23}$$

or

$$2.30 \times 10^{10} \mu_{\rm e}^{-4/3} \mathcal{M}^{4/3} R^{-4} R^3 = 4.00 \times 10^{-11} \mathcal{M}^2 R^{-1} \tag{2A.1.24}$$

thus giving

$$5.75 \times 10^{20} \mu_{\rm e}^{-4/3} = \mathcal{M}^{2/3}. \tag{2A.1.25}$$

For helium and the heavier elements

$$\mu_{\rm e} \simeq 2 \tag{2A.1.26}$$

and so

$$\mathcal{M} = 3.44 \times 10^{30}\ \text{kg} \tag{2A.1.27}$$

$$= 1.72 \mathcal{M}_\odot. \tag{2A.1.28}$$

A more precise calculation gives

$$\mathcal{M} = 1.44 \mathcal{M}_\odot. \tag{2A.1.29}$$

Other values between about 1 and 1.4 solar masses may result when due allowance is made for actual compositions. This mass limit, first calculated by Chandrasekhar and now named after him, is the largest mass which may be supported against gravity by the pressure in a relativistically electron-degenerate material. It is thus also the largest mass possible for a white dwarf star (*2, and §§2A.5, 3A.6). Since degeneracy pressure increases with increasing density, the larger the mass of a white dwarf, the smaller its size. The Chandrasekhar limit represents the point at which the radius has reduced to zero (figure 2A.3).

The collapse of a condensed object over the Chandrasekhar limit may be halted by baryon degeneracy pressure (equation 2P.1.74), resulting in the formation of a neutron star (§3A.5). However, there is a mass limit for neutron stars in a similar fashion to that of white dwarfs, and its value is estimated to lie between 1.5 and 2 solar masses. Above this latter limit, the collapse may continue indefinitely, and the object will become a black hole (§3A.9). Neutron stars of very low masses ($< 0.1\mathcal{M}_\odot$) are unstable and must undergo oscillations.

These various mass limits translate into size limits of

$$R_* < 15\,000 \text{ km} \qquad \text{for white dwarfs}$$
$$10 \text{ km} < R_* < 200 \text{ km} \quad \text{for neutron stars}$$

and mean density ranges of

$$2 \times 10^7 \text{ to } 4 \times 10^{10} \text{ kg m}^{-3} \quad \text{for white dwarfs}$$
$$10^{13} \text{ to } 5 \times 10^{16} \text{ kg m}^{-3} \qquad \text{for neutron stars}$$

the central densities reaching perhaps 10^{12} kg m^{-3} in massive white dwarfs, and 10^{18} kg m^{-3} in massive neutron stars.

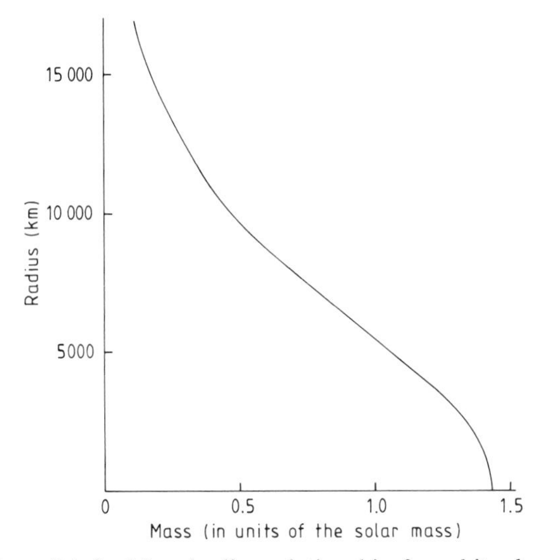

Figure 2A.3 Mass/radius relationship for white dwarfs.

These figures are quite outside the range of human experiences. A couple of anecdotal examples of their implications are therefore presented to try to help to place them into context. If an ordinary matchbox (roughly 30 ml in volume) were to be filled with average white dwarf matter, its weight on Earth would be about 30 tonnes, while a similar matchbox filled with average neutron star material would weigh some twenty million tonnes. The material from the *centre* of a neutron star would give a weight of thirty thousand million tonnes, compared with the actual normal weight of a filled matchbox of about fifteen grams! The gravitational forces at the surfaces of these objects are such that a 'mountain' requiring the same effort to climb as does Everest on the Earth would be 0.1 m high on a white dwarf, and 1 μm high on a neutron star.

2A.2 ENERGY SOURCES

We have already used the idea that the energy sources for most stars are nuclear fusion reactions, and have shown that other sources such as potential energy (equation 2P.5.29), and chemical energy (§2P.3) are insufficient for the purpose. Here, the details of these reactions, their variations from one star to another, and within one star during its lifetime are examined.

The problem of the sources of stars' energies in general, and of the Sun's energy in particular, has exercised astrophysicists and, before them, astronomers for many centuries. Potential and chemical energy release have already been discussed. Other suggestions have included a deluge of meteorites hitting the surface, and a variety of electrical phenomena. Even the possibility that the solar heat and radiation were effects within our own atmosphere, induced by the astronomical object that we call the Sun, was considered. Some, at least, of these were realistic in terms of the time scales of a few thousand to a few million years which were variously thought necessary in the past to permit the existence of mankind and of the Earth. However, as the properties of radioactive elements became understood in the first decades of this century, it quickly became apparent that geological time scales were at least hundreds of millions, and perhaps thousands of millions of years long. Fossils of creatures 500 or 600 million years old could be found which appeared to require conditions on the Earth not very different from those at present. The Sun must also therefore have been reasonably similar to its present form over the same length of time. Indeed, we now place the ages of the Earth and of other bodies in the solar system at about 4.55 thousand million years, with at least a comparable age for the Sun in its present form. None of the previously hypothesised energy sources are capable of supplying the solar energy for more than a very small fraction of such a period.

As these problems over the time scales for 'conventional' energy sources became apparent, many scientists came to believe that nuclear energy in some form must be the true source of the Sun's energy. However, no plausible suggestion for the actual mechanism was found until 1939. Then Hans Bethe showed that a cycle of reactions could occur at the conditions expected for the solar interior, which would lead to the conversion of hydrogen to helium. This, as already seen (§2P.2) would release about 6×10^{14} J of energy for every kg of hydrogen so converted. Bethe's process became known as the C–N–O cycle since it involves carbon, nitrogen, and oxygen in addition to hydrogen and helium (see below). It is now thought that the C–N–O cycle is relatively unimportant for stars of less than two solar masses, and that another route for the conversion of hydrogen into helium, the proton–proton chain (see below), produces the energy of most stars. In either case, the production of the energy currently radiated away by the Sun, 4×10^{26} W, requires the conversion of about 6×10^{11} kg of hydrogen into helium each second. If the initial Sun were composed solely of hydrogen, then its mass of 2×10^{30} kg would place an upper limit on its lifetime of about 3×10^{18} s (10^{11} years). In fact, even at formation, the Sun had only about 74% of its mass in the form of hydrogen (§1A.2), and only a fraction of this will be subject to the temperatures and pressures which enable the nuclear reactions to occur. Thus a more realistic figure for the lifetime of the Sun in more or less its present form is a factor of ten down on this maximum estimate; that is about 10^{10} years. Even so, this is more than adequate to cover the present age of the solar system of 4.55×10^9 years. Indeed, we may confidently expect a further five thousand million years to pass before the Sun undergoes any radical changes in its nature (*3).

The details of the two sets of reactions which convert hydrogen into helium are shown in figures 2A.4 and 2A.5. In the Sun, the mainstream reactions, producing about 91% of the energy, are the first pathway through the proton–proton chain:

$$p^+ + p^+ \rightarrow {}^2_1H + e^+ + \nu_e \qquad (2A.2.1)$$

$$ {}^2_1H + p^+ \rightarrow {}^3_2He + \gamma \qquad (2A.2.2)$$

$$ {}^3_2He + {}^3_2He \rightarrow {}^4_2He + p^+ + p^+ \qquad (2A.2.3)$$

(the relative probabilities of the reactions for solar conditions are shown on the figures). The total energy released by the reactions, including that coming from the subsequent annihilation of the positrons, is 26.8 MeV (4.30×10^{-12} J). The neutrinos which are produced in some of the reactions, however, can carry off differing amounts of energy. This is lost to the Sun since very few indeed of these neutrinos subsequently interact with any of the solar material (§§2P.2, 4A.4). The net gains of energy from the various pathways are therefore

26.0–26.8 MeV	Basic proton–proton pathway
24.9–25.4 MeV	Basic pathway starting with one p–e–p reaction
25.5–26.8 MeV	Pathway via ${}^7_4Be$ and ${}^7_3Li$
12.4–26.8 MeV	Pathway via ${}^7_4Be$ and ${}^8_5B$
21.8–26.8 MeV	Basic C–N–O cycle
21.3–26.8 MeV	C–N–O cycle via ${}^{17}_9F$.

The mean energies contributed to the Sun are thus 26.2 MeV for the proton–proton cycle, and 25.0 MeV for the C–N–O cycle.

The rates at which these reactions occur depend mainly upon the temperature, and are given to a good degree of approximation by

$$N_{pp} = \frac{5.63 \times 10^{14} X^2 \rho}{T^{2/3} \exp(3381/T^{1/3})} \ (\mathrm{kg^{-1}\,s^{-1}}) \qquad (2A.2.4)$$

$$N_{CNO} = \frac{1.80 \times 10^{36} X Z_N \rho}{T^{2/3} \exp(15231/T^{1/3})} \ (\mathrm{kg^{-1}\,s^{-1}}) \qquad (2A.2.5)$$

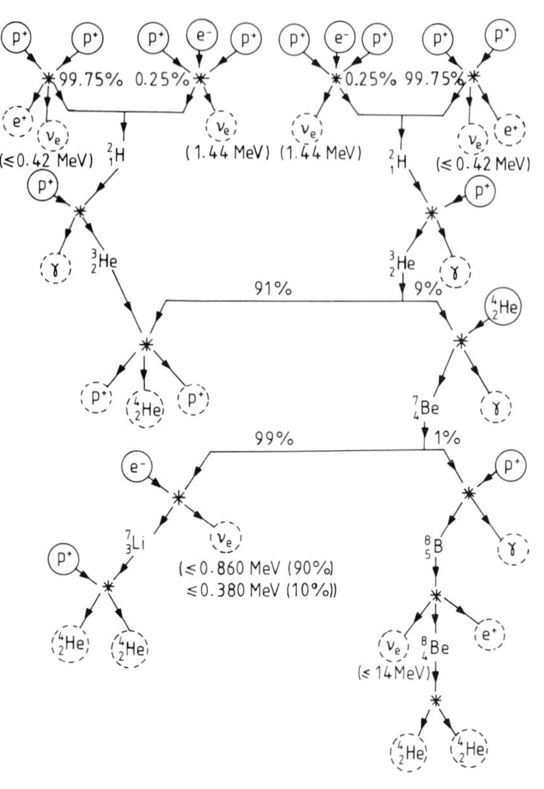

Figure 2A.4 The proton–proton chain and its variants. Particles consumed in the reactions are encircled with a solid line, final products with a broken line. Neutrino energies are shown in parentheses.

where N_{pp} and N_{CNO} are the number of conversions of hydrogen to helium per second, for the proton–proton and C–N–O reactions respectively, and Z_{N} is the mass fraction of $^{14}_{7}\mathrm{N}$. The corresponding energy generation rates are therefore

$$E_{\mathrm{pp}} = \frac{2.36 \times 10^{3} X^{2} \rho}{T^{2/3} \exp(3381/T^{1/3})} \ (\mathrm{W\,kg^{-1}}) \tag{2A.2.6}$$

$$E_{\mathrm{CNO}} = \frac{7.21 \times 10^{24} X Z_{\mathrm{N}} \rho}{T^{2/3} \exp(15231/T^{1/3})} \ (\mathrm{W\,kg^{-1}}) \tag{2A.2.7}$$

where E_{pp} and E_{CNO} are the energy generation rates for the proton–proton and C–N–O reactions, respectively. The ratio of the contributions from these two reaction routes to the total energy of the star is thus

$$E_{\mathrm{pp}}/E_{\mathrm{CNO}} = 3.27 \times 10^{-22} \frac{X}{Z_{\mathrm{N}}} \exp(11850/T^{1/3}). \tag{2A.2.8}$$

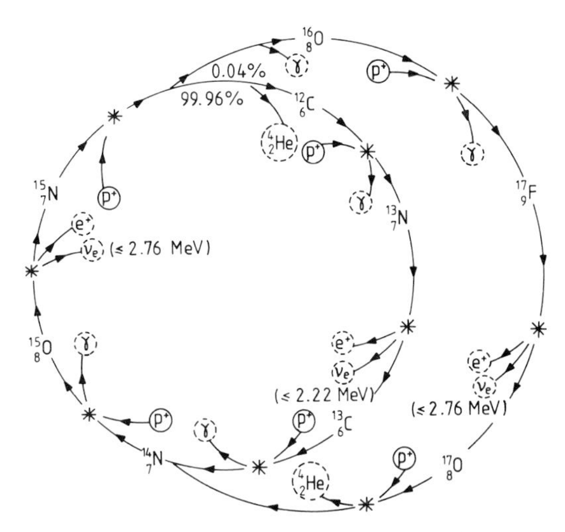

Figure 2A.5 The CNO cycle. Particles consumed in the reactions are encircled with a solid line, final products with a broken line. Neutrino energies are shown in parentheses.

For the centre of the Sun, we have

$$X \simeq 0.360 \tag{2A.2.9}$$

$$Z_{\mathrm{N}} \simeq 0.0116 \tag{2A.2.10}$$

$$T \simeq 1.55 \times 10^7 \mathrm{~K} \tag{2A.2.11}$$

giving

$$(E_{\mathrm{pp}}/E_{\mathrm{CNO}})_{\mathrm{Sun}} = 4.4\,. \tag{2A.2.12}$$

The Sun's energy, however, is mostly generated within a region about 0.2 $R_\odot$ across, and the temperature towards the edge of this energy-generating zone falls towards 9.5×10^6 K, when

$$(E_{\mathrm{pp}}/E_{\mathrm{CNO}})_{\mathrm{Sun}} = 2 \times 10^4\,. \tag{2A.2.13}$$

The actual proportion of the solar energy generated by the C–N–O cycle is therefore estimated at about 5%. The very strong dependence upon temperature of equation (2A.2.8) means that normally either the proton–proton chain or the C–N–O cycle predominates (figure 2A.6). The proton–proton chain is dominant for temperatures under about 16×10^6 K, which occur in stars with masses lower than about 1.2 $\mathcal{M}_\odot$. The C–N–O cycle dominates at temperatures over 20×10^6 K (stars with masses greater than about 2.0 $\mathcal{M}_\odot$). The total energy generated is also very sensitive to temperature (equations 2A.2.6, 2A.2.7), and so only a small range of central temperatures of 10^6 to 3.5×10^7 K is required for stars in their hydrogen-burning phase, even though the luminosity range for those stars is very large; there being a factor of 10^9 or more between the faintest and the brightest.

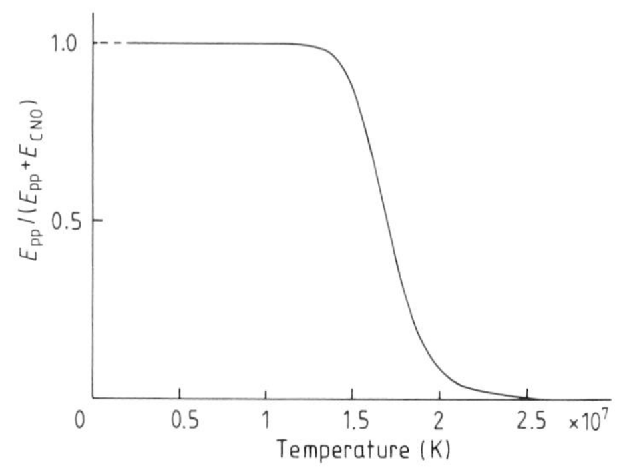

Figure 2A.6 Proportion of energy produced by the proton–proton chain and CNO cycle as a function of temperature (equation 2A.2.8) for the composition of the centre of the Sun.

The source of most stars' energy thus seemed well established two decades ago. A problem, however, has since arisen which casts some doubt on our insight into processes occurring inside the Sun, and by implication inside other stars as well. This is the solar neutrino problem (§4A.4). In essence, the problem is that too few neutrinos are observed to be coming from the Sun; the discrepancy being by about a factor of three and well outside the error limits of the observations. Amongst the possible explanations for the problem are:

(*a*) Non-zero rest mass for the neutrino. This would imply the possibility of interchange between the three types of neutrino (electron neutrino, muon neutrino and tauon neutrino) and reduce the expected neutrino flux by a factor of three since only the electron neutrinos are detected.

(*b*) Significant mixing of the material at the centre of the Sun. More massive stars are expected to have convective cores (§8A.2), but to date solar models predict radiative transport of energy out to 75% of the solar radius, so that the centre should remain unmixed.

(*c*) Initial helium abundance by mass much less than 24%.

(*d*) Sun much younger than 4.55×10^9 years.

(*e*) Solar composition initially inhomogeneous and/or the surface abundances altered by recent accretion of interstellar material.

(*f*) Errors in the nuclear and neutrino interaction cross sections.

(*g*) Variations in the rates of the nuclear reactions with time, perhaps due to transient convective mixing. The energy presently radiated by the Sun has been produced over the last ten million years, and therefore represents a long average energy generation rate. The neutrinos reflect the instantaneous energy generation rate since they take only eight minutes to reach the Earth.

(*h*) Magnetic fields, black holes etc at the centre of the Sun.

At the time of writing (1985), which of these possibilities, if any, provides the explanation for the solar neutrino problem remains to be established. Should any except the first turn out to be correct, then the implications for stellar modelling and for astrophysics in general may be quite extensive.

Stars burning hydrogen at their centres form more than 90% of the observed stars and are called main sequence stars (Chapter 3A). Now we shall see (§2A.4) that such stars exhibit a relationship between their luminosities and masses, and that it takes the form (equation 2A.4.1)

$$L \propto \mathcal{M}^n \tag{2A.2.14}$$

where the index n takes the values

3.9	$\mathcal{M}_* < 7\mathcal{M}_\odot$
3.0	$7\mathcal{M}_\odot < \mathcal{M}_* < 25\mathcal{M}_\odot$
2.7	$\mathcal{M}_* > 25\mathcal{M}_\odot$.

For the Sun, the length of this hydrogen-burning stage, $T_H(*)$ is about 10^{10} years, so that for other stars, assuming similar structures and processes, it is

$$T_H(*) = T_H(\odot)(\mathcal{M}_*/\mathcal{M}_\odot)(L_\odot/L_*) \tag{2A.2.15}$$

whence, using equation (2A.2.14), we have

$$T_H(*) = 10^{10}(\mathcal{M}_*/\mathcal{M}_\odot)^{1-n} \quad \text{(years).} \tag{2A.2.16}$$

The more massive stars therefore have much shorter main sequence lifetimes than those of lower masses (figure 2A.7).

Most of the non-main sequence stars are white dwarfs (§§2A.5, 3A.6). Their energy output arises from the radiation of stored thermal energy, with little of significance in the way of nuclear reactions taking place. The small remaining fraction, perhaps 1% to 2% of all stars, may have alternative nuclear reactions to the conversion of hydrogen to helium as their energy sources. That this must be the case is apparent from the increasing abundance of the heavier elements with time (figure 1A.12). Reactions to produce those elements must therefore be occurring somewhere, and stellar interiors are prime candidates for the sites of many of the required processes, the elements being redistributed into the interstellar medium by stellar winds, novae, supernovae etc. The processes hypothesised for the production of the elements give a guide to, but not a precise explanation of, their observed abundances (figure 1A.11). Their study is the concern of the subject entitled nucleosynthesis.

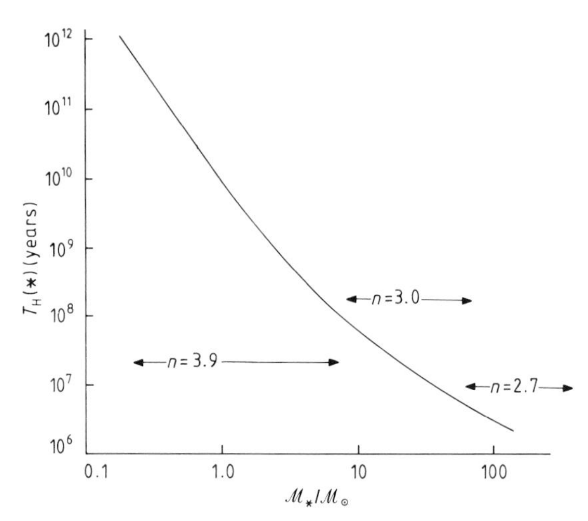

Figure 2A.7 Lengths of the hydrogen-burning stages of stars ('main sequence'). *n* is the index in the empirical mass/luminosity relation.

The first stage of nucleosynthesis, as just discussed, is the formation of helium from hydrogen. The Sun has currently reduced its initial 75% hydrogen by mass to about 37% at its centre by this process. In a further 5×10^9 years or so, all the hydrogen at the centre of the Sun will have been converted to helium. All other main sequence stars will similarly run out of hydrogen at their centres in greater or lesser times and with slight differences of detail according to whether or not convection is occurring at the centre and causing mixing. The formation of this almost pure helium core marks the end of the star's main sequence lifetime. Hydrogen to helium conversion will still continue around the boundary of the core, and the core will tend towards becoming isothermal. In so doing, it must slowly contract in order to maintain the pressure balance through the increase of density towards its centre. The necessity for this collapse is reinforced by the loss of pressure consequent upon the conversion of four protons and two electrons into one helium nucleus (equation 2P.1.15). The collapse releases gravitational energy, half of which goes into heating the core (equation 2P.5.21). Eventually the central temperature rises to near 10^8 K, and the central density to near $10^8\ \mathrm{kg\,m^{-3}}$, providing the conditions required for nuclear reactions commencing with helium to start. The large increase in temperature which is required for the initiation of helium-burning is a consequence of the instability of the ${}^{8}_{4}\mathrm{Be}$ nucleus. This decays into two ${}^{4}_{2}\mathrm{He}$ nuclei in about 2×10^{-16} s (figure 2A.4). Thus the simple combination of two helium nuclei into beryllium results in the almost instantaneous fission back to helium again:

$${}^{4}_{2}\mathrm{He} + {}^{4}_{2}\mathrm{He} \rightarrow {}^{8}_{4}\mathrm{Be} \rightarrow {}^{4}_{2}\mathrm{He} + {}^{4}_{2}\mathrm{He}. \qquad (2A.2.17)$$

Only when the temperature and pressure are high enough for a third helium nucleus to interact with the beryllium before it decays can a stable higher mass product result. The ${}^{12}_{6}\mathrm{C}$ nucleus formed in this way is in an excited state (*4), and it emits 7.656 MeV in the form of a gamma ray in order to revert to its ground state. The overall sequence, which is known as the triple-alpha reaction, since helium nuclei are the alpha particles of radioactivity, is thus

$${}^{4}_{2}\mathrm{He} + {}^{4}_{2}\mathrm{He} \rightarrow {}^{8}_{4}\mathrm{Be} \qquad (-0.092\ \mathrm{MeV}) \qquad (2A.2.18)$$

$${}^{8}_{4}\mathrm{Be} + {}^{4}_{2}\mathrm{He} \rightarrow {}^{12}_{6}\mathrm{C}^{*} \qquad (-\ 0.286\ \mathrm{MeV}) \qquad (2A.2.19)$$

$${}^{12}_{6}\mathrm{C}^{*} \rightarrow {}^{12}_{6}\mathrm{C} + \gamma \qquad (+\ 7.656\ \mathrm{MeV}). \qquad (2A.2.20)$$

The first two reactions are slightly endothermic, and require an input of energy from the thermal motions of the particles. At 10^8 K, the mean thermal energy of a helium nucleus is about 0.01 MeV, so that only a few nuclei in the high-energy tail of the velocity distribution will have sufficient energy to initiate the second reaction of the sequence. The reaction requires what amounts to a triple collision of helium nuclei, and so the reaction rate varies as the square of the density, as well as being very sensitive to temperature:

$$N_{3\alpha} = \frac{9.96 \times 10^{28} Y^2 \rho^2}{T^2 \exp(4.405 \times 10^9/T)}\ (\mathrm{kg^{-1}\,s^{-1}}) \qquad (2A.2.21)$$

for $30 \times 10^6\ \mathrm{K} < T < 2000 \times 10^6$ K. The net energy released by the sequence (equations (2A.2.18) to (2A.2.20)) is 7.278 MeV, and so the energy generation rate is

$$E_{3\alpha} = \frac{1.16 \times 10^{17} Y^2 \rho^2}{T^2 \exp(4.405 \times 10^9/T)}\ (\mathrm{W\,kg^{-1}}). \qquad (2A.2.22)$$

Near 10^8 K, this gives a dependence of the energy generation rate upon temperature of about T^{40}; doubling the temperature therefore increases the energy produced by a factor of about 10^{12}. This results in a comparatively restricted temperature range for the triple-alpha helium-burning regions of stars of 150×10^6 to 250×10^6 K. The energy released by the conversion of 1 kg of helium into carbon is about 6×10^{13} J, or about a tenth of that from the conversion of 1 kg of hydrogen to helium. This lower energy release is a consequence of the decreasing change in the binding energies per nucleon as higher mass nuclei are formed (figure 2P.3), and it results in the consumption of helium at a very much greater rate than the previous consumption of the hydrogen. A core containing significant quantities of ${}^{12}_{6}\mathrm{C}$ is therefore formed in about a thousandth of the main sequence lifetime. Heavier elements can then be built up from the carbon as helium nuclei are successively added in the alpha-capture process.

Alpha capture starts at temperatures only a little over those required for the triple-alpha reactions and produces $^{16}_{8}O$:

$$^{12}_{6}C + ^{4}_{2}He \rightarrow ^{16}_{8}O + \gamma \qquad (+7.161\ MeV). \qquad (2A.2.23)$$

Further additions of helium nuclei produce a sequence of elements as the temperature rises towards 6×10^8 K:

$$^{16}_{8}O + ^{4}_{2}He \rightarrow ^{20}_{10}Ne + \gamma \qquad (+4.730\ MeV) \qquad (2A.2.24)$$

$$^{20}_{10}Ne + ^{4}_{2}He \rightarrow ^{24}_{12}Mg + \gamma \qquad (+9.317\ MeV) \qquad (2A.2.25)$$

$$^{24}_{12}Mg + ^{4}_{2}He \rightarrow ^{28}_{14}Si + \gamma \qquad (+9.981\ MeV). \qquad (2A.2.26)$$

The total energy released as 1 kg of carbon and helium is converted to silicon is only about 10^{14} J, or about 15% of the energy from the conversion of 1 kg of hydrogen to helium. Alpha capture can also start from nuclei other than $^{12}_{6}C$, and thereby produce other sequences of elements. In particular the intermediate products of the C–N–O cycle (figure 2A.5) produce

$$^{14}_{7}N + ^{4}_{2}He \rightarrow ^{18}_{9}F + \gamma \qquad (+4.416\ MeV) \qquad (2A.2.27)$$

$$^{18}_{9}F \xrightarrow{EC} ^{18}_{8}O + e^{+} + \nu_e(*5) \qquad (+1.65\ MeV) \qquad (2A.2.28)$$

$$^{18}_{8}O + ^{4}_{2}He \rightarrow ^{22}_{10}Ne + \gamma \qquad (+9.667\ MeV) \qquad (2A.2.29)$$

and the important neutron-releasing reactions (see the *r* and *s* processes below)

$$^{13}_{6}C + ^{4}_{2}He \rightarrow ^{16}_{8}O + n \qquad (+2.214\ MeV) \qquad (2A.2.30)$$

$$^{18}_{8}O + ^{4}_{2}He \rightarrow ^{21}_{10}Ne + n \qquad (-0.699\ MeV) \qquad (2A.2.31)$$

$$^{22}_{10}Ne + ^{4}_{2}He \rightarrow ^{25}_{12}Mg + n \qquad (-0.481\ MeV). \qquad (2A.2.32)$$

Above a temperature of about 6×10^8 K, the carbon nuclei can react together directly, and the oxygen nuclei similarly react directly with each other for temperatures over about 10^9 K:

$$^{12}_{6}C + ^{12}_{6}C \rightarrow ^{24}_{12}Mg + \gamma \qquad (+13.930\ MeV) \qquad (2A.2.33)$$

$$^{16}_{8}O + ^{16}_{8}O \rightarrow ^{32}_{16}S + \gamma \qquad (+16.539\ MeV). \qquad (2A.2.34)$$

Alternative paths for some of these reactions can also result in the production of $^{23}_{11}Na$, $^{23}_{12}Mg$, $^{31}_{15}P$, and $^{31}_{16}S$, together with additional protons and neutrons. The direct burning of carbon and oxygen can occur explosively over only a few seconds, or take up to 10^{-5} of the star's main sequence lifetime.

Nuclei above silicon can be formed by alpha capture:

$$^{28}_{14}Si + ^{4}_{2}He \rightarrow ^{32}_{16}S + \gamma \qquad (+6.948 \text{ MeV}) \qquad (2A.2.35)$$

$$^{32}_{16}S + ^{4}_{2}He \rightarrow ^{36}_{18}Ar + \gamma \qquad (+6.645 \text{ MeV}) \qquad (2A.2.36)$$

but the mutual repulsion of the two positively charged nuclei grows as the atomic number increases, and so other processes start to take over. There are a very great many possible such reactions, and their probable durations range from seconds to years at most. All the exothermic reactions stop at the iron-group nuclei, but higher elements can be built up by some of the reactions by utilising the particles' thermal energies. The most important groups of these reactions are the *e*, *r*, *s*, and *p* processes.

The *s* process occurs at the lowest temperatures, somewhat over 10^8 K being adequate. It is the capture of neutrons by nuclei at a rate *s*lower than the radioactive decay times of the products. The neutrons are produced in many reactions (equations 2A.2.30, 2A.2.31, 2A.2.32 etc) and can react much more easily than the protons with the nuclei, through the lack of electrostatic repulsion between the particles. In the *s* process, the capture of a neutron is followed by β decay, so that the net effect is to produce nuclei of higher atomic number. For example:

$$^{31}_{15}P + n \rightarrow ^{32}_{15}P \qquad (2A.2.37)$$

$$^{32}_{15}P \xrightarrow{\beta^-} ^{32}_{16}S + e^- + \tilde{\nu}_e \text{ (14.3 days)} \qquad (2A.2.38)$$

$$^{32}_{16}S + n \rightarrow ^{33}_{16}S \text{ (stable)} \qquad (2A.2.39)$$

$$^{33}_{16}S + n \rightarrow ^{34}_{16}S \text{ (stable)} \qquad (2A.2.40)$$

$$^{34}_{16}S + n \rightarrow ^{35}_{16}S \qquad (2A.2.41)$$

$$^{35}_{16}S \xrightarrow{\beta^-} ^{35}_{17}Cl + e^- + \tilde{\nu}_e \text{ (88 days)} \qquad (2A.2.42)$$

$$^{35}_{17}Cl + n \rightarrow ^{36}_{17}Cl \qquad (2A.2.43)$$

$$^{36}_{17}Cl \xrightarrow{\beta^-} ^{36}_{18}Ar + e^- + \tilde{\nu}_e \text{ } (3 \times 10^5 \text{ yr}) \qquad (2A.2.44)$$

and so on. The *s* process is thought to occur after the initiation of helium burning and when there is convective mixing of the hydrogen and helium layers.

The *r* process requires temperatures over 10^{10} K. It is related to the *s* process and is just neutron capture at a rate which is more *r*apid than the products' decay times. It results in neutron-rich nuclei, and the decay to radioactively stable end products occurs only after the synthesis has finished. It requires a very high neutron flux and is thought to occur for a few seconds during supernova explosions (§§3A.5, 8A.2).

The *p* process appears to be of less significance than the preceding reactions. It is the direct capture of *p*rotons by nuclei. It requires temperatures up to about 3×10^9 K, and probably produces the low-abundance proton-rich nuclei. It also may occur in supernovae, particularly in the outer hydrogen-rich layers as the supernova shock wave passes through.

Above temperatures of about 3×10^9 K, the *e* process may determine abundances. This process is simply the condition that a reaction and its inverse have reached *e*quilibrium. The rate of formation of a nucleus is balanced by its rate of destruction. The abundances of the elements involved then depend only upon the relative rates of the reactions. The 25% or so of the material in the Universe which is in the form of helium is thought to be due to an *e* process during the early stages of the 'Big Bang'. There is then equilibrium between the formation and destruction of neutrons which leads to 12% of the matter being in the form of neutrons and most of the rest in the form of protons. The later incorporation of these neutrons into heavy hydrogen nuclei and then into helium nuclei gives the observed helium ratio (*6).

The reactions which have just been outlined give the impression that stellar energy sources are very diverse. It is worth re-emphasising therefore that the vast majority of stars are simply converting hydrogen into helium. Not all the subsequent stages occur in all stars, and in any case they occur very rapidly when compared with the main sequence lifetime. The relative durations are (very approximately)

Reactions	*Relative duration*
hydrogen to helium	1
helium to carbon, oxygen etc, via the triple-alpha and alpha-capture processes	10^{-3} to 10^{-2}
direct carbon and oxygen burning	few seconds to 10^{-5}
s process	10^{-7} to 10^{-3}
p process	tens of seconds
e process	few seconds
r process	tens of seconds.

Only stars with initial masses greater than about twenty times that of the Sun appear likely to go through the whole sequence of reactions up to the iron-group elements, and these form only 10^{-7} of the total number of stars (figure 1A.10). With a stellar population of 2×10^{11} for the Galaxy, we would thus expect some 2×10^4 stars to have masses greater than 20 $\mathcal{M}_\odot$. Their main sequence lifetimes of about 3×10^7 years would then suggest that, in the whole Galaxy, only for a few seconds once per century would the final stages of nucleosynthesis be occurring in any star (*7).

Nucleosynthesis within stars can explain most of the features of the abundances of the elements (figure 1A.11). The main spheres of influence are thought to be

Mass number	*Reaction*
12–20	triple-alpha and alpha capture
16–35	direct carbon and oxygen burning
25–65	*e* process
> 60	*r*, *s*, and *p* processes.

These various reactions suffice to explain the observed element abundances almost in their entirety. There are, however, two major areas of discrepancy. First, helium appears overabundant by a factor of five to ten compared with the amount which could have been produced inside stars since the origin of the Universe. Despite one or two alternative hypotheses, as already discussed this is almost certainly due to the primordial helium produced in the early stages of the Universe. The second anomaly is that some of the light elements such as ${}^{6}_{3}Li$, ${}^{7}_{3}Li$, ${}^{9}_{4}Be$, ${}^{10}_{5}B$, ${}^{11}_{5}B$, etc exist in far greater abundance than might be expected, since they are destroyed rather than produced in nucleosynthetic reactions. This may be due to the production of these light nuclei by cosmic ray spallation reactions (§7A.4) in interstellar space. Heavy nuclei are fragmented in these collisions, and the rate of the reaction appears to be sufficient to account for the observed abundances. The interstellar medium and the surface layers of stars are thus enriched in these elements compared with the interiors of stars. There remain one or two other smaller anomalies, but despite these, there can be little doubt of the overall validity of the nucleosynthetic hypotheses. The generation of stars' energies by these reactions may therefore be regarded as equally well established.

2A.3 ENERGY TRANSFER

Most of the energy released by the nuclear reactions inside stars initially takes the form of very high energy gamma rays. This energy, together with that going into the particles' velocities, must percolate to the surface of the star before it may be radiated away. If the gamma rays were able to travel directly to the surface, like the neutrinos, then the energy would be lost to the star within a few seconds of its production. The photon, however, is scattered many times, usually losing a small fraction of its energy to the scattering particle, and is then finally absorbed completely, long before it reaches the surface. The thermal energy stored in the solar material amounts to about 1.4×10^{41} J. This would take about 10^{7} years to radiate away at the Sun's present luminosity. This time interval represents a rough estimate of the time required for a packet of energy released at the centre of the Sun to reach the surface, τ. Similar figures apply to other stars and are given by

$$\tau \simeq 10^7(\mathcal{M}_*/\mathcal{M}_\odot)(R_\odot/R_*)(L_\odot/L_*). \qquad \text{(years)} \qquad (2A.3.1)$$

The energy must thus undergo many interactions on its journey to the surface of the star in order to be delayed to such an extent. As we have seen (Chapter 2P), two forms of energy transfer are of potential significance inside most stars: radiative and convective transfer (*8). If we consider momentarily the energy only in the form of a photon, then its motion is a series of short straight-line steps each of which terminates in a scattering event. The direction of motion of the photon after each scattering will be randomised. Such a situation, known as a random walk, is well understood mathematically, and the average distance of a photon from its point of origin is

$$D = (n)^{1/2}\bar{d} \qquad (2A.3.2)$$

where n is the number of steps and d is the mean step length. Thus, travelling at 3×10^8 m s^{-1} for 10^7 years, we have for the photon

$$n\bar{d} \simeq 10^{23} \text{ m} \qquad (2A.3.3)$$

while the solar radius gives

$$D = 7 \times 10^8 \text{ m}. \qquad (2A.3.4)$$

Thus

$$n \simeq 10^{28} \qquad (2A.3.5)$$

would be a typical number of scattering events for such a photon, and we would have

$$\bar{d} \simeq 10^{-5} \text{ m}. \qquad (2A.3.6)$$

Of course, very few photons would remain unabsorbed on such a journey, but the calculations do give a 'feel' for the situation. The two types of energy transfer are now considered more rigorously.

Radiative energy transfer

In regions of radiative energy transfer, the temperature gradient of the material is relatively gentle (equation 2P.4.11, etc), so that portions of the material becoming displaced from their positions will be denser than their surroundings, if rising, and so will sink back again. Convective motions are thus inhibited. Conduction is very slow except in degenerate matter, and so in these conditions significant energy transfer must be due to radiation alone. Now a photon is essentially a packet of energy moving at up to

$3 \times 10^8\ \mathrm{m\,s^{-1}}$, so the actual movement of the energy needs no further consideration. The main concern of this section is therefore how the movement of the photons is affected by their interaction with matter.

A photon may undergo two types of interaction with matter: *scattering*, in which the photon's direction of motion is changed, but there is only a small or zero change in its energy; and *absorption*, in which the whole or a significant fraction of the photon's energy is converted into other forms of energy. In particular, the excitation energies of electrons in atoms, ions and molecules, the rotational and vibrational energies of molecules, and the thermal motions of all types of particles are important in this context. In most cases the absorbed energy will eventually be re-emitted as radiation, but it is likely to have suffered a time delay, to have its direction randomised, and is often changed very considerably in its energy.

The two types of interaction result in two quite different types of behaviour within the star. Scattering destroys the information carried by the photon on the structural properties of its source, while absorption destroys all the remaining information (such as temperature) about the source that the photon may be carrying. In a terrestrial situation, a thick fog or cloud layer scatters the solar radiation so that it is no longer possible to determine the size, shape or position of the Sun. The majority of the solar photons, however, still penetrate to ground level, allowing night and day to be distinguished, and a spectral analysis of the light would still enable the solar temperature, composition, etc to be determined. By contrast a very thick pall of smoke from a major fire or volcanic eruption results in the absorption of the solar photons, leading to the phrase beloved of novelists: 'Midday became as black as night'. The illumination is then entirely governed by the local conditions, and whether or not the Sun has risen has no measurable effect. Conversely, measurements of the local conditions will fail to provide any information about the properties of the Sun. These two situations are described physically by two lengths; the mean free path, and the destruction length. The mean free path is simply the average distance between successive scattering events, and has previously (equation 2A.3.2) been denoted by $\bar{d}$. It is closely related to optical depth (§3P.2), and corresponds to the physical distance at which the optical depth is unity:

$$\bar{d} \simeq 1/\varkappa \tag{2A.3.7}$$

where $\varkappa$ is the opacity (§3P.2).

The destruction length is at least as long as the mean free path, and since photons are often scattered many times before being absorbed, it is usually much greater. It is the distance over which the photon has a 50% probability of being absorbed, and its energy returned to the thermal pool. Denoting the destruction length by $\bar{\Lambda}$, the probability of destruction of the photon, P_d, at its next interaction is then approximately

$$P_\mathrm{d} \simeq \bar{d}/\bar{\Lambda}. \tag{2A.3.8}$$

When P_d is close to unity, the radiation field is very closely coupled to the local properties of the matter. The photon's energy is deposited into the particles at almost every interaction. Photons subsequently emitted by the material are then characteristic of the properties of the material at that point. Thus the radiation field will be described by Planck's equation (equation 2P.3.2 etc), the matter will obey the appropriate pressure law (equations 2P.1.15, 2P.1.74, 2P.1.76, 2P.1.77), Boltzmann's and Saha's equations (3P.4.10, 3P.4.1) etc and the temperature will have the *same* value in all the equations. In such a situation, we say that the radiation field is thermalised to the local conditions, and the conditions are those of thermodynamic equilibrium (TE).

When P_d is small, the radiation field at any point will contain contributions from many sources, some at considerable physical distances. The sources may be quite inhomogeneous, and the actual radiation field may then differ markedly from that to be expected from the local conditions in the material. If the properties of the matter are described by the various physical equations using a single value for the temperature, and the radiation field is described by Planck's equation, but with a *different* value for the temperature, then we have the conditions known as local thermodynamic equilibrium (LTE). This is an approximation which has been of great value in the past, and is still of some use in the interpretation of stellar spectra and in the subsequent modelling of the conditions in the outer layers of stars (Chapters 3P, 3A, and 4A).

Deep inside the star, P_d may approach unity and in any case $\bar{d}$ is very small because of the high density. The radiation field is thus very closely coupled to the matter, and conditions are very well described by the assumption of TE. In such a situation every interaction is balanced by its inverse, and the black body equations are valid. Closer to the star's surface, however, P_d can become small, and the interaction volume (the sphere of radius $\bar{\Lambda}$ centred on the region of interest) may extend beyond the surface of the star. Some of the photons will then escape completely from the star, and we will have a probability of escape, P_e, for photons from that point. The depth at which

$$P_e = P_d \tag{2A.3.9}$$

is known as the thermalisation depth and marks the point below which an assumption of TE may be useful. It is also the greatest depth from which photons observed outside the star are likely to have originated.

Thus we see that when energy is transferred by radiation, the photons are constantly being thermalised to the local conditions in the matter. Since the temperature of the matter is decreasing from tens of millions of degrees at the star's centre to thousands of degrees at its surface, the radiation field must likewise adjust its spectral distribution to those of the corresponding black bodies. Only very close to the surface, where photons may escape directly into space, may significant departures from TE occur (*9).

Convective energy transfer

When convection is occurring, most of the energy will be transported outwards by the physical movement of the material inside the star. Some energy is still transported by radiation, however, and deep inside the star the matter and radiation field remain closely coupled. The convective flow is thought to be turbulent, with eddies on many size scales, moving and interacting in very complex patterns. No precise quantitative theory exists to describe this behaviour, and this is a very serious limitation to stellar modelling calculations. Only the semi-empirical mixing length theory is available, but this does provide some insight into what may be happening in the regions of stars undergoing convective transfer of energy.

We have already seen that the criterion for convection to occur (equation 2P.4.9) is

$$\frac{\mathrm{d}\log_e T}{\mathrm{d}\log_e P} > \frac{\gamma - 1}{\gamma} \tag{2A.3.10}$$

and this is generally known as the Schwarzschild criterion. In practice convection will not become significant until the temperature gradient is somewhat steeper than suggested by this equation since the viscosity of the material slows down the motions of the material. Nonetheless, Schwarzschild's criterion remains a useful measure of when convection is likely to occur. The ratio of the specific heats, γ, has a value of $\frac{5}{3}$ for a perfect monatomic gas. The gas in a real star, however, is not perfect because of the effects of excitation and ionisation. Furthermore, radiation pressure may need to be taken into account in some cases. Thus we may more generally write equation (2A.3.10) as

$$\frac{\mathrm{d}\log_e T}{\mathrm{d}\log_e P} > \frac{\Gamma - 1}{\Gamma} \tag{2A.3.11}$$

where Γ replaces γ and takes the values $\frac{5}{3}$ for a perfect monatomic gas, $\frac{4}{3}$ for pure radiation pressure and ~ 1.1 for regions where hydrogen is partially ionised. Thus we get

$$\frac{\Gamma - 1}{\Gamma} = 0.4 \quad \text{for a perfect monatomic gas} \tag{2A.3.12}$$

$$= 0.25 \quad \text{for pure radiation pressure} \tag{2A.3.13}$$

$$\simeq 0.1 \quad \text{for hydrogen ionisation regions} \tag{2A.3.14}$$

and so we see that convection is most likely to occur in regions where the hydrogen is partially ionised. Thus, in the hottest stars hydrogen is completely ionised throughout most of their outer regions, and energy transport is then by radiation. In cool stars, significant ionisation can still be occurring at great depths, and so they have very deep convective outer layers which essentially determine the structure of the star. In solar-type stars, as already mentioned (§2P.4), convection transports the energy for about the outer 25% of the star's radius. The central regions of stars may also become convective, this time through the existence of a very steep temperature gradient arising from the high sensitivity of the energy-generating reactions to the conditions (equations 2A.2.6, 2A.2.7, 2A.2.22 etc). Thus a main sequence star with a mass about five times that of the Sun is likely to have convective energy transport over about 7% of its radius. In more massive stars the convective regions may extend out to 25 or 30% of the radius. The central condensation inside stars means that over 50% of the mass of the star is involved in these central convection zones. Unlike less massive stars, therefore, the central regions of larger stars are well mixed and of uniform composition. This affects the main sequence and subsequent behaviour patterns of the stars somewhat (§8A.2).

Prandtl mixing length theory

The physical behaviour of the material within a region of convective energy transport is described semi-empirically by the mixing length theory due to Prandtl. This assumes that the convective elements move a characteristic length, the mixing length, releasing all their excess energy only at the top of this motion (if moving upwards), or absorbing all their energy deficit only at the bottom of their motion (if moving downwards). The transfer of energy thus arises through both the upward and downward motions of the material. The convective energy transfer per unit volume may be estimated as

$$C_P \rho \delta T \tag{2A.3.15}$$

where δT is the temperature difference between the moving element and its surroundings, C_P and ρ are the specific heat at constant pressure and the density for the material at the top (or bottom) of its motion. This energy is carried over a distance equal to the mixing length, l. However, considering many such elements, at any given instant they will be at all stages of travel along this distance, and so on average will be at a distance $l/2$ from their points of origin. The temperature differential is therefore on average

$$\delta T = \left(\left.\frac{\mathrm{d}T}{\mathrm{d}r}\right|_{\text{star}} - \left.\frac{\mathrm{d}T}{\mathrm{d}r}\right|_{\text{element}} \right) \frac{l}{2} \tag{2A.3.16}$$

and so the rate of energy transport, or the net convective flux, $\mathscr{F}_\mathrm{c}$, is

$$\mathscr{F}_\mathrm{c} = \bar{v} C_P \rho \left(\left.\frac{\mathrm{d}T}{\mathrm{d}r}\right|_{\text{star}} - \left.\frac{\mathrm{d}T}{\mathrm{d}r}\right|_{\text{element}} \right) \frac{l}{2} \tag{2A.3.17}$$

where $\bar{v}$ is the mean velocity of the convective elements. Now, neither $\bar{v}$ nor l can be determined with any precision from theory, but semi-quantitative arguments lead to estimates that

$$0.1H < l < 2H \tag{2A.3.18}$$

where H is the pressure scale height, i.e. the distance over which the pressure changes by a factor of e, and is given by

$$H = \frac{kT}{\mu m_{\mathrm{H}} g} \tag{2A.3.19}$$

where g is the local gravitational acceleration; and

$$\bar{v} \simeq \left(\frac{P}{2\rho HT}\right)^{1/2} \left(\left.\frac{\mathrm{d}T}{\mathrm{d}r}\right|_{\mathrm{star}} - \left.\frac{\mathrm{d}T}{\mathrm{d}r}\right|_{\mathrm{element}}\right)^{1/2} \frac{l}{2} \tag{2A.3.20}$$

so that

$$\mathscr{F}_{\mathrm{c}} \simeq C_P \rho \left(\frac{P}{2\rho HT}\right)^{1/2} \left(\left.\frac{\mathrm{d}T}{\mathrm{d}r}\right|_{\mathrm{star}} - \left.\frac{\mathrm{d}T}{\mathrm{d}r}\right|_{\mathrm{element}}\right)^{3/2} \left(\frac{l}{2}\right)^2. \tag{2A.3.21}$$

Now

$$\frac{\mathrm{d}T}{\mathrm{d}r} = \frac{\mathrm{d}T}{\mathrm{d}P} \frac{\mathrm{d}P}{\mathrm{d}r} \tag{2A.3.22}$$

and the equation of hydrostatic equilibrium (equation 2P.5.12) gives

$$\frac{\mathrm{d}P}{\mathrm{d}r} = -g\rho. \tag{2A.3.23}$$

Strict hydrostatic equilibrium is not valid in the convective regions of stars, but equation (2A.3.23) is still a good approximation, and so

$$\frac{\mathrm{d}T}{\mathrm{d}r} = g\rho \frac{\mathrm{d}T}{\mathrm{d}P} \tag{2A.3.24}$$

(the negative sign disappearing because P increases as r decreases). This gives

$$\frac{\mathrm{d}T}{\mathrm{d}r} = g\rho \frac{T}{P} \frac{\mathrm{d}\log_e T}{\mathrm{d}\log_e P} \tag{2A.3.25}$$

and from the perfect gas law (equation 2P.1.15)

$$\frac{\mathrm{d}T}{\mathrm{d}r} = \frac{g\mu m_{\mathrm{H}}}{k} \frac{\mathrm{d}\log_e T}{\mathrm{d}\log_e P}. \tag{2A.3.26}$$

Thus from equation (2A.3.19), we have finally

$$\frac{\mathrm{d}T}{\mathrm{d}r} = \frac{T}{H} \frac{\mathrm{d}\log_e T}{\mathrm{d}\log_e P}. \tag{2A.3.27}$$

As we have already seen (equation 2A.3.11), the element's temperature gradient is

$$\left.\frac{\mathrm{d}\log_e T}{\mathrm{d}\log_e P}\right|_{\mathrm{element}} = \frac{\Gamma - 1}{\Gamma} \tag{2A.3.28}$$

so that the convective energy flux (equation 2A.3.21) becomes

$$\mathscr{F}_{\mathrm{c}} \simeq C_P\rho\left(\frac{P}{2\rho HT}\right)^{1/2}\left(\frac{T}{H}\right)^{3/2}\left(\left.\frac{\mathrm{d}\log_e T}{\mathrm{d}\log_e P}\right|_{\mathrm{star}} - \frac{\Gamma-1}{\Gamma}\right)^{3/2}\left(\frac{l}{2}\right)^2 \tag{2A.3.29}$$

$$= \left(\frac{gH}{32}\right)^{1/2} C_P\rho T\left(\left.\frac{\mathrm{d}\log_e T}{\mathrm{d}\log_e P}\right|_{\mathrm{star}} - \frac{\Gamma-1}{\Gamma}\right)^{3/2}\left(\frac{l}{H}\right)^2. \tag{2A.3.30}$$

The situation is further complicated in practice by radiative energy loss or gain from the moving element, leading to non-adiabatic changes within it, and by ionisation and excitation changes in the atoms and ions of the material. The latter is partially taken into account by the use of Γ instead of γ (see equation 2A.3.11), while a measure of the effects of the former is the convective efficiency, Δ_{c}, defined by

$$\Delta_{\mathrm{c}} = \left(\left.\frac{\mathrm{d}\log_e T}{\mathrm{d}\log_e P}\right|_{\mathrm{star}} - \left.\frac{\mathrm{d}\log_e T}{\mathrm{d}\log_e P}\right|_{\mathrm{element}}\right) \times \left(\left.\frac{\mathrm{d}\log_e T}{\mathrm{d}\log_e P}\right|_{\mathrm{element}} - \left.\frac{\mathrm{d}\log_e T}{\mathrm{d}\log_e P}\right|_{\mathrm{adiabatic}}\right)^{-1} \tag{2A.3.31}$$

and which is effectively the ratio of the energy transported convectively to that lost during the element's motion.

The total energy flux is the sum of the convective and radiative fluxes, and must also be given by the Stefan–Boltzmann equation

$$\mathscr{F} = \mathscr{F}_{\mathrm{c}} + \mathscr{F}_{\mathrm{R}} = \sigma T_{\mathrm{eff}}^{\ 4}. \tag{2A.3.32}$$

In convective regions, $\mathscr{F}_{\mathrm{c}}$ will typically be 95% or more of $\mathscr{F}$, while in radiative regions mass motions will generally account for less than 0.001% of the flux.

2A.4 LUMINOSITY

However it may have originated or been transported, the energy inside a star must eventually reach the surface. It is then lost to the star, mostly in the form of near infrared, visible and ultraviolet electromagnetic radiation. The details of this process are considered further in §§3P.2 and 3A. Here we are simply concerned with the total energy emission of the star, or, as it is commonly called, its *luminosity*.

A relationship between a star's mass and its luminosity has already been mentioned (equations 1A.1.16, 1A.1.17). The existence of such a relationship arises from the operation of the Vogt–Russell theorem. This states that *the mass, composition and age of an isolated star uniquely determine its luminosity and effective surface temperature.* While not strictly proven, this theorem has not been found to have any significant deviations. But as we have seen (§1A.2), over 90% of stars are formed essentially of hydrogen and helium in the ratio of about three to one by mass, the only variations occurring in the very small proportion of the mass of the star which is in the form of the heavier elements. Furthermore, for a substantial part of the star's life its age has only a marginal effect upon its structure; whilst burning hydrogen to helium at its centre, only the composition of these central regions changes, and that very slowly. Thus the only significant variable which remains to affect a star's luminosity is its mass. We find therefore a relationship between the two, and it takes the form

$$L \propto \mathcal{M}^n. \qquad (2A.4.1)$$

The value of the index n varies slightly with the mass, and in practice takes the values

$$n = 3.9 \qquad \mathcal{M}_* < 7\mathcal{M}_\odot \qquad (2A.4.2)$$

$$= 3.0 \qquad 7\mathcal{M}_\odot < \mathcal{M}_* < 25\mathcal{M}_\odot \qquad (2A.4.3)$$

$$= 2.7 \qquad 25\mathcal{M}_\odot < \mathcal{M}_*. \qquad (2A.4.4)$$

Thus for most stars ('solar'-type in the sense of Chapter 1A), there is a direct relationship between mass and luminosity which is almost independent of composition and age (figure 2A.8), and which has the form

$$L_* = 2.9 \times 10^{-89}\mathcal{M}_*^{3.9} \text{ (W)} \qquad (2A.4.5)$$

($\mathcal{M}_*$ in kg) or

$$L_*/L_\odot = (\mathcal{M}_*/\mathcal{M}_\odot)^{3.9}. \qquad (2A.4.6)$$

White dwarfs, giants etc (Chapter 3A), which may amount to 10% of the stars by number, do not fit the relation, and as we have just seen, it is also slightly different for the very rare high-mass stars. The range of stellar masses (§1A.1) gives a range for stellar luminosities excluding white dwarfs, giants etc, of 10^{19} to 10^{34} W ($3 \times 10^{-8}\ L_\odot$ to $3 \times 10^7\ L_\odot$), with a mean value of 10^{25} W ($0.02\ L_\odot$).

Since surface temperature is also uniquely determined by mass, composition and age according to the Vogt–Russell theorem, a similar argument to the above leads to a mass/surface temperature relationship, and this is shown in figure 2A.9. An equation of the form

$$T_* = 5870\mathcal{M}_*^{0.5} \text{ (K)} \qquad (2A.4.7)$$

where $\mathcal{M}_*$ is in units of the solar mass fits the data quite well.

These two relationships lead to the majority of the stars being restricted to a narrow band when plotted on a luminosity/temperature diagram. Such a diagram is often called an HR diagram after Hertzsprung and Russell who plotted the first version of it. The band containing most of the stars is called the main sequence (§3A.4).

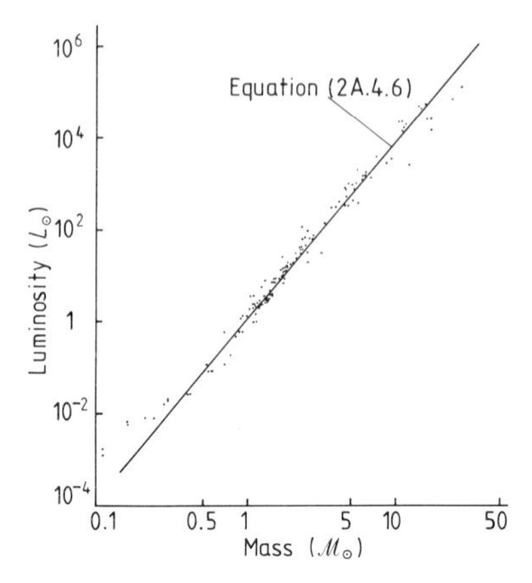

Figure 2A.8 Mass/luminosity relationship for main sequence stars.

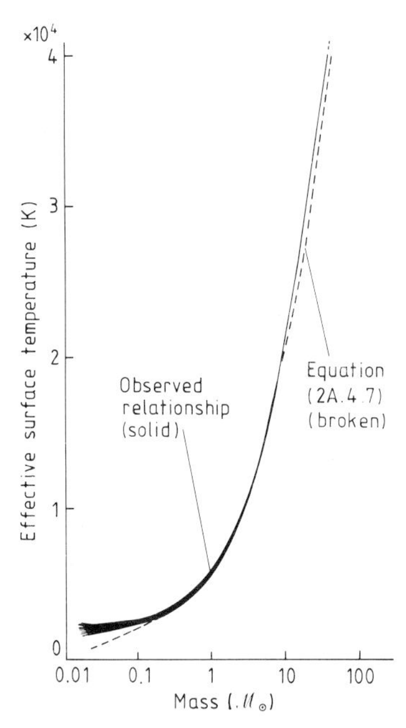

Figure 2A.9 Mass/surface temperature relationship for main sequence stars.

Non-main sequence stars have luminosities which are very different from those suggested by equation (2A.4.6) for their masses. Their properties are discussed in more detail in Chapter 3A. Here, just a summary of the observations is presented.

Star type (Chapter 3A)	*Luminosity range* ($L_\odot$)
White dwarfs	10^{-4}–10^{-2}
Giants	10–10^3
Supergiants	10^3–10^6

For a given mass, white dwarfs will be one to two hundred times fainter than main sequence stars. Giants and supergiants will be of comparable luminosities to massive main sequence stars, but of very different temperatures and sizes.

2A.5 SIZES AND DENSITIES

In the previous section relationships were found between mass, luminosity and temperature for main sequence stars. Since the luminosity of a star is given by

$$L_* = 4\pi R_*^2 \sigma T_{\rm eff}^4 \qquad (2A.5.1)$$

there is also a mass/radius relationship for these stars. Using equations (2A.4.5) and (2A.4.7), we have

$$R_* = 3.7 \times 10^{-19} \mathscr{M}_*^{0.9}\ (\text{m}) \qquad (2A.5.2)$$

or

$$R_*/R_\odot = (\mathscr{M}_*/\mathscr{M}_\odot)^{0.9} \qquad (2A.5.3)$$

and this is shown on figure 2A.10 along with the observed relation. The range of sizes for main sequence stars is thus from about 10^7 m to 5×10^{10} m (0.015 to 70 $R_\odot$), the mean value being about 3×10^8 m ($0.4R_\odot$). Direct measurement of the angular sizes of stars is possible in a very few cases by techniques such as amplitude interferometry, occultation, infrared photometry etc (see *Astrophysical Techniques*). If the star's distance (Appendix I) is also known, then its actual radius may be found. The results of these techniques generally confirm those of this more indirect analysis, showing that deviations of the star's spectrum from that of a black body are not of significance in this context.

From equation (2A.5.2) we may find the mean density for main sequence stars:

$$\bar{\rho}_* = 4.7 \times 10^{54} \mathscr{M}_*^{-1.7}\ (\text{kg m}^{-3}). \qquad (2A.5.4)$$

The range is then from about 1 $\mathrm{kg\,m^{-3}}$ to $10^7\ \mathrm{kg\,m^{-3}}$, with a mean value of about $10^4\ \mathrm{kg\,m^{-3}}$ (N.B. the lower the mass of the star, the higher its density).

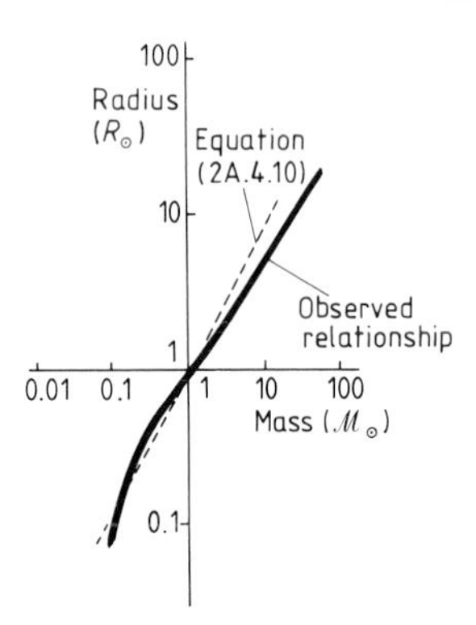

Figure 2A.10 Mass/radius relationship for main sequence stars.

In white dwarfs the relationship of density with mass is reversed; the higher the mass, the higher the density, with the radius decreasing as the mass increases. The Chandrasekhar mass limit of 1.44 $\mathcal{M}_\odot$ (equation 2A.1.29) corresponds to zero radius for the star. For a pure-iron electron-degenerate sphere, we have the relationship between mass and radius shown in figure 2A.11. The largest observed white dwarfs are about 1.4×10^7 m (0.02 $R_\odot$) in radius, corresponding to a mass of 10^{29} kg and a mean

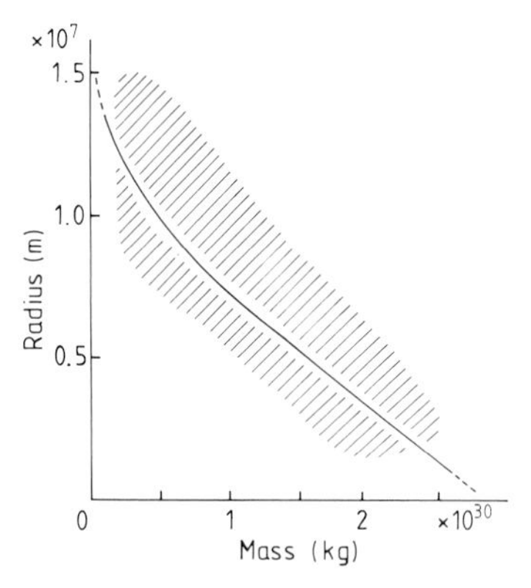

Figure 2A.11 Mass/radius relationship for an electron-degenerate iron sphere (solid line) and for white dwarfs (shaded area).

density of 10^7 kg m^{-3}. The smallest are about 10^6 m in radius with masses of perhaps 2.5×10^{30} kg and densities up to 6×10^{11} kg m^{-3}.

Giants are a much more diverse group of stars as evidenced by their many subdivisions: subgiants, red giants, giants, supergiants, bright giants etc. Furthermore, the diversity arises from stars in different stages of their evolution (§8A.2), so that the same star will occupy different spots within the giant region of the HR diagram at different stages of its life. The largest stars of all are the red giants. They have radii up to a thousand times that of the Sun. Their masses, though, are probably only in the region of twenty times that of the Sun, so that their mean densities are very low, down to as little as 10^{-5} kg m^{-3}. The hottest supergiants by contrast may be only slightly larger than main sequence stars of similar masses.

The mean densities just calculated are very different from the surface or central densities of the stars. For the Sun, for example, the mean density is 1.4×10^3 kg m^{-3}, compared with a central density of 1.6×10^5 kg m^{-3}, and a density at the visible surface of 3.4×10^{-7} kg m^{-3}. Similar figures will apply to other stars.

PROBLEMS

2A.1 By assuming a linear variation from the centre to the surface for the density of the solar material, and that the density effectively reduces to zero at the surface, estimate values for the solar central density and pressure.

$$\mathcal{M}_\odot = 2 \times 10^{30} \text{ kg}$$
$$R_\odot = 7 \times 10^8 \text{ m}$$
$$G = 6.67 \times 10^{-11} \text{ J m}^2 \text{ kg}^{-2}.$$

2A.2 Calculate the temperature at which the proton–proton chain and the C–N–O cycle might be expected to produce equal quantities of energy in a newly formed star, and compare this with the figure for a star with the present Sun's central composition (figure 2A.6).

2A.3 By following a similar line of reasoning to that outlined in equations (2A.1.20) to (2A.1.29), and by taking the limiting radius of a neutron star to be that of the appropriate Schwarzschild black hole (§3A.9), determine an upper limit to the mass of a neutron star.

$$g = 2$$
$$m_n = 1.675 \times 10^{-27}$$
$$h = 6.625 \times 10^{-34} \text{ J s}$$
$$G = 6.67 \times 10^{-11} \text{ J m}^2 \text{ kg}^{-2}.$$

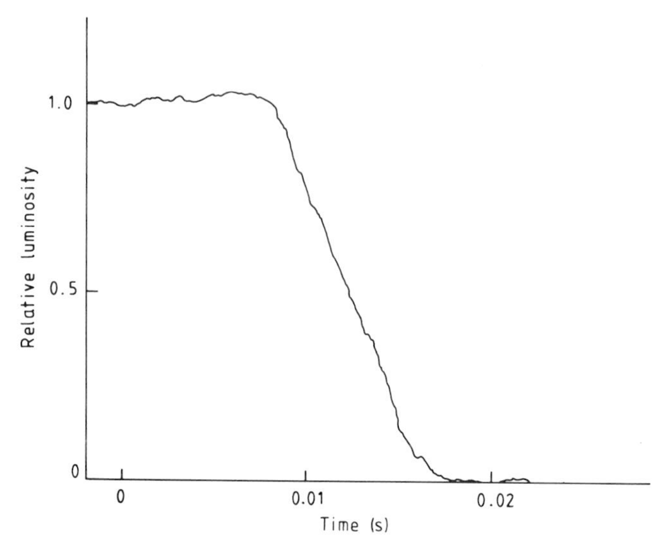

Figure 2A.12 Occultation of Antares.

2A.4 Figure 2A.12 shows the change in the brightness of Antares as it is occulted by the Moon. Given the lunar synodic period of 29.531 days, estimate the angular diameter of the star. Compare your value with that of 0.040″ determined by interferometry. What is the physical size of the star if its distance is 130 pc?

3P Physical Background

SUMMARY

Formation of spectrum lines, transition probabilities, forbidden and allowed transitions, Zeeman effect, radiative transfer, the equation of transfer, optical depth, the source function, gravity darkening, atomic and ionic populations.

Provides background material for Chapters 3A, 6P, and 6A in particular.

See also spectral classification (§3A.2), line profiles and broadening mechanisms (§3A.2), surface temperature (§3A.3), HR diagram (§3A.4), variable stars (§3A.5), abundance determinations (§§1A.2, 3A.6), stellar rotation (§3A.7), stellar and interstellar magnetic fields (§§3A.8, 4A.3, 4A.5, 7A.4), limb darkening (§4A.2), sunspots (§4A.3), coronal spectrum (§4A.5), masers (§6P.1), interstellar gaseous nebulae (§§6A.2, 6A.5, 6A.6), interstellar gas (§7A.3).

3P.1 SPECTROSCOPY

Spectrum line formation

Spectral lines are formed when an electron moves from one level (*1) within an atom, ion or molecule to another, or a molecule changes its vibrational or rotational state. Such a change of level is often called a transition. An individual transition produces no significant effect, but many atoms or ions undergoing the same transition nearly simultaneously will change the radiative flux at the appropriate wavelength noticeably. This then appears as a line because the spectrum is formed of a series of overlapping monochromatic images of the entrance aperture of the spectrograph, and most spectrographs use linear slits for this purpose (*Astrophysical Techniques*, Chapter 4).

An absorption line results when the electron increases its energy during the transition, and an emission line when it decreases its energy. Emission and absorption lines are thus just the inverses of each other, and a more general expression of this is found in Kirchhoff's law,

$$j_\nu(T)/\varkappa_\nu(T) = \text{constant} \tag{3P.1.1}$$

where $j_\nu(T)$ is the emission at frequency ν per unit frequency interval and per unit mass for a material at a temperature T (usually called the *emission coefficient*), and $\varkappa_\nu(T)$ is the absorption at frequency ν per unit frequency interval per unit mass for the same material at temperature T (usually called the *absorption coefficient*). For material in thermodynamic equilibrium (TE) we have

$$j_\nu(T)/\varkappa_\nu(T) = \mu^2 B_\nu(T). \tag{3P.1.2}$$

For material not in TE, the ratio is still constant but no longer necessarily given by the black body intensity. The ratio of emission and absorption coefficients is often called the *source function* and determining its value is usually the critical problem in modelling the processes of radiative transfer in stellar atmospheres (§3P.2).

The energy of each level is not unique, but has a certain spread about its nominal value. The photons absorbed or emitted by a transition may therefore have a small spread in their energies. This results in an intrinsic breadth for the spectrum line and is known as the natural line width. The line profile is given by

$$I(\nu) = I(\nu_0)\frac{(\gamma/4\pi)^2}{(\nu-\nu_0)^2 + (\gamma/4\pi)^2} \tag{3P.1.3}$$

where γ is the damping constant and has the shape shown in figure 3P.1, often known as the Lorentz or damping line profile. This is rarely observed, being usually far smaller in its effect than other astrophysical processes which broaden spectrum lines (§3A.2), and so is usually disregarded. The *shape* produced by equation (3P.1.3) occurs for other broadening mechanisms however, and these may often be of significance.

The pattern of allowed energies for electrons can be very complex. It is most usefully displayed on an energy level diagram (often called a Grotrian diagram). An example is shown in figure 3P.2. The details of the organisation of this diagram and of the notation for the levels is beyond the scope of this book; further details may be found in references in the bibliography.

Even if the spectrum of an element is examined extremely carefully, only a few of the potentially available lines will be found. There are two main reasons for this. First, the higher energy levels represent physically large separations of the electron from the nucleus. In most situations, therefore,

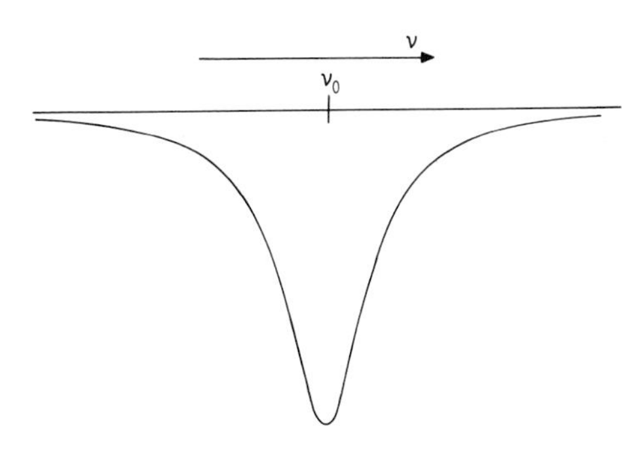

Figure 3P.1 The Lorentz or damping spectrum line profile.

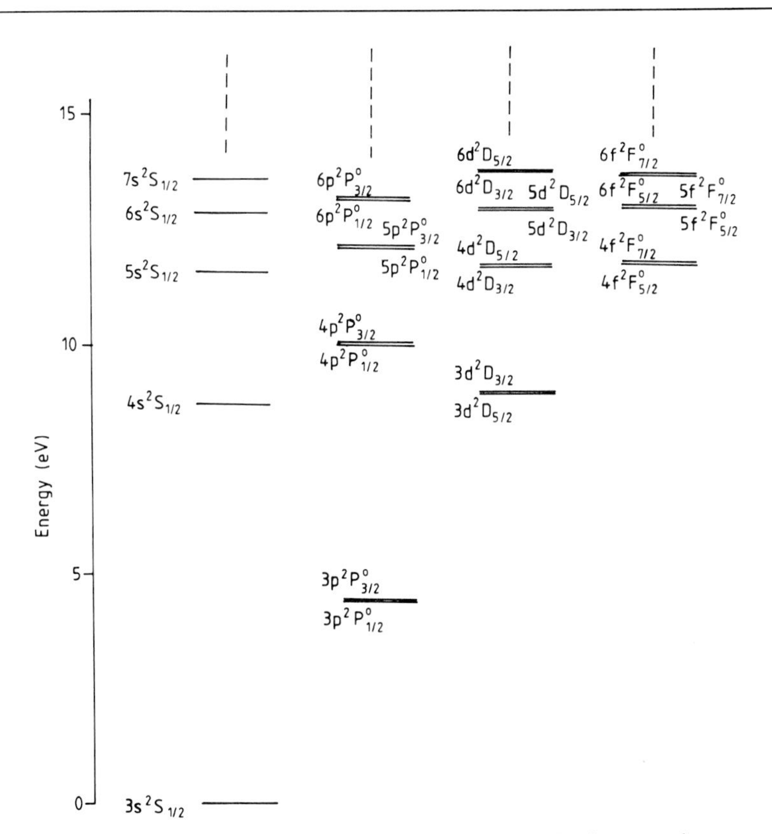

Figure 3P.2 Partial Grotrian diagram of singly ionised magnesium.

these levels will be destroyed by interactions with other atoms, ions and electrons. In interstellar clouds (Chapter 7A) the pressure and temperature may be low enough to permit the existence of many more levels than normal, and some of the observed radio lines arise from transitions amongst these outer levels. Even there, however, interactions with other particles eventually place a limit on the energy of a bound electron. The second reason is much more fundamental; there are restrictions on the transitions which are possible. The quantum numbers which describe the properties of the electrons and of their configurations in the atom can only change in certain ways during a transition. The permitted transitions are governed by conservation laws on the changes in the quantum numbers, known as selection rules. The details are complex, and will not be discussed here. The interested reader is again referred to the literature in the bibliography for further reading. The net result of the selection rules, however, is that transitions are forbidden between many of the energy levels shown in a diagram such as figure 3P.2. The allowed transitions form the normal spectrum of the element.

The details of the allowed transitions of an element are complex but clear-cut. Unfortunately, the transitions 'forbidden' by the selection rules are only partially forbidden! A more correct description is to say that the transition probability (see below) is small but not zero. The lines which result from these other transitions are known as forbidden lines and are normally several orders of magnitude fainter than those from the allowed transitions. They are thus normally lost in the noise level of the spectrum. Special circumstances, however, may lead to the forbidden lines becoming detectable, and sometimes to their dominating the spectrum. Interstellar nebulae (Chapters 6A, 7A) often have spectra comprising little else other than very strong forbidden emission lines. This arises because some of their elements' levels have no allowed transitions to the lowest energy (ground state) level. Excited electrons may cascade downwards to such a level and then become stuck. In the absence of collisions etc to provide alternative paths to the ground state, most of the atoms of that element will eventually make their way into this metastable state. The forbidden transition is then the only way out for the electrons, and so the only observed lines are forbidden ones. Until this situation was understood, the spectra of nebulae were such a puzzle that a new element, nebulium, was invented to account for them!

The transition probability referred to above is discussed in detail later, when it will be seen to be the inverse of the lifetime of an electron in that level. Normal transitions are known as electric dipole transitions, and their lifetimes are typically 10^{-8} s. Forbidden transitions may be permitted by various interactions within the atom and their lifetimes range from 10^{-3} s to many years. A half-forbidden type of transition resulting in what are known as intercombination lines has a lifetime of about 10^{-6} s. These various transition types are often symbolised in the literature using square brackets:

allowed transition: no brackets;
  e.g. O I 630.0304 (*2)
intercombination line: one bracket;
  e.g. K III 348.111]
forbidden line: two brackets;
  e.g. [O II 731.99].

(The precise positioning of the brackets may vary.) The arabic numerals gives the line's wavelength in nanometres.

In addition to these various transitions which produce spectrum lines, continuous absorption or emission is possible. There is a maximum energy for any bound electron known as the ionisation energy. If the electron absorbs sufficient energy to take it above this limit, then it will be lost to the atom and move off as a free electron, leaving behind an ionised atom. The reverse of ionisation is called recapture, and occurs when a free electron rejoins an ion with the emission of radiation.

Multi-electron atoms can be ionised as many times as they have electrons, the ions having ionisation energies in the same manner as the neutral atom. On a Grotrian diagram the ionisation limit may be plotted along with the quantised levels (figure 3P.3). For this discussion, energies above this limit

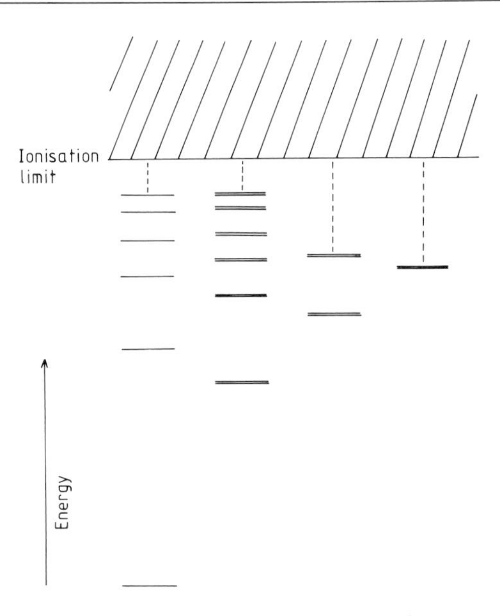

Figure 3P.3 Schematic Grotrian diagram.

may be regarded as unquantised, and so the electron may take any energy (see the section on electron-degenerate material in §2P.1 for a more precise analysis of the situation). Transitions from a bound level to above the ionisation limit, or vice versa, may therefore be of any energy equal to or more than the ionisation energy from that level. The resulting absorption or emission is thus continuous. The highest probability of a transition occurs close to the limit. The spectrum hence contains an emission or absorption band with a sharp edge and its greatest intensity at the wavelength corresponding to the ionisation energy. The best known such absorption band is the Balmer discontinuity in stellar spectra arising from ionisation from the first excited level of the hydrogen atom. Similar discontinuities exist for the Lyman and Paschen series, and for other atoms and ions (figure 3P.4). The most important ionisation edges from an astrophysicist's point of view are in the ultraviolet and occur from the ground states of the lighter neutral and ionised elements. Transitions such as these are sometimes called bound–free transitions for an absorption, and free–bound for a recombination, since the electron is going from being bound to the atom or ion to being free from it. Similarly, transitions producing spectrum lines may be called bound–bound transitions, and those taking place entirely above

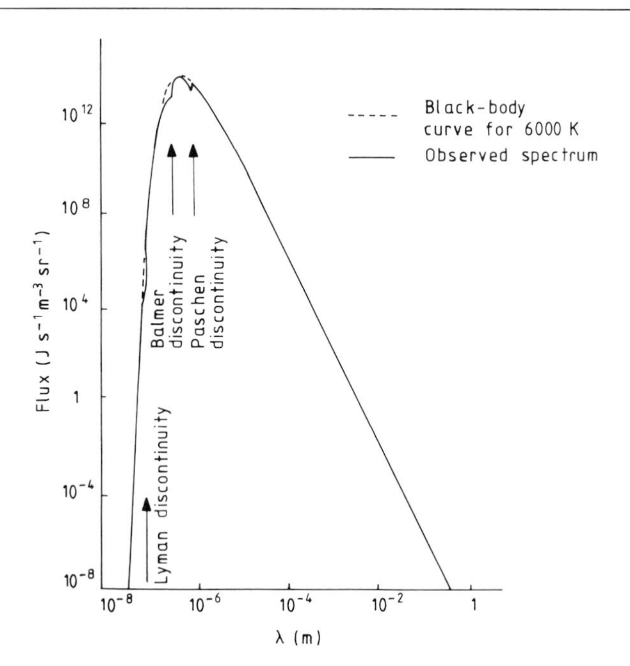

Figure 3P4. Schematic continuous spectrum of a solar-type star, showing ionisation edges.

the ionisation limit (see synchrotron radiation, Chapter 6P) free–free transitions.

A special case of considerable importance involves the negative hydrogen ion. This is a neutral hydrogen atom which has acquired a second electron. That this is possible at all is due to the positively charged nucleus not being totally screened by the single electron normally present and so a second electron may be attracted and held. The ion has only one bound state and this has an ionisation energy of 0.75 eV (1.2×10^{-19} J). Thus any radiation with a wavelength less than 1.65 μm can ionise it. Although the ion is rare under solar surface conditions, its interaction probability is so large that these few ions cause most of the opacity of the outer layers of the Sun. Thus by Kirchhoff's law (equation 3P.1.1), most of the continuum radiation that we receive from the Sun is due to the formation of the ion. The sharp edge to the visible disc of the Sun is likewise due to the H^- ion; its number density is very highly sensitive to temperature. At high temperatures it is too unstable to exist in significant quantities, while at lower temperatures the ionised metals supplying the extra electrons recombine so that there are too few electrons for its significant formation.

Apparent continuum absorption and emission may also be derived from molecules. Their spectra appear to consist of a series of broad bands, reminiscent of an atomic spectrum, but with the lines broadened by a factor of 100 to 1000. Higher resolution, however, will show that the bands comprise very large numbers of individual lines of normal widths, which overlap on the lower resolution spectra. The lines are produced by electron transitions within the molecule in a manner directly analogous to atomic transitions. The greater complexity of the molecular spectrum is just a consequence of the increased complexity of energy levels possible with the greater number of degrees of freedom in a molecule.

Molecules can also produce spectrum lines by two other processes. The molecule can rotate around its centre of mass, and the atoms within it may vibrate about their mean positions. The rotational and vibrational energies are quantised. Changes in those energies can only occur in the same discrete steps as changes in the energy of a bound electron. If the charge axis of the molecule is *not* coincident with the rotational axis, then a change in the rotational energy will correspond to the absorption or emission of radiation. The rotational energies are generally low, since molecules are easily dissociated, and so the resulting spectrum lines are in the far infrared, microwave and radio regions of the spectrum. These transitions have recently come into prominence through their use in the study of molecules in the interstellar medium (Chapter 7A). Some molecules which would be expected to be common, such as H_2, N_2, O_2 etc are symmetrical, however. Their rotational and charge axes therefore coincide, and the rotational transition to the first excited level is forbidden. Their rotational spectrum lines are hence very weak or non-existent. Thus despite their probable high

abundance they are much more difficult to detect than very much rarer asymmetrical molecules. Changes in the frequency of vibration of atoms in a molecule result in vibrational transitions, the energies of these placing most of the resulting spectrum lines in the near infrared.

An absorption line in a spectrum results from the subtraction of energy at that point of the spectrum. This energy is not destroyed, but must reappear in some other form, generally being returned to the thermal pool of energy. When there are large numbers of lines, sufficient energy may be returned in this way to affect the temperature of the outer layers of the star to a significant extent. This phenomenon is called line blanketing. It also occurs on planets where it is known as the greenhouse effect; shortwave radiation from the Sun penetrates through to the planet's surface, but the longwave energy radiated by the planet is blocked by molecular absorptions in the atmosphere. It can have a very considerable effect; the mean surface temperature of the Earth is increased by about 40 K, and that of Venus by some 400 K in this way.

A further process which can occur in both atoms and ions and which produces lines in the x-ray region of the spectrum is the Auger process. An x-ray photon is absorbed by a transition of one of the inner electrons of a multi-electron atom, ion or molecule (transitions previously discussed were all for the outer, or valence, electrons of the particle). One of the outer electrons then drops down to fill the vacancy, but instead of emitting radiation, the energy of the transition goes into ionising one of the other outer electrons. This is a special case of autoionisation; when two electrons are simultaneously in excited states, one may return to the ground state, the released energy causing the other excited electron to be ejected and the particle ionised.

We have already mentioned that normally the forbidden lines are far weaker than the allowed lines. Furthermore, a brief glance at a spectrum will show that even amongst the allowed lines there is a very considerable variation in intensity. This variation is a consequence of the differing likelihoods of the various transitions, and these are measured by three coefficients known as the Einstein transition probabilities after their discoverer.

Consider first an electron in an excited level. Even in the absence of any external disturbing processes, such a situation is normally unstable, and the electron will drop down spontaneously to a lower level. The mean time interval for such an electron to remain in the excited level is called the lifetime of that level. The reciprocal of this lifetime is the Einstein transition probability for spontaneous emission, and is denoted by A_{21} (where the upper and lower levels are 2 and 1 respectively). For normal (electric dipole) transitions, A_{21} is given by

$$A_{21} = \frac{2\pi e^2}{m_e c^3 \varepsilon_0} f_{21} \nu^2 \qquad (3P.1.4)$$

where f_{21} is the oscillator strength for emission. It is related to the oscillator strength for absorption by

$$f_{21} = (g_1/g_2) f_{12}. \quad (3P.1.5)$$

f_{12} is the ratio of the number of classical oscillators, N_c, with resonant frequency equal to that of the spectrum line to the number of atoms (etc) producing the actual absorption, N_1, which are required to produce an equal degree of absorption:

$$f_{12} = N_c/N_1. \quad (3P.1.6)$$

g_i is the statistical weight of the ith state, level or term.

The sum of all the oscillator strengths for all the transitions from any given level (emission transitions being given negative oscillator strengths) is equal to the number of valence electrons in the atom, these being the only electrons to take part in most optical transitions. The oscillator strength can thus also be regarded as the effective number of available, optically active electrons for the level. For simple atoms, f_{12} can be calculated; thus for hydrogen it is given by

$$f_{12} = \frac{2^5}{3\sqrt{3}\pi} \frac{g}{n_1^5 n_2^3} \left(\frac{1}{n_1^2} - \frac{1}{n_2^2} \right)^{-3} \quad (3P.1.7)$$

where n_i is the principal quantum number of the ith level, and g is the Gaunt factor which corrects the classical to the quantum mechanical formula, and which is usually close to unity. For other atoms f_{12} must be determined experimentally, and values are tabulated in many places (see Appendix III).

The intensity of the spontaneous emission is then given simply by

$$I(\nu_{21}) = N_2 A_{21} h \nu_{21} \quad (3P.1.8)$$

where N_2 is the number or column density (as appropriate) of atoms or ions with electrons excited to level 2.

Transitions in the opposite sense—absorptions—can only occur if radiation of the correct frequency is illuminating the particle. For a thin layer of the substance, the change in the intensity of radiation passing through it, $\Delta I(\nu_{12})$, is then

$$\Delta I(\nu_{12}) = -N_1 B_{12} I(\nu_{12}) h \nu_{12} \quad (3P.1.9)$$

(for $|\Delta I(\nu_{12})| \ll |I(\nu_{12})|$) where B_{12} is the Einstein absorption transition probability and N_1 is the number of atoms with electrons in level 1. Note:

$$\nu_{12} = \nu_{21}. \quad (3P.1.10)$$

B_{12} is thus the probability that an electron in level 1 of the atom will absorb a photon and be boosted to level 2, in unit time, and when illuminated by radiation of the requisite frequency which has unit intensity. It is related to the spontaneous transition probability by

$$B_{12} = \frac{c^2 g_2}{2h\nu^3 g_1} A_{21} \tag{3P.1.11}$$

or, from equation (3P.1.4),

$$B_{12} = \frac{\pi e^2}{h m_e \varepsilon_0 c} f_{12} \nu^{-1} . \tag{3P.1.12}$$

The third of the transition probabilities is for stimulated emission, and is denoted by B_{21}. Stimulated emission is at first sight a more obscure process than spontaneous emission or absorption, but in fact it is simply the inverse of absorption. In a classical analogy, an oscillator illuminated by radiation of its resonant frequency would absorb energy from it for phase differences between its own oscillations and those of the wave of 0 to π, and contribute energy to it for phase differences of π to 2π, the latter situation being the equivalent of stimulated emission. The physical process is thus that when an excited atom, ion or molecule is illuminated by radiation of the same frequency as one of the possible downward transitions from that excited state, that transition can be induced to occur by the radiation. The newly emitted radiation will be of the same frequency, phase, direction and polarisation as the stimulating photon. Since this process is just the inverse of absorption we have

$$g_2 B_{21} = g_1 B_{12} \tag{3P.1.13}$$

or

$$B_{21} = \frac{\pi e^2}{\varepsilon_0 m_e c h} \frac{g_1}{g_2} f_{12} \nu^{-1} \tag{3P.1.14}$$

and the intensity of the stimulated emission, by analogy with equation (3P.1.9), is

$$\Delta I(\nu_{12}) = N_2 B_{21} I(\nu_{12}) h\nu_{12} . \tag{3P.1.15}$$

From equation (3P.1.4) we see that the spontaneous emission probability is proportional to ν^2, while equation (3P.1.14) gives the stimulated emission probability proportional to ν^{-1}. Thus

$$A_{21}/B_{21} \propto \nu^3 \tag{3P.1.16}$$

and so at high frequencies (infrared, visible, ultraviolet etc) stimulated emission is usually negligible. Stimulated emission is the process occurring in masers and lasers (Chapter 6P), and the greater probability of its occurrence at lower frequencies is reflected in the observation of masers in interstellar gas clouds (Chapters 6A, 7A), but not the shorter wavelength lasers.

Knowledge of the transition probabilities, or equivalently of the oscillator strengths, is essential to many aspects of astrophysics, and much work has been invested in their experimental determination (see the Bibliography for appropriate references). The results, however, are often of low reliability due to experimental problems and the difficulties in understanding all the processes and interactions which may be taking place. Thus much of the uncertainty in stellar element abundance determinations (§§1A.2, 3A.6), for example, is due to the lack of accurate transition probabilities for the lines which are observed in stellar spectra.

Zeeman effect

It was earlier remarked (Note 1) that many states are degenerate in the strict physical sense (as opposed to the looser astrophysical usage—see §2P.1). That is, they have identical energies. Such degeneracy may be removed if the particle is in the presence of a magnetic field. Transitions then occurring will produce photons of different energies from the different pairs of states involved. In a spectrum, the lines will be observed to split into two, three or more components when the material producing them is in a magnetic field. This effect is called the Zeeman effect and applies to both emission and absorption lines, in the latter case strictly being called the inverse Zeeman effect. A similar effect, the Stark effect, arises for electric fields. The latter, however, are rarely of significance in astrophysics since the very high conductivity of plasmas causes any charge differences within them to be smoothed out very quickly.

In the normal Zeeman effect, the spectrum lines split into three components (figure 3P.5). One of these is at the normal wavelength of the line, the others are shifted to longer and shorter wavelengths by an amount

$$\Delta\lambda = (e/4\pi m_e c^2)\lambda^2 gH \tag{3P.1.17}$$

where H is the field strength, g is called the Landé factor, and is a correction factor whose value is usually close to unity. In the optical region, therefore, the splitting is by about 0.01 nm per tesla (10^{-11} m T^{-1}).

For the situation sketched in figure 3P.5, the line splits into three components, each being triply degenerate. The radiation absorbed or emitted for each component is polarised. When the line of sight is perpendicular to the magnetic field direction, all the components are linearly polarised, the directions for the outer components being perpendicular to that for the central component. When the line of sight is along the magnetic field direction, then only the two outer components are seen, and these are circularly polarised in opposite directions. At other orientations, a mixture of the two limiting cases is observed.

If the Landé factors for the two levels differ, then the energy degeneracy of the transitions is lifted. Each allowed transition can then produce a separate spectrum line. The resulting effect upon the spectrum in this case is called the anomalous Zeeman effect, since unlike the normal Zeeman effect it cannot be explained by classical physics. At very high field strengths (half a tesla or more) some of the components start to combine, a phenomenon known as the Paschen–Back effect. At moderate spectral resolutions, therefore, the spectrum may look as though the normal Zeeman effect is operating; each of the three components in this case, however, is a close multiplet.

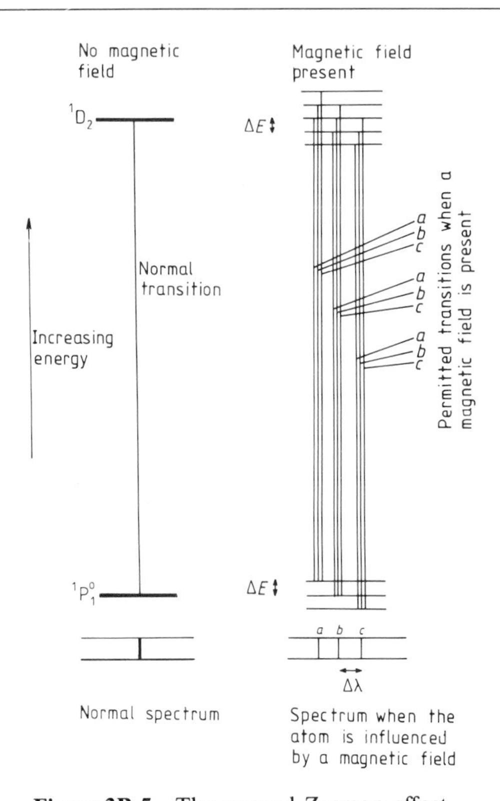

Figure 3P.5 The normal Zeeman effect.

At extreme field strengths ($>10^3$ T) such as might be found on white dwarfs and neutron stars, the quadratic Zeeman effect becomes important. The lines are then all displaced to higher frequencies by an amount

$$\frac{\varepsilon_0 h^3 n^4}{8\pi^2 m_e^3 c^2 e^2 \mu_0}(1 + M^2)H^2 \tag{3P.1.18}$$

where n is the principal quantum number and M is the projected component of the atom's angular momentum (in units of $h/2\pi$) on to the magnetic field.

3P.2 RADIATIVE TRANSFER

The details of the interaction of individual atoms and photons which we have just discussed need to be combined with macroscopic considerations in order to see how a real stellar spectrum is formed. The equation of transfer is the device for describing these macroscopic effects. It is deceptively simple in concept, and we may state it as:

The radiative energy emerging from a medium is equal to that entering it plus any energy emitted by the medium and minus any energy absorbed by the medium.

However, the practical realisation of this concept may become very complex indeed.

This macroscopic description is based upon absorption and emission coefficients. If we consider radiation passing through a layer of material (figure 3P.6), then the fraction of the energy absorbed from the beam will depend upon the thickness, density and composition of the layer. For a thin layer, we will have direct proportionality

$$dI_\nu \propto \rho I_\nu \, dr \qquad (3P.2.1)$$

where ρ is the density of the material, I_ν is the incident radiation intensity at frequency ν, dr is the thickness of the layer and dI_ν is the change in the intensity of the radiation (negative for absorption). Thus

$$dI_\nu = -\varkappa_\nu \rho I_\nu \, dr \qquad (3P.2.2)$$

where $\varkappa_\nu$ is the proportionality constant for radiation of frequency ν, and

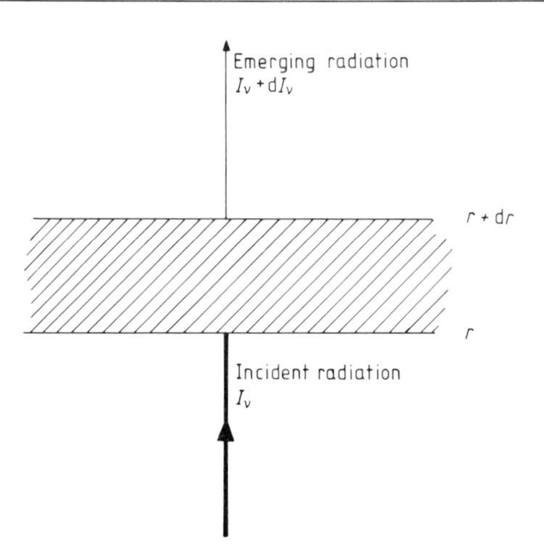

Figure 3P.6 Absorption of radiation.

depends upon the properties of the material. $\varkappa_\nu$ is called the mass absorption coefficient, and is the fractional absorption per unit mass of the material. It varies with frequency, sometimes very rapidly, resulting in the absorption lines and other features of the spectrum.

A useful related quantity is the optical depth, previously mentioned in §2A.3. This is a measure of the fractional change in the radiation intensity for a thick layer of the material. It may be found from equation (3P.2.2) by integration.

$$\int_{I_\nu(1)}^{I_\nu(2)} \frac{\mathrm{d}I_\nu}{I_\nu} = -\int_{r_1}^{r_2} \varkappa_\nu \rho \, \mathrm{d}r \tag{3P.2.3}$$

where I_ν (1) is the intensity at position 1, etc. Therefore,

$$\log_e [I_\nu(2)/I_\nu(1)] = -\int_{r_1}^{r_2} \varkappa_\nu \rho \, \mathrm{d}r. \tag{3P.2.4}$$

If we set r_2 equal to the top (emergent) surface of the layer, and r_1 to correspond to a physical depth of x, then we have

$$I_\nu(0) = I_\nu(x) \exp\left(-\int_x^0 \varkappa_\nu \rho \, \mathrm{d}r\right). \tag{3P.2.5}$$

We now define the modulus of the exponent as the 'optical' depth, τ_ν, (though the formalism applies at all wavelengths)

$$\tau_\nu = \int_x^0 \varkappa_\nu \rho \, \mathrm{d}r \tag{3P.2.6}$$

and its value may be found when the detailed behaviours of $\varkappa_\nu$ and ρ are known between depth x and the surface. For a deep layer of material, therefore, the radiation emerging at the surface is related to the intensity at an optical depth τ_ν by

$$I_\nu(0) = I_\nu(\tau_\nu) \mathrm{e}^{-\tau_\nu}. \tag{3P.2.7}$$

Apart from the simplicity of the relation, the great virtue of optical depth is that the apparently visible surface of a star or other semi-transparent object occurs at an optical depth of unity. Thus the solar photosphere corresponds to unit optical depth integrated over the visible spectrum. To an extraterrestrial being with eyes sensitive to 100 nm radiation, the visible surface of the Earth would be 100 km above the solid surface and so on. An optical depth of unity means that 1/e (36.8%) of the radiation from that point emerges at the surface. For other optical depths we have the following emergent intensities:

τ_ν	$I_\nu(0)/I_\nu(\tau_\nu)$ (%)
0	100
0.1	90.5
0.5	60.7
1.0	36.8
1.5	22.3
2.0	13.5
3.0	5.0
5.0	0.7

Emission of radiation on a macroscopic scale may be considered in a similar manner to its absorption. The intensity of the emitted radiation will be proportional to the mass of the material, and the proportionality constant for a particular frequency is called the mass emission constant, j_ν. Thus

$$\mathrm{d}I_\nu = j_\nu \rho \,\mathrm{d}r. \qquad (3P.2.8)$$

With these definitions, we may now return to the equation of transfer. The change of intensity of a beam of radiation is the sum of the absorption and emission contributions from the material through which the radiation is passing. So, from equations (3P.2.2) and (3P.2.8), we have

$$\mathrm{d}I_\nu = -\varkappa_\nu \rho I_\nu \,\mathrm{d}r + j_\nu \rho \,\mathrm{d}r \qquad (3P.2.9)$$

and this is the equation of transfer. It may more commonly be encountered in several slightly different forms, however:

$$\mathrm{d}I_\nu/\mathrm{d}r = -\varkappa_\nu \rho I_\nu + j_\nu \rho \qquad (3P.2.10)$$

or

$$\mathrm{d}I_\nu/\mathrm{d}\tau_\nu = I_\nu - j_\nu/\varkappa_\nu \qquad (3P.2.11)$$

where

$$\mathrm{d}\tau_\nu = -\varkappa_\nu \rho \,\mathrm{d}r. \qquad (3P.2.12)$$

The ratio of the emission and absorption coefficients is called the source function, and is usually denoted by S_ν. Most actual problems involving the equation of transfer reduce to that of determining the behaviour of the source function so that the equation may be solved. If the radiation is emerging at an angle θ to the normal to the surface, then the physical lengths of the paths are increased by a factor $(\cos\theta)^{-1}$. Conventionally, $\cos\theta$ is denoted by μ. Thus we get the final and most generally used form of the equation of transfer,

$$\mu(\mathrm{d}I_\nu/\mathrm{d}\tau_\nu) = I_\nu - S_\nu. \qquad (3P.2.13)$$

If S_ν is known, then equation (3P.2.13) may be solved through the use of an integrating factor of $\exp(-\tau_\nu/\mu)$:

$$\frac{d}{d\tau_\nu}(I_\nu e^{-\tau_\nu/\mu}) = e^{-\tau_\nu/\mu}\frac{dI_\nu}{d\tau_\nu} - \frac{I_\nu}{\mu}e^{-\tau_\nu/\mu}. \qquad (3P.2.14)$$

Thus

$$\frac{d}{d\tau_\nu}(I_\nu e^{-\tau_\nu/\mu}) = -e^{-\tau_\nu/\mu}\frac{S_\nu}{\mu} \qquad (3P.2.15)$$

and so upon integration between levels 1 and 2

$$I_\nu(2,\mu)\,e^{-{}_2\tau_\nu/\mu} - I_\nu(1,\mu)\,e^{-{}_1\tau_\nu/\mu} = \int_{{}_1\tau_\nu}^{{}_2\tau_\nu} -e^{-\tau_\nu/\mu}\frac{S_\nu}{\mu}\,d\tau_\nu. \qquad (3P.2.16)$$

For a star, the optical depth of its centre is effectively infinity for an external observer. So setting ${}_1\tau_\nu$ to infinity and ${}_2\tau_\nu$ to zero, the emergent radiation intensity is given by

$$I_\nu(0,\mu) = \int_0^\infty \mu^{-1}\,e^{-\tau_\nu/\mu}S_\nu\,d\tau_\nu. \qquad (3P.2.17)$$

Equation (3P.2.17) is the formal solution of the equation of transfer, and it may become an actual solution when S_ν is known as a function of τ_ν. We may see, however, that the emerging radiation amounts to the integral of the source function along the line of sight, weighted by $\exp(-\tau_\nu/\mu)$. Since the weighting factor decreases very rapidly with increasing optical depth, in most situations the form of the emerging radiation is dominated by the values of the source function close to the surface.

3P.3 EFFECTIVE SURFACE GRAVITY

A remarkable result due to von Zeipel and hence named after him is that:

The brightness of a point on the surface of a star is directly proportional to the local effective acceleration.

Thus rotating stars, or stars in binary systems, etc, where the effective acceleration is a combination of that due to the actual gravitational force plus centrifugal and/or tidal effects, must vary in their brightness over their surfaces. For pure rotation, and ignoring any effect of the change from sphericity for the star which may be occurring, we have

$$I(\varphi) \propto (G\mathcal{M}_*/R_*^2) - R_*\omega_*^2\cos^2\varphi \qquad (3P.3.1)$$

where $I(\varphi)$is the surface brightness at the stellar 'latitude', φ, and $\mathcal{M}_*$, R_*, and ω_* are the star's mass, radius and angular rotational velocity, respectively. Currently (1985) no star has been studied directly with sufficient angular resolution for the effect to be detected. Even for the Sun, where the angular resolution is sufficient, the rotational velocity is too low. Thus, from equation (3P.3.1),

$$I(\varphi) \propto 273 - 5.5 \times 10^{-3} \cos^2 \varphi \qquad (3P.3.2)$$

so that the poles will be brighter than the equator by only 0.002%; too little for detection by present day techniques. The effect, however, must be taken into account when studying spectrum line profiles in detail from rapidly rotating stars, and when interpreting the light curves of close binary systems. In reality the details of gravitational darkening will be complicated by the differential rotation occurring in most stars (see §4A.1 for example) and by the variation in gravity due to the non-spherical shape of the stars.

3P.4 ATOMIC AND IONIC POPULATIONS

In TE and LTE (§2A.3), the properties of the gas may be described by the use of a single temperature in the equations. Some of the equations have already been mentioned (Chapter 2P), but here two important additional aspects of the material's behaviour are discussed. In a large assemblage of atoms some will be ionised and some neutral. Furthermore, within a given type of atom or ion, some will be in the ground state, whilst others are excited to various degrees. These phenomena are described in TE and LTE by the use of the Saha and Boltzmann equations.

The Saha equation gives the proportions of a species of atom which are neutral, ionised, doubly ionised, triply ionised, etc. It takes the form

$$\frac{N_{I+1}}{N_I} = \frac{2U_{I+1}}{N_e U_I} \left(\frac{2\pi m_e kT}{h^2}\right)^{3/2} e^{-\chi_I/kT} \qquad (3P.4.1)$$

where N_I and N_{I+1} are the number densities of the ions of a single atomic species in the Ith and $(I + 1)$th stages of ionisation. The neutral ion corresponds to $I = 0$. N_e is the free electron number density, and U_I and U_{I+1} are called the partition functions. They are obtained for a particular ion from the statistical weights (the number of degenerate states forming the level) and the energies of its levels:

$$U_i = g_{i0} + g_{i1}\, e^{-E_{i1}/kT} + g_{i2}\, e^{-E_{i2}/kT} + \cdots. \qquad (3P.4.2)$$

Here, g_{in} and E_{in} are the statistical weight and excitation energy of the nth level of the ith ion respectively. The partition functions take account of the ions with electrons in excited levels. Often only the first few terms of equation (3P.4.2), or even just the first term, are sufficient to determine the partition function with sufficient accuracy for most purposes. m_e is the mass of the electron, k is Boltzmann's constant, h is Planck's constant, T is the gas temperature and χ_I is the ionisation potential of the Ith ion.

With N_e in m^{-3}, T in K, and χ_I in eV (their customary units), Saha's equation becomes

$$\frac{N_{I+1}}{N_I} = 4.829 \times 10^{21} \frac{U_{I+1}}{U_I} N_e^{-1} T^{3/2} \exp(-1.160 \times 10^4 \chi_I T^{-1}). \qquad (3P.4.3)$$

By starting from the neutral atom, the equation may be used to determine all the atomic and ionic populations for any atom at any temperature. Thus in the line-forming regions of the Sun, where

$$T = 6390 \text{ K} \qquad (3P.4.4)$$

$$N_e = 6.398 \times 10^{19} \text{ m}^{-3} \qquad (3P.4.5)$$

we find for hydrogen

$$N_1/N_0 = 3.7 \times 10^{-4} \qquad (3P.4.6)$$

so that only 0.037% of the hydrogen in the solar surface layers is ionised. By contrast, for calcium we find

$$N_1/N_0 = 1.2 \times 10^3 \qquad (3P.4.7)$$

$$N_2/N_1 = 8.5 \times 10^{-3}. \qquad (3P.4.8)$$

Thus the bulk of calcium is in the form of the singly ionised state, and the absorptions from its ground state (often called resonance transitions) give rise to the extremely strong Fraunhofer H and K lines.

It is also useful to have Saha's equation in terms of the electron pressure rather than number density for some purposes, and it then becomes

$$\frac{N_{I+1}}{N_I} = 0.06667 \frac{U_{I+1}}{U_I} P_e^{-1} T^{5/2} \exp(-1.160 \times 10^4 \chi_I T^{-1}) \qquad (3P.4.9)$$

where P_e is the pressure of the free electrons in N m^{-2}.

Within a given type of atom or ion, the relative populations of its ground and excited states are given by Boltzmann's equation:

$$\frac{N_b}{N_a} = \frac{g_b}{g_a} e^{-(E_b - E_a)/kT} \qquad (3P.4.10)$$

where N_a and N_b are the number densities of a single type of atom or ion which has electrons excited to levels a and b respectively, g_a and g_b are the statistical weights of levels a and b, and E_a and E_b are the excitation energies of levels a and b. If we refer all the populations to that of the ground state, then we find

$$\mathcal{N}_i = \frac{g_i}{g_0} e^{-1.160 \times 10^4 E_i T^{-1}} (\text{m}^{-3}) \qquad (3P.4.11)$$

where $\mathcal{N}_i$ is the relative population of the ith level, g_0 and g_i are the statistical weights of the ground state and of the ith level respectively, and E_i is the excitation energy of the ith level in eV. Thus, again for solar surface conditions, we find that only one hydrogen atom in 250 000 is excited above the ground state. A temperature of some 85 000 K is needed for there to be equal populations in the ground and first excited states. However, by then only one hydrogen atom in 100 000 will remain unionised! We may thus see that the variation over stellar spectra (§3A.2) in the strength of the Balmer lines (which originate from the first excited level of hydrogen) is governed by the effects of both excitation and ionisation. At low temperatures too few atoms are excited to produce strong lines, while the majority of the atoms are ionised at higher temperatures, again leaving few of them available for line production. The balance of these effects occurs for a stellar surface temperature of about 10 000 K, and the Balmer lines then achieve their greatest strengths. Similar considerations for other atoms and ions result in similar variations in their line strengths, their peak intensities generally occurring at different temperatures from that for hydrogen (figure 3A.4).

The Saha and Boltzmann equations are only strictly valid in TE, and approximately so in LTE. Often, however, they may still be useful even when the actual conditions do not approximate to these states. Once departures from LTE become large, however, then each individual reaction must be considered separately in order to determine the overall properties of the material. Non-LTE stellar atmospheric models and synthetic spectra (§3A.6) are becoming more widespread. But they require vastly more computer time than simpler models, so that LTE and the use of the Saha and Boltzmann equations etc are likely to remain useful assumptions for some time to come, and in any case they may provide first approximations from which the more precise analyses may start.

3A Stars: Observations and Observed Types and Properties

SUMMARY

Apparent and absolute magnitudes, colour indices, spectral and luminosity classifications, radial velocities of stars, line broadening mechanisms (pressure, Doppler, magnetic), Fraunhofer lines, stellar surface temperatures, the HR diagram, photometric variable stars, classes (Cepheids, explosive variables, pulsars, T Tauri stars), non-'solar' type stellar compositions, stellar rotation and magnetic fields, black holes.

See also stellar masses (§1A.1), abundances (§1A.2), physics of gases (§2P.1), degeneracy pressure (§2P.1), radiation laws (§2P.3), mass limits (§2A.1), stellar luminosities, sizes and densities (§§2A.4, 2A.5), spectrum line formation (§3P.1), Zeeman effect (§3P.1), equation of transfer (§3P.2), effective surface gravity (§3P.3), atomic and ionic populations (§3P.4), solar constant, luminosity, mass, composition and radius (§4A.1), stellar winds (§5A.2), synchrotron radiation (§6P.2), supernova remnants (§6A.6), and stellar formation and evolution (§§8A.1, 8A.2).

INTRODUCTION

The main basic observations of a star are a simple measurement of its brightness at a particular wavelength (photometry), and the measurement of the changes in its brightness with wavelength (spectroscopy). We start, therefore, by considering these techniques in some detail, before going on to look at the observed and derived properties of stars.

3A.1 PHOTOMETRY

It is obvious that repeated measurements of the luminosity of a star will reveal whether or not it is a variable. Were this all that photometry could reveal about the nature of stars, then it would still be a very useful observing technique. However, with the addition of only a very slight complication to the basic process, the technique can be made to give vastly more, and more fundamental, information on both variable and non-variable stars. The additional complication is simply to measure the star's brightness at two or more different and precisely defined wavelengths. Before considering this further, however, we take a look at the basic process.

Apparent magnitude

The brightness of a star *as it is seen in the sky* is measured by its apparent magnitude. This is a logarithmic scale, defined in its present form by Pogson in 1856 by the equation

$$m_1 - m_2 = -2.5 \log_{10}(E_1/E_2) \tag{3A.1.1}$$

where E_1 and E_2 are the energies per unit area and per unit time at the surface of the Earth from the stars 1 and 2. m_1 and m_2 are called the apparent magnitudes of the same two stars.

The zero of the scale is set from the apparent magnitudes assigned to a series of standard stars near the north pole (the North Polar Sequence), and is such that stars of magnitude 6 are just visible to the naked eye from a good observing site. The precise form of the resulting scale traces its roots back over two millennia to one of the earliest star catalogues, that due to Hipparchus. As a result of this historical antecedent, the scale is an awkward one; it assigns a numerically smaller number to the magnitude of a brighter object, and the energy changes by a factor of 2.51189... ($=10^{0.4}$) for a difference of one magnitude. The apparent magnitudes in visible radiation of some well known objects are listed below (the magnitudes outside the visible region may be very different from these figures in both relative and absolute senses).

Sun	−26.7
Full Moon	−12.7
Venus (maximum)	− 4.3

Jupiter (opposition)	− 2.6
Mars (opposition)	− 2.02
Sirius A	− 1.45
Betelgeuse	− 0.73
Mercury (maximum)	− 0.2
Saturn (opposition)	+ 0.7
Polaris (slightly variable)	+ 2.3
M31 (integrated)	+ 3.5
Ganymede (opposition)	+ 4.6
Io (opposition)	+ 4.9
Uranus (opposition)	+ 5.5
Faintest object visible to the unaided eye	+ 6.0
Neptune (opposition)	+ 7.9
Faintest object visible in a 300 mm telescope	+14
Pluto (opposition)	+14.9
Faintest object visible in a 1 m telescope	+16.5
Faintest object visible in a 5 m telescope	+20
Faintest object detectable by modern techniques using a 5 m telescope on the surface of the Earth	+24
Projected limit for the Hubble space telescope	+29

Absolute magnitude

Apparent magnitude is a function of both the intrinsic luminosity of the object and its distance. To obtain a measure that reflects the intrinsic luminosity alone, we define the absolute magnitude as

The apparent magnitude of the object if it were a distance of 10 pc (*Appendix I*) *from the Earth.*

It is normally denoted by an upper-case M to distinguish it from the lower-case m used for apparent magnitude. The two measures are related by

$$M = m + 5 - 5 \log_{10} D \qquad (3A.1.2)$$

where D is the actual distance of the object in parsecs. The absolute magnitudes of some of the previously listed objects, together with some other examples, are listed below

Sun	+ 4.8
Full Moon	+32
Venus (maximum)	+29
Jupiter (opposition)	+26
Sirius	+ 1.4
Betelgeuse	− 6
Polaris	− 4.6
M31 (integrated)	−21
Supernova (maximum)	−19
Nova (maximum)	− 8
Brightest supergiants	− 8
Faintest observed white dwarfs	+16

Spectral band

The examples listed above to illustrate apparent and absolute magnitudes are all *visual* magnitudes: the magnitudes estimated by eye or measured using a detector with a similar response to that of the eye. Since the peak sensitivity of the eye occurs at a wavelength of about 550 nm, the magnitudes correspond to the yellow-green luminosities of the objects. Other detectors can have different sensitivity ranges from that of the eye, however; the basic photomultiplier has a peak sensitivity near 400 nm, while the basic charge-coupled device (CCD) peaks near 800 nm (see *Astrophysical Techniques* for further details of these and other detectors and of the techniques of photometry). Thus when such devices are used in photometers, they will generally give somewhat different results from these visual measures. In order to provide consistent and reproducible photometric measures, the wavelengths over which the energy is detected must be specified. This is usually accomplished via the use of filters of known properties to define the observed spectral region. Many hundreds of such filters have been and are being used, but the most widespread system in common use is called the *UBV* system, and was first defined by Johnson and Morgan in 1953.

The *UBV* system is based upon three filters with peak transmissions at 365, 440 and 550 nm, and bandwidths of about 90 nm. The labels of the pass bands of the system stand for *u*ltraviolet, *b*lue and *v*isual regions of the spectrum. The *V* region's longwave cutoff is provided by the decreasing sensitivity of a photomultiplier with a caesium antimonide photocathode rather than by the filter, and so such a detector must be used in the photometer if it is to provide results consistent with those obtained by other workers. The measured *V* magnitudes, however, then correspond quite well to those obtained by eye. The *B* magnitudes similarly correspond to early photographically determined measurements made on unsensitised emulsions. The *U* region's shortwave cutoff is due to atmospheric absorption, and so it can give very variable results except from the most favoured observing sites.

The *UBV* photometric system is a wide-band system. Most other filter systems are medium- or narrow-band, the filter widths being from a few tens to a few nanometres, and they tend to have more specialised applications. There is, however, another magnitude 'system' with a bandwidth even larger than those of the *UBV* filters, and this gives the *bolometric* magnitude of the object. It is based upon the total energy emitted by the star at all wavelengths. Since the limited range of detectors and atmospheric absorption means that it cannot be measured directly, its value is estimated by various methods, and its difference from the *V* magnitude calculated. This difference is called the bolometric correction, BC,

$$\mathrm{BC} = M_{\mathrm{Bol}} - M_V = m_{\mathrm{Bol}} - m_V \qquad (3A.1.3)$$

where M_{Bol} and M_V are the absolute bolometric and *V* magnitudes respectively, and m_{Bol} and m_V are the apparent bolometric and *V* magnitudes respectively. The bolometric correction is tabulated below for various values of effective temperature (table 3A.1).

Table 3A.1 Bolometric corrections.

Temperature (K)	Bolometric correction Main sequence	Giants	Supergiants
3 000	−2.0	−1.8	−1.6
4 000	−0.72	−0.8	−0.8
5 000	−0.17	−0.2	−0.25
6 000	−0.03	–	−0.10
7 000	−0.03	–	−0.10
8 000	−0.09	–	−0.18
9 000	−0.22	–	−0.25
10 000	−0.42	–	−0.35
15 000	−1.40	–	−0.90
20 000	−2.0	–	−1.7
25 000	−2.5	–	−2.3
30 000	−3.0	–	−3.0
35 000	−3.5	–	–
40 000	−4.0	–	–

The absolute bolometric magnitude is related to the luminosity of the star, L_*, by

$$M_{\mathrm{Bol}} = 71.193 - 2.5 \log_{10} L_* \qquad (3A.1.4)$$

or

$$L_* = 3 \times 10^{28} \times 10^{-0.4 M_{\mathrm{Bol}}} \qquad (3A.1.5)$$

for L_* in watts.

Colour indices

We may now return to the 'slight complication' which is required to turn photometry into one of the most useful astronomical techniques. This simply consists of comparing the star's brightnesses at two different wavelengths. For a pure black body energy distribution (figures 2P.6, 2P.7) the result depends only upon the temperature. Stellar spectra approximate quite well to black body distributions, but not perfectly. Careful choice of the measured spectral regions, however, still enables the temperature of the star to be estimated. The B and V regions are two such appropriate choices. The comparison of the brightness is by means of the difference between the two magnitudes, and is called the *colour index*. Thus we have

$$B - V = m_B - m_V = M_B - M_V \qquad (3A.1.6)$$

and from equation (3A.1.1) we then find

$$B - V = -2.5 \log_{10}(E_B/E_V). \qquad (3A.1.7)$$

The star's effective surface temperature is then given by the semi-empirical relation

$$T \simeq \frac{8540}{(B-V) + 0.865} \quad (\mathrm{K}) \qquad (3A.1.8)$$

for temperatures up to about 10 000 K. The relationship for higher temperatures is shown on figure 3A.1.

Since stellar spectra are not truly those of black bodies (figure 3P.4, for example), some of the deviations between the two may be investigated by appropriate choices of other colour indices. Thus the Balmer discontinuity (§3P.1) is bracketed by the U and B regions. The value of the $U - B$ colour index hence largely reflects the depth of the Balmer discontinuity in the particular star's spectrum. This peaks in stars with temperatures near 10 000 K (see discussion in §3P.4). The $U - B$ colour index thus also varies with temperature, but in a much more complex manner than the $B - V$ index (figure 3A.2). Under appropriate conditions UBV photometry can provide estimates of stellar distances, absolute magnitudes, spectral types etc in addition to the surface temperatures.

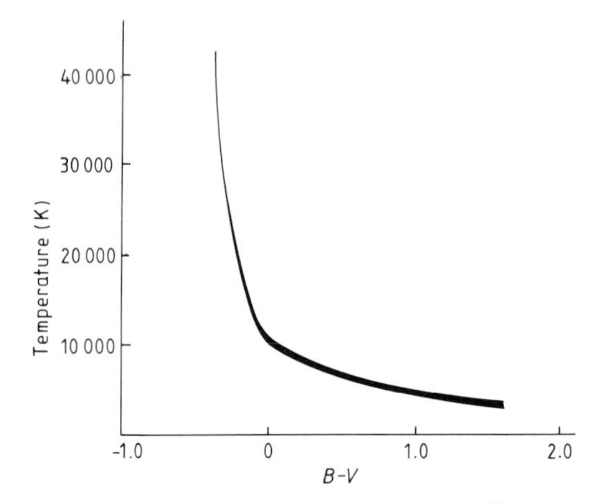

Figure 3A.1 Observed relationship between a star's effective temperature and its $B-V$ colour index.

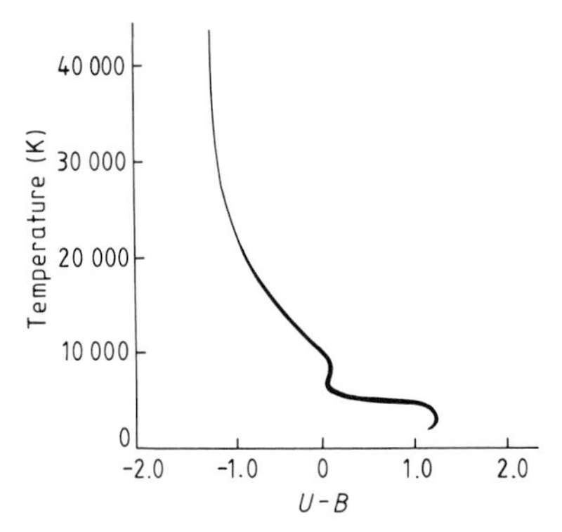

Figure 3A.2 Observed relationship between a star's effective temperature and its $U-B$ colour index.

3A.2 SPECTROSCOPY

Spectral and luminosity classifications

One of the most pervasive astronomical uses of spectroscopy is in the determination of the spectral types and luminosity classes of the stars. These are based upon eye estimates of the features visible on medium dispersion spectrograms. An experienced worker can determine a great deal of information about the star for a small investment of his time and effort, so that it is likely to remain a popular method for studying the stars for the foreseeable future.

It was realised very early in the history of stellar spectroscopy that stellar spectra showed certain recurring patterns which enabled them to be usefully grouped together. One of these early grouping systems was based upon the intensities of the hydrogen lines relative to the other lines in the spectrum. The strongest hydrogen line spectra were group A, those slightly weaker, group B, and so on. It was also thought that the evolution of stars caused the spectra to change in the sense of decreasing hydrogen line intensity and of increasing complexity, as the star aged. Thus the least complex spectra, types A, B, etc, were thought to come from young or early stars, while the more complex spectra came from older or later stars. These ideas are now known to be incorrect, and the underlying reason for the different appearances of stellar spectra is the variation in the surface temperatures of the stars. Thus a more useful classification would be based upon a monotonic temperature sequence of the spectra. Somewhat unfortunately, this was accomplished by rearranging and adapting the older system, so that it is now rather untidy and unnecessarily difficult to use. It seems unlikely to be rationalised now, however, so the student must perforce become familiar with it as it stands.

The basis of the modern system of classification was laid down in Harvard's Henry Draper star catalogue, and so is sometimes known as the Harvard classification system. More commonly it is called the MKK system after the later development of it by Morgan, Keenan and Kelman at Yerkes. There are thirteen groups of spectra, of which seven, labelled by the letters

O B A F G K M

form the core. The O-type stars are the hottest (40 000 K or more, though some white dwarfs may have temperatures up to 100 000 K; see below), and the M-type the coolest (2500 K). In another hangover from the earlier system, the first three or four groups are often called the early-type stars, while the last two or three are called the late-type stars. This is now just a convenient convention however, and no longer has any evolutionary significance. A more precise relationship between spectral type and temperature is shown in figure 3A.3. A useful mnemonic for the order of the classes is

*O*h *B*e *A* *F*ine *G*irl, *K*iss *M*e.

Each of the major classes is subdivided into ten, with arabic numerals denoting the subdivisions. Thus the Sun, for example, is of spectral class G2, while Sirius is A1, and Betelgeuse M2.

In addition to the spectral class, we may add a luminosity parameter. This

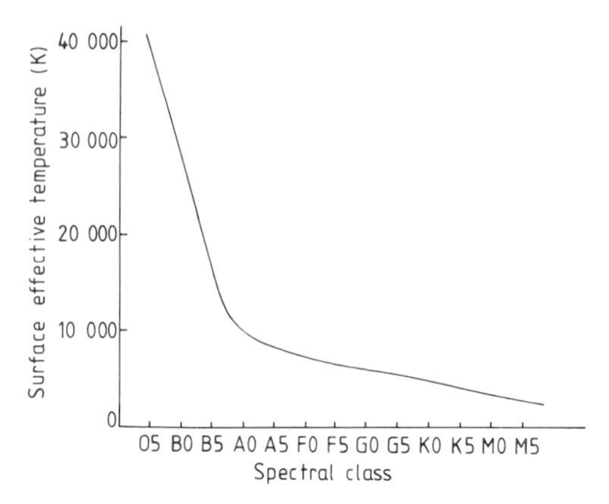

Figure 3A.3 Relationship between spectral class and surface effective temperature for main sequence stars.

is based upon the fact that the widths of spectral lines increase as the gas pressure in the regions producing them increases (see the discussion on pressure broadening later in this section). The physically largest stars have the lowest densities (§2A.5), and hence the lowest surface pressures. Thus the spectral lines originating from large and hence very luminous stars generally have smaller widths than those from smaller stars. The net effect is to change the intensity ratios between some pairs of lines in the spectra from stars of similar spectral classes, but with different luminosities. The luminosity class is added as a roman numeral after the spectral class, with the following groupings:

Luminosity class	Star type	Line width
I	Supergiants	Narrow
II	Bright giants	↓
III	Giants	
IV	Subgiants	↓
V	Main sequence (or dwarfs)	
VI	Subdwarfs	
VII	White dwarfs	Very broad

Some workers also distinguish a class, 0, which we might call super-supergiants! A more complete classification for the earlier examples is therefore: the Sun, G2 V; Sirius, A1 V; Betelgeuse, M2 I.

The classification of a spectrum is actually based upon the intensity ratios of pairs of lines which are especially sensitive to temperature or luminosity.

The lines used vary over the classification system, some only being used to distinguish one or two subclasses, others being of wider use. The reader is referred to specialist works on the topic, such as *An Atlas of Representative Stellar Spectra* by Y Yamashita, K Nariai and Y Norimoto (University of Tokyo Press, 1977), for further details of the techniques of spectral classification.

The main features of the spectra are listed below (table 3A.2), and the variations in the strengths of the main spectral lines are shown in figure 3A.4.

Table 3A.2 Characteristics of spectral classes.

Spectral class	Major characteristics of the spectrum
O4–B0	Few lines, most of those actually visible being from highly ionised Si, N, etc. H Balmer and ionised He lines visible.
B0–B5	Balmer lines strengthening, neutral He and lower stages of ionisation of Si, N etc now producing lines.
B5–A0	Neutral He lines disappear, H Balmer lines peak in intensity at A0, Ca II *K* line appears.
A0–F0	Balmer lines weakening, numerous lines due to singly-ionised metals appear.
F0–G0	Spectra becoming more complex, Balmer and ionised metal lines weakening, neutral metal lines strengthening.
G0–K0	Ca II *H* and *K* lines peak in their intensities, Balmer lines continue to weaken, neutral metal lines continue to increase in intensity.
K0–M0	Balmer lines still just visible, many lines due to neutral metals, a few TiO bands appear.
M0–M8	TiO bands dominate the spectrum.

Thus we have a relatively straightforward system for classifying stellar spectra, and thereby rapidly making estimates of important physical parameters such as temperature, size, luminosity, density, pressure etc. Unfortunately we now have to add a considerable number of special cases which complicate the classification greatly.

First there are six remaining major classes:

R N S W P Q.

R and N are sometimes combined into the single class C. They are rare variable giants with temperatures similar to stars of classes G and K. Their spectra are characterised by strong bands of cyanogen (CN), molecular carbon, and other carbon compounds (hence their alternative designation: C for *C*arbon stars).

There are other abundance peculiarities, notably the presence of technetium, all of whose isotopes are radioactive and none with half lives longer than 2.5×10^6 years. Hence they must have been formed recently within the star. The relationship of these stars to more normal types is not clear, although they are generally believed to be in an advanced stage of evolution. S-type stars are generally long-period variables with many spectral characteristics in common with the N-type stars, though probably of rather lower temperature. Their defining characteristics are the presence of strong ZrO bands in their spectra, and the absence or very weak presence of TiO bands.

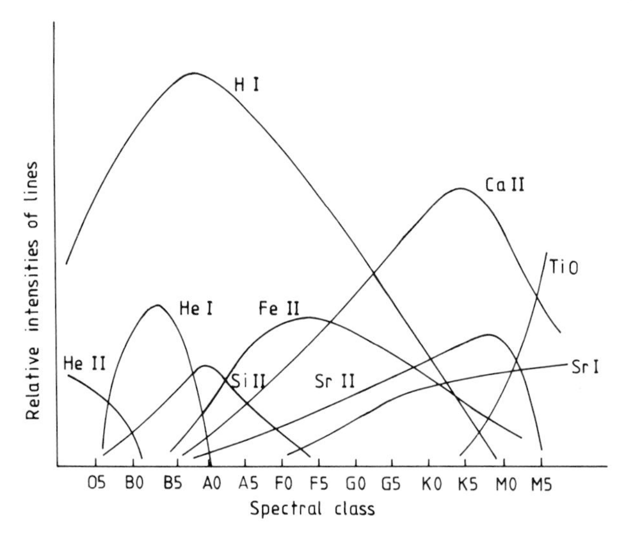

Figure 3A.4 Schematic diagram showing the variation of the relative intensities of the spectral lines of some atoms and ions with spectral class.

The stars of class W are known as the Wolf–Rayet stars. They are discussed in more detail in §3A.6. They are very hot, and exhibit abundance peculiarities; hydrogen lines are weak or absent and helium lines very strong. There are two subdivisions; the WC stars which have excess carbon and oxygen, and the WN stars with excess nitrogen. They are interpreted as possibly being the central cores of evolved stars whose overlying layers have been stripped away by an intense stellar wind (Chapter 5A). A sequence of intermediate stellar types, labelled Of, O(f), and O((f)) have been identified linking the Wolf–Rayet stars to the O and early B-type stars.

The remaining two classes, P and Q, are separate from the main groupings. Class P are nebular spectra; the spectrum consists almost entirely of emission lines, often due to forbidden transitions (§3P.1) with almost no continuum, and this is interpreted as arising in low-pressure gaseous nebulae. Class Q denote novae (§3A.5) during their outbursts. In practice, the last two classes are rarely used; their spectra are more fully described, or they are compared with well known examples of the type. However, including these last two types, the full arrangement of the major spectral classes is indicated in the schematic diagram below:

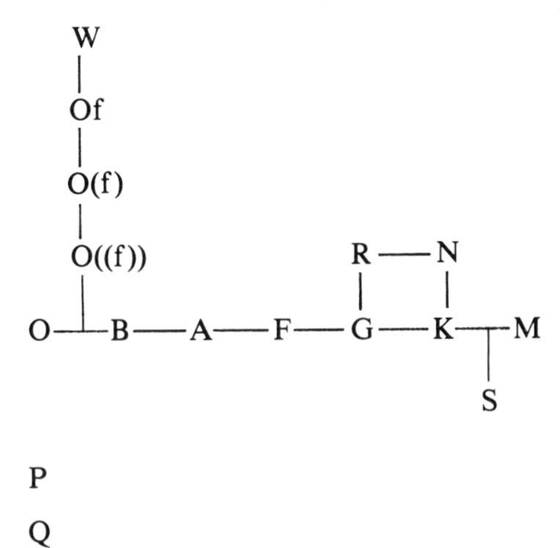

The decimal subdivision of the major classes also has a few vagaries. Class O and M stars at one time were subdivided by the use of lower case letters,

Oa Ob Ma etc

and this may still sometimes be encountered in the literature. The Of stars (above) are also a remnant of this system. O3 is the hottest and M8 the coolest of the classes currently in use. The gap between O9 and B0 is much larger than average, and so O9.5 is often used and represents a whole division rather than half of one.

Subdivisions of the luminosity class are used for the white dwarfs. They are subdivided into

DA	Hydrogen-rich
DB	Helium-rich
DC or BC	Continuum spectrum (lines less than 5% of the continuum intensity)
DO	He II strong
DQ	Carbon-rich
DZ	Metal lines only.

An older system replaces the last three of these classes with

DF	Calcium-rich
DP	Magnetic stars

The only other important variation is that luminosity class I is very often subdivided into classes Ia and Ib.

Special features or additional details of the spectra may be indicated by the addition of prefixes and suffices to the main spectral classes. The usage of these varies somewhat, and there is some redundancy in the notation, but the main terms likely to be encountered are:

Prefix	
c or sg	Very narrow spectrum lines; a supergiant; luminosity classes 0, I, Ia, Ib etc.
g	Giant; luminosity class III.
d	Dwarf; main sequence; luminosity class V.
sd	Subdwarf; luminosity class VI.
w, wd or D	White dwarf; luminosity class VII.
Suffix	
e	Emission lines present in the spectrum. The Greek letter of the last Balmer line in emission may also be added.
n	Very broad or nebulous lines.
s	Sharp lines (but less so than supergiants).
v	Variable spectrum.
p, pec	Peculiar spectrum. Most of the features are standard to the main spectral class, but there are some non-standard features as well.
m	Metallic lines present when not expected or stronger than expected.
k	Interstellar absorption lines in the spectrum.
a, ab, b	Suffix to the *luminosity* class: a, the brightest and b, the faintest subdivision.

Often several of the prefixes and suffices may be combined to indicate all the features of the spectrum. A few examples are

γ Vel	WC8 + O7v	(composite spectrum)
γ Cas	B0ev IV	
α^2 CMa	wA5	(Sirius B)
α Boo	K2p III	
α Ori	M2ev I	(Betelgeuse).

Radial velocities

After spectral and luminosity classification, the commonest parameter of a star to be obtained from its spectrum is its radial velocity. By this term, unless otherwise stated, an astronomer means the velocity of the star radial to the Earth, and not that radial to the star or other objects concerned. The convention is used that the radial velocity is positive when directed away from the Earth and negative when directed towards the Earth. It is usually necessary to correct the measured velocities for that of the Earth in its orbit, to produce the heliocentric velocities. The corrected velocity is given by

$$V_H = V_G + v[\sin \delta_a \sin \delta_* + \cos \delta_a \cos \delta_* \cos(\alpha_* - \alpha_a)] \quad (3A.2.1)$$

where V_G is the object's geocentric velocity, V_H is the object's heliocentric velocity, v is the Earth's orbital velocity (29.78 km s^{-1} on average), α_* and δ_* are the right ascension and declination of the object and α_a and δ_a are the right ascension and declination of the instantaneous apex of the Earth's orbital motion, given by

$$\alpha_a = \tan^{-1}(-\cot \varepsilon \cot \lambda_\odot) \quad (3A.2.2)$$

$$\delta_a = \sin^{-1}(-\sin \varepsilon \cos \lambda_\odot) \quad (3A.2.3)$$

where ε is the obliquity of the ecliptic (23° 27′) and $\lambda_\odot$ is the celestial longitude of the Sun at the instant of the observation.

The velocity itself, of course, is obtainable from the Doppler formula

$$\Delta\nu/\nu = \Delta\lambda/\lambda = v/c \quad (3A.2.4)$$

or, for high velocities, from the relativistic version

$$\frac{\Delta\nu}{\nu} = \frac{\Delta\lambda}{\lambda} = \frac{1-(v/c)}{[1-v^2/c^2]^{1/2}} - 1 \quad (3A.2.5)$$

where $\Delta\lambda$ and $\Delta\nu$ are the wavelength or frequency shifts

$$\Delta\lambda = \lambda_L - \lambda_o \quad (3A.2.6)$$

$$\Delta\nu = \nu_L - \nu_o. \quad (3A.2.7)$$

The subscript 'o' denotes an observed value and the subscript 'L' denotes a laboratory (unshifted) value; while v is the velocity of the object along the line of sight. There is also a redshift for the lines from material moving across the line of sight at very high velocities—the transverse Doppler shift. With the exception of the intriguing system known as SS433 (§3A.9), however, this is not generally of significance in stellar astrophysics.

Spectrophotometry

The greatest return of information from a spectrum comes from the detailed study of the line strengths and shapes (profiles). This is known as spectrophotometry and provides information on (amongst other matters) the star's surface structure, temperature, pressure, electron, ion and total densities, turbulent motions, mass motions, element abundances, rotation, magnetic fields, etc.

Spectrum lines have an intrinsic shape which arises from the natural line width (Lorentz profile; §3P.1). The shape is given by a formula of the type (see equation (3P.1.17))

$$S(\nu) = A/[(\nu - \nu_0)^2 + B] \tag{3A.2.8}$$

where $S(\nu)$ is the line strength at frequency ν, ν_0 is the frequency of the line centre, and A and B are constants. Such a shape is often called a *damping* profile from its close association with the line shape produced by a classical damped oscillator. The natural line width is very small, typically 10^{-5} nm, compared with observed line widths of 10^{-4} to 1 nm in normal stars. We have already mentioned that lines are broader in white dwarfs than supergiants through the effects of the gas pressure within the line-producing regions of the two types of object. This and other broadening mechanisms dominate the actual shapes of the lines. Potentially, therefore, the lines contain information on all the effects influencing their shapes, and if these can be deciphered, then the various physical details mentioned above are determinable. In order to accomplish this, a thorough understanding of the action and results of the important line broadening mechanisms is required.

Pressure broadening

The energy levels of atoms and ions are perturbed by the nearby presence of other atoms, ions, electrons etc. Any transition occurring will therefore have a somewhat different wavelength from 'normal', and the spectrum line resulting from many such transitions will be broadened. This process is known as pressure broadening, since its effects increase with the pressure.

Pressure broadening of lines from hydrogen and hydrogen-like ions can be described accurately by quantum theory with the assumptions that electron interactions are instantaneous impacts, while the ions produce a quasi-static electric field. Other atoms and ions cannot be modelled in this way, and predictions of their behaviour have to be based upon simpler approaches. In either case the resulting line profile is very close to the Lorentz profile of the natural line shape (equations (3P.1.17), (3A.2.8), figure 3P.1.4). However, the width of the line is at least tens of times greater than the natural line width, even at low pressures. Increasing pressure decreases the separations of the interacting particles, so that their mutual interference increases, and the spectrum line width increases with pressure. As discussed earlier in this section, this effect is the basis of the luminosity classification of stars.

Doppler broadening

Motion of absorbing or emitting particles along the line of sight induces changes in the observed wavelength through the Doppler shift (equations 3A.2.4, 3A.2.5). If there are a range of motions of the particles, then the resulting lines will be broadened by the ensuing range of the Doppler shifts.

The important mechanisms leading to Doppler broadening are thermal motions, macroscopic motions (turbulence, convection), rotation, and expansion or contraction. We briefly consider each of these in turn below.

Thermal motions

The root mean square velocity of a particle in a perfect gas at temperature T is given by

$$\bar{v} = (3kT/m)^{1/2} \tag{3A.2.9}$$

where m is the particle's mass and the velocity distribution is Maxwellian (equation 2P.1.46). The resulting line profile for emission lines and faint absorption lines is then given by

$$I(\nu) = I(\nu_0)\left(\frac{mc^2}{2\pi kT}\right)^{1/2}\nu_0^{-1}\exp\left[\frac{-mc^2}{2kT}\left(\frac{\nu-\nu_0}{\nu}\right)^2\right] \tag{3A.2.10}$$

where ν is the normal frequency of the transition. The shape of the line is shown in figure 3A.5, and is known as a Gaussian profile. The width of the line at half its maximum intensity is then

$$\Delta\lambda_{1/2} = \frac{2\lambda_0}{c}\left[\frac{2kT}{m}\log_e 2\right]^{1/2}. \tag{3A.2.11}$$

The line width is thus proportional to the square root of the temperature. Typically, for say an oxygen atom or ion at a temperature of 10 000 K, an optical spectrum line would have a width due to thermal broadening of 0.01 nm—a thousand or so times the natural line width. Strong absorption lines of course have their shapes distorted as their centres reach saturation (§3A.6)

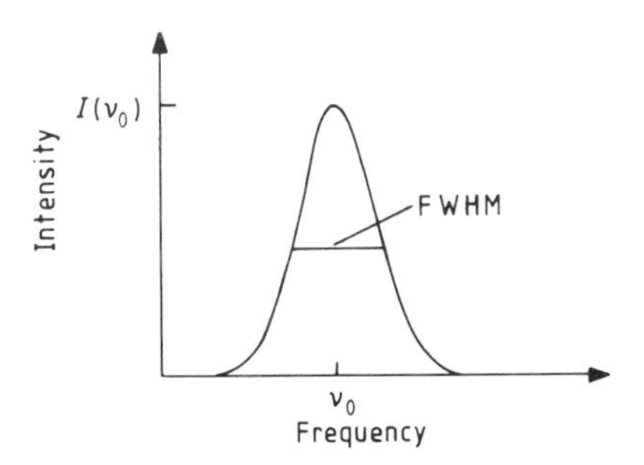

Figure 3A.5 The Gaussian profile.

Turbulence

A Gaussian line profile also results from random turbulent motion. The effects of the various individual velocities may be characterised by a single mean velocity, V. The line profile is then

$$I(\nu) = I(\nu_0)\frac{1}{\sqrt{\pi}}\frac{c}{V}\nu^{-1}\exp\left[-\frac{c^2}{V^2}\left(\frac{\nu-\nu_0}{\nu}\right)^2\right]. \tag{3A.2.12}$$

When thermal and turbulent broadening are both present, their effects simply combine into a third Gaussian profile given by

$$I(\nu) = I(\nu_0)\frac{c}{\sqrt{\pi}(2kT/m + V^2)^{½}}\nu^{-1}\exp\left[\frac{-c^2}{(2kT/m + V^2)^{½}}\left(\frac{\nu-\nu_0}{\nu}\right)^2\right]. \tag{3A.2.13}$$

The width at half maximum is then

$$\Delta\lambda_{½} = \frac{2\lambda_0}{c}\left[\left(\frac{2kT}{m} + V^2\right)\log_e 2\right]^{½}. \tag{3A.2.14}$$

The oxygen lines considered previously would thus be doubled in width by the presence of turbulent motions of about 5 $\mathrm{km\,s^{-1}}$.

Rotation

The profile of a line widened purely by rotation is quite different from any of the cases considered so far. If we make the assumptions that the object is angularly unresolved with the line-producing regions thin compared with the object's size, and that it rotates as a solid body without significant limb darkening, gravity darkening, etc, then a simple calculation gives an idea of the line profile. Taking the object's rotational axis to be perpendicular to the line of sight, we may see from figure 3A.6 that the component of the velocity along the line of sight for a point on the surface at position (θ, φ) is

$$V\cos\theta\cos\varphi \tag{3A.2.15}$$

where V is the equatorial rotational velocity. However, we may also see that the distance of this point from the rotational axis in the plane perpendicular to the line of sight is

$$\cos\theta\cos\varphi. \tag{3A.2.16}$$

Thus the radial velocity (in the astronomical sense), and hence also the Doppler shift, for any point on the surface is simply proportional to the projected distance of that point from the rotational axis. The line profile is thus given by

$$I(\nu) = I(\nu_0)\left[1 - \frac{c^2(\nu-\nu_0)^2}{V^2\nu^2}\right]^{½} \tag{3A.2.17}$$

and this is semicircular in shape (figure 3A.7). The width at half maximum is thus

$$\Delta\lambda_{½} = (\surd 3/2)(\lambda_0 V/c). \tag{3A.2.18}$$

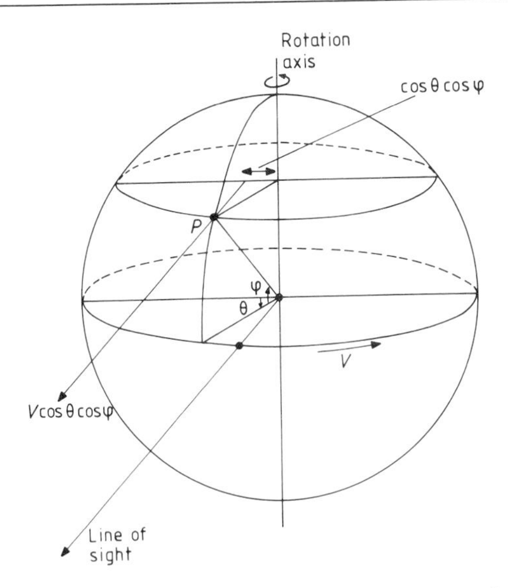

Figure 3A.6 Coordinate system for stellar rotation.

For an optical line and a value of V of 100 km s^{-1}, this gives a half width of 0.15 nm. Limb darkening, inclination of the rotational axis to the line of sight, etc will tend to reduce the actual width. But nonetheless, amongst the hotter stars which have the higher rotational velocities (§3A.7), rotational broadening is often the predominant broadening mechanism.

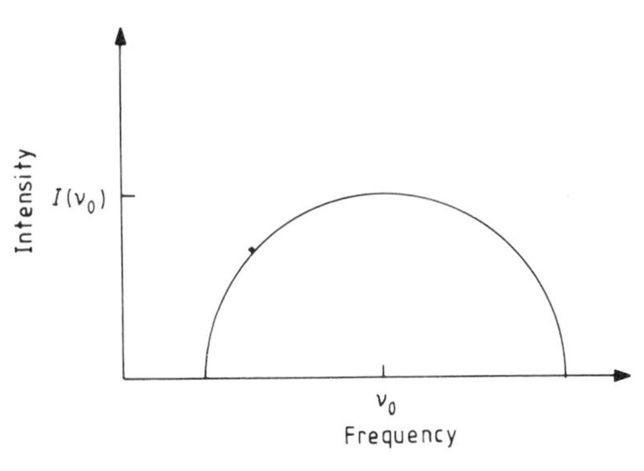

Figure 3A.7 Rotational line profile.

Expansion and contraction

Spherically symmetric expansion or contraction of a star is a rare phenomenon, Cepheids and novae being probably the best known examples. More commonly a steady outflow of material may occur, as in a stellar wind. Then changing conditions within the flow may lead to some spectrum lines being formed only over a restricted region, and this has a similar effect on the line profiles.

Emission line profiles in an ideal case (no limb darkening etc) are given by equation (3A.2.17) with V now being the velocity of expansion. But, unless the line-producing region is large compared with the star, obscuration of its rear half by the star will leave only the sections contributing to the higher frequencies visible. Thus the profile becomes asymmetric (figure 3A.8) towards the higher frequencies and has a width at half maximum of

$$\Delta\lambda_{1/2} = (\sqrt{3}/4)(\lambda_0 V/c). \qquad (3A.2.19)$$

Generally the region producing the line cannot be regarded as thin in comparison with its radius, and the velocity may therefore vary across the region. This and other complications in practice usually alter the shape of the line profile significantly from its ideal quadri-circular form, but an asymmetry towards the higher frequencies usually remains for expanding regions. Absorption line profiles are displaced to higher frequencies when produced by an extended expanding envelope around the star, and in ideal cases are roughly triangular in shape (see problem 3A.2 and its answer). The much rarer situation of contraction would similarly lead to an asymmetry or displacement to lower frequencies.

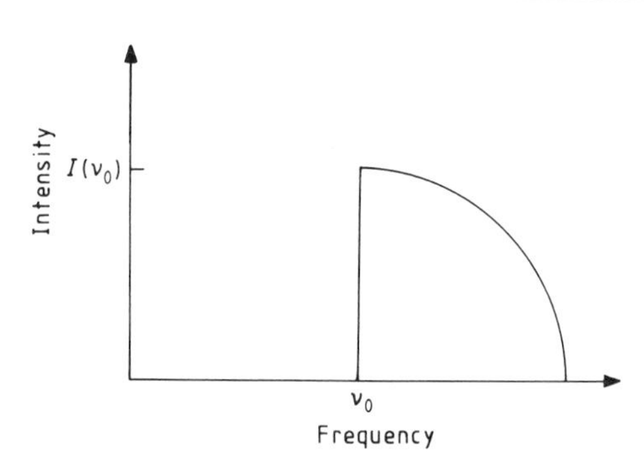

Figure 3A.8 Expansion line profile.

A frequently occurring line profile is called the P Cygni profile after one of the stars in which it figures prominently. The line has two components: an emission line close to the laboratory wavelength of the transition (after correction for any radial motion of the star as a whole), and an absorption line displaced to higher frequencies. Four classes of such lines have been distinguished by Beals, and these are shown in figure 3A.9. The usual interpretation of the profile is via an outwardly expanding and accelerating envelope or shell around the star. The emission component originates in the hot, slowly moving inner portions of the shell, while the absorption component arises further away from the star in cooler, more rapidly expanding regions. Infall would produce an inverse P Cygni profile with the absorption to lower frequencies. Although uncommon, such lines have been found in the YY Orionis stars (§3A.5).

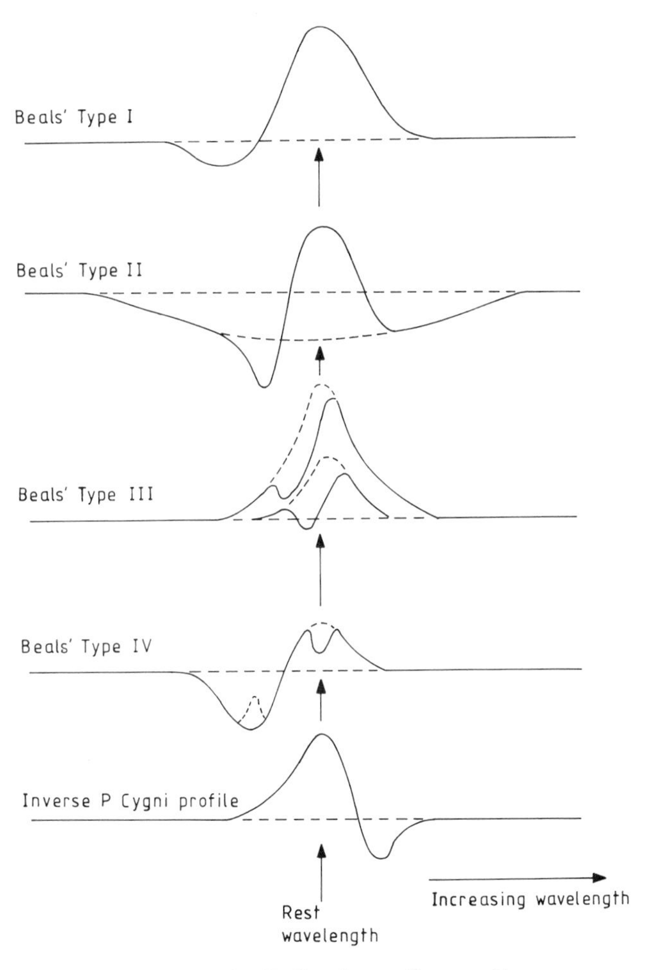

Figure 3A.9 P Cygni type line profiles.

Other effects
Magnetic fields split lines into two or more components via the Zeeman effect (§3P.1). Apart from the Sun, the splitting cannot be observed to produce separate components in stellar spectra; the components blend together and the effect is just to broaden the line. If the magnetic field itself is being studied, then the imbalance of the various polarisations in the wings of the line (the instrumental profile) is known. Its effects can be removed from Chapter 5). In other cases, the star's lines merely appear broader than for a comparable non-magnetic star.

Other effects which can contribute to line broadening in particular instances include isotope shifts, hyperfine splitting, and the degradation of the data by the telescope/spectroscope combination. The latter effect can often be reduced in importance if the effect of the instruments on an ideal line (the instrumental profile) is known. Its effects can be removed from the data by various mathematical computer-based processes.

Only rarely will a line profile be the result of the operation of just one of the various line broadening mechanisms. Two commonly occurring situations have rotation and expansion combined or Gaussian and Lorentz profiles combined. The former case can produce a wide range of line shapes, including the double emission often to be found in Be stars etc. The latter occurs in many stars and results in a Voigt profile. The Gaussian and Lorentzian components are separable out from such a profile, and suitable tables for accomplishing this may be found in *Astrophysical Quantities* by C W Allen (Athlone Press, 3rd edition, 1973), amongst other places.

Fraunhofer lines

Fraunhofer in his early work on the solar spectrum (*circa* 1814) labelled the nine most prominent lines with the letters A to K (missing out I and J). Several of the lines are still commonly referred to by this notation, so that it is useful to be aware of them. Including ultraviolet lines added since Fraunhofer's day, the full list is given in table 3A.3. The use of this notation that is most likely to be encountered is for the G band, the sodium D lines and the calcium H and K lines. The other symbols are used only rarely now.

3A.3 SURFACE TEMPERATURE

Both spectral type and colour index depend ultimately upon the star's surface temperature. This is most conveniently defined as the effective temperature: the temperature of a black body emitting the same total energy per unit area as does the star. This definition is chosen because the outer layers of the star are not in TE, and often not even approximating to LTE.

Table 3A.3 Fraunhofer lines.

Fraunhofer line		Wavelength (nm)	Identification
A		760	O_2 (telluric)
B		710	O_2 (telluric)
C		656.28	Hα
D	*D*1	589.59	Na I
	*D*2	589.00	
E	*b*1	518.36	Mg I
	*b*2	517.27	
	*b*3	516.73	
F		486.13	Hβ
G		430–432	CH
H		396.85	Ca II
K		393.37	Ca II
L		382.04	Fe I
M		373.49	Fe I
N		368.12	Fe I

Different measures of the temperature will therefore lead to different results. Thus, in the Sun, for example (Chapter 4A), the effective temperature is 5778 K, but excitation and ionisation temperatures of 5600 to 6200 K are found, and the brightness temperature varies from 5000 K at the limb to 6300 K at the centre of the disc. In the chromosphere, there are temperatures between 4000 and 10 000 K, while in the corona the kinetic temperature can reach 10^6 K. The effective temperature, being based upon the total energy emission, rather than one aspect of it, is regarded as the most generally representative parameter. It cannot always be used, however; for example, the excitation temperature must be used when studying the excitation of atoms, and so on.

The colour index is closely related to the effective temperature, since the spectral regions involved in the measurements are broad and often well separated. Spectral type is defined in terms of spectrum line intensities, and this may lead to a less direct relation with the effective temperature. Thus we have already seen how the excitation and ionisation temperatures may differ from each other and from the effective temperature. Additionally, however, lines of differing strengths will generally originate in different portions of the stellar atmosphere, depending upon where unit optical depth in the line is reached. Thus, on the Sun, Hα absorption arises at a height of some 3000 km above the normal photosphere. Nonetheless, the averaging effect of looking at the whole disc of the star and at many lines ensures that spectral type is also reasonably unequivocally related to the star's effective temperature. Figures 3A.1 and 3A.3 show the actual relationships in the two cases.

From those diagrams we may see that stellar temperatures range from about 2500 K to 50 000 K. The mean stellar mass of 0.35 $\mathcal{M}_{\odot}$ (§1A.1) corresponds for main sequence stars to a B—V colour index of + 1.56, and a spectral type of M3 V. Thus the mean stellar temperature is about 2700 K.

A very few exceptional stars, such as the nuclei of some 'planetary' nebulae, may have temperatures as high as 200 000 K. The reality of these temperatures, however, is not firmly established, since the methods used to determine them contain large uncertainties.

3A.4 HR DIAGRAM

The two most fundamental properties of stars revealed by photometry and spectroscopy are the absolute magnitude (or luminosity) and the effective temperature (or colour index or spectral type). In 1913, Russell (as did Hertzsprung, independently and a little earlier) plotted those stars for which these two parameters were known at the time on a single diagram with absolute magnitude as the ordinate and spectral type as the abscissa. Such a diagram (and its variants using luminosity, effective temperature or colour index) has since become known as the Hertzsprung–Russell diagram, or more commonly as the HR diagram. When temperature is used, it is plotted conventionally as increasing towards the left to match the way in which the diagram was first drawn using spectral types. It is one of the most useful ways of summarising many of the properties of stars, and enables determinations of stellar ages and distances to be made in some cases. Figure 3A.10 shows the main features of the diagram, and includes the luminosity classes as well as the basic spectral class.

The terminology for the main types of stars which has been used in the preceding chapters derives from their positions on this diagram. About 92% of the observed stars lie within a broad band descending from the top left to the bottom right, with over half of these being of class M. This band is called *the main sequence*, and stars on it are called main sequence stars (also known as dwarfs and luminosity class V stars). Of the remainder, most (7.5% of the total) lie towards the bottom left-hand corner of the diagram. These stars are called the white dwarfs (or luminosity class VII stars) since their low luminosities and high temperatures imply small sizes. The 0.5% or so of the stars not falling into these two classes are mostly to be found above the main sequence. They are generally physically large stars and hence are called the giants. Their various subclasses are shown on Figure 3A.10 (luminosity classes IV to I).

This true distribution of the stars on the HR diagram may be obscured by the manner in which it is often plotted. Stars on it may be selected down to some limiting apparent magnitude, since this is the natural way in which the observations are obtained. The brighter stars are then, however, preferentially selected, since they are visible out to much greater distances than the fainter ones. The impression is thereby gained that the stars at the top end of the main sequence, and in the giant region, are relatively common, instead of the actual situation in which they form less than 1% of the total. The true distribution can only be found by plotting all the stars out to a given distance, irrespective of their apparent magnitudes, as has been done here.

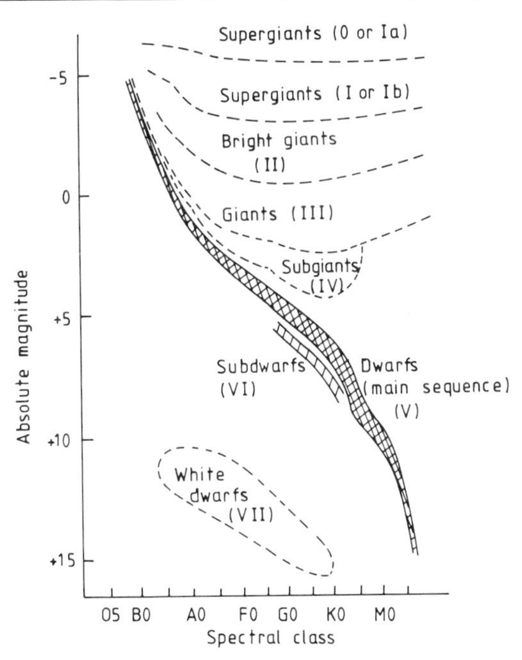

Figure 3A.10 Hertzsprung–Russell diagram showing luminosity classes.

3A.5 VARIABLE STARS

Ninety-nine per cent or more of the stars do not change their luminosities by significant amounts except over time scales which are appreciable fractions of their lifetimes. The remainder may change their luminosities by amounts from 0.01 to 20 stellar magnitudes, with time scales from seconds to decades. They are called variable stars, or, more commonly, just variables. Partly through human nature which develops its greatest curiosity for the exceptional, and partly because these stars can reveal so much about stars in general, astronomers have probably devoted more time to the study of these stars than to all the other stars put together! There is thus a very large observational data set and a rather smaller theoretical understanding of this very diverse group of stars. The classification system (see below) imposes some order on the situation, but may still result in stars with physically unrelated phenomena being put together, and many of the classes have still to have their underlying mechanisms determined.

Nomenclature

Variables are labelled in a variety of ways, and many of them may have several designations. The extension to the Bayer system (which labels the brighter stars on each constellation with capital letters from A to Q), is the most widespread and useful system. This labels the variables in each constellation with capital letters starting at R for the brightest. Letters from R to Z suffice for only nine variables in each constellation, and so thereafter double capitals are used in a rather complex manner. RR to RZ is used for the next nine variables discovered, then SS to SZ for the next eight, TT to TZ for the next seven, and so on to ZZ, giving another 54 labels in all. After that the sequences AA to AZ, BB to BZ, CC to CZ . . . QQ to QZ are used. A total of 334 variables in each constellation may thus be identified, the format of the label being the letters and the constellation abreviation; T Tau, RR Lyr, UX Ari etc. When more than 334 variables are found the simpler practice of the letter V followed by a number starting from 335 is used. Thus we have V1057 Cyg, V439 Ori etc. Pulsars and other special groups of variables may use other systems of labelling; a common practice is to use a label to indicate the variable type and combine it with an approximate right ascension and declination. Thus we have the pulsar PSR 1919 +21, and the x-ray source A 1909 +04 etc.

Classification

The variables discussed in this section are all photometric variables; their brightnesses as seen from the Earth vary. Other properties of the star may also change, and lead to spectroscopic, polarimetric, magnetic variables etc. Often, but not always, such stars are additionally photometric variables, and so they may be considered here, or their properties may be dealt with in other sections.

We may divide the variables into two major groups on the physical basis of their variations: the extrinsic and the intrinsic variables. In the first of these the star's emission is constant, but the radiation received on Earth varies through some additional process. Examples are the eclipsing binaries

(§1A.1) and the pulsars (see later in this section). With intrinsic variables, the luminosity of the star itself is changing. The list of variable types is very long and a brief summary is tabulated below. Their more significant representatives are then discussed in more detail in the next subsection of this chapter. Figure 3A.11 shows the approximate positions of those variables from the list which have reasonably well defined properties.

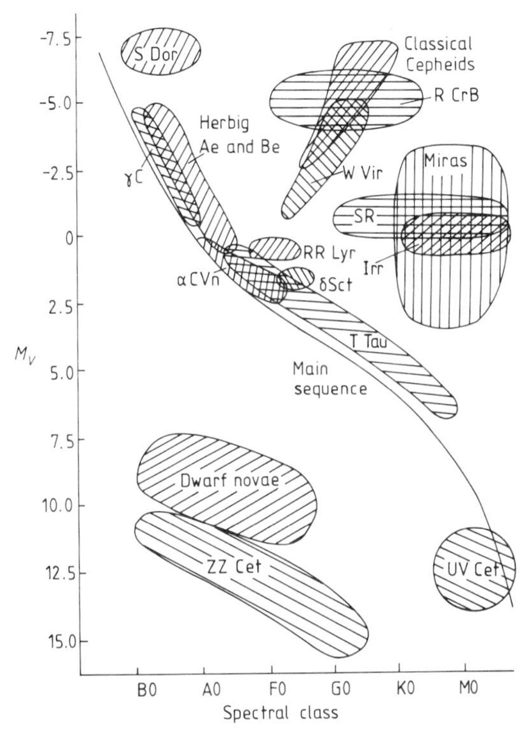

Figure 3A.11 Approximate positions of variables on the HR diagram (only those variables with reasonably well defined properties are plotted).

IAU classes of variable stars

I INTRINSIC VARIABLES

A Pulsating variables

Cep, Cδ, CW	Long period Cepheids, W Virginis stars; periods 1 to 100 days, amplitudes 0.1 to 2 magnitudes, temperatures 5000 to 12 000 K.†

† Stars discussed in detail in the next subsection.

RR, RRab, RRc, RRs	RR Lyrae stars or short period Cepheids; periods 0.05 to 2 days, amplitudes less than 1 magnitude, temperatures around 10 000 K.†
δ Sct	δ Scuti type stars. Some resemblance to RRs stars, but the form of the light curve highly variable. Amplitude up to 0.1 magnitude, temperature about 8000 K.
ZZ	ZZ Ceti stars; variable white dwarfs. Periods of a few minutes sometimes with several superimposed cycles, amplitudes small.
αCV	α Canum Venaticorum variables; spectral class A stars with many peculiarities such as abundance anomalies, magnetic fields etc. Also known as Ap stars or magnetic variables. Periods 0.5 to 30 days, amplitudes usually less than 0.1 magnitudes.
M	Mira variables; late-type giants with very long and rather variable 'periods' of 100 to 1000 days. Amplitudes usually more than 2.5 magnitudes. Lower amplitude examples may be classified with the SR stars.
SR, SRa, SRb, SRc, SRd	Semi-regular variables; giants and supergiants, mostly of late spectral class and with unstable periods from 20 to 2000 days. Amplitudes less than 2 magnitudes; the form of the light curve between the different types is very changeable.
L, Lb, Lc	Irregular variables with long characteristic time scales for their variations. Usually giants of late spectral class.
SD	S Doradus variables; high luminosity variables of early spectral class and sometimes with large amplitudes.
γC	Irregular variables of early B spectral class, time scales of minutes to years and usually small amplitudes. The variation may be due to obscuration of the star by a thick and clumpy shell, so that they might also be classed with the extrinsic variables.

B Eruptive variables

N, Na, Nb, Nc	Novae; amplitudes 6 to 10 magnitudes, the initial rise taking a few days and the subsequent decline several months. May be periodic on time scales of 10 000 to 100 000 years.†

† Stars discussed in detail in the next subsection.

Nr	Recurrent novae; periods of a decade or so, amplitude about 5 magnitudes. †
UG, Z Cam	Dwarf novae; periods of 10 to 200 days, amplitudes of 2 to 6 magnitudes. †
SN	Supernovae; amplitudes up to 20 magnitudes, initial rise over about a month, followed by a decline taking a year or more (†, also §§6A.6, 8A.2).
Nl	Nova-like variables; slow and irregular variations of up to one magnitude. Low amplitude rapid flickering. †
MXB	X-ray bursters; a very rapid (<1 s) rise in x-ray brightness followed by a slower (10 s) fall back to normal. †
RCB	R Coronae Borealis stars; 'anti-novae' or stars with a rapid decrease in brightness by 1 to 10 magnitudes followed by a slower return to normal. The drops in brightness recur in an irregular fashion and with varying amplitudes. Temperatures 4000 to 8000 K (see §7A.2).
Z And	Symbiotic stars; a very diverse group whose spectra exhibit features characteristic of two or more different stars.
AM Her	Polars; binaries with periodic variations in phase with the orbital period and also longer more irregular variations. The defining characteristic is their strong and variable polarisations.

C Irregular variables

Ia, In, Ina, InB, L	A very heterogeneous group, usually associated with nebulosity, sometimes also called the Orion variables, or the T Orionis stars. Time scales for their variations around one day, with amplitudes up to 3 magnitudes.
Is, Isa, Isb	Similar to the Orion variables, but not associated with nebulosity.
InT	T Tauri stars; irregular variables associated with nebulosity. Spectral types A to M, but with possible higher temperature analogues known as the Herbig Ae and Be stars. Time scales of days to months and amplitudes of several magnitudes. †

† Stars discussed in detail in the next subsection.

UV, UVn	UV Ceti or flare stars; spectral classes K and M. Very rapid (tens of seconds) brightening by 1 to 6 magnitudes followed by a return to normal over tens of minutes. The variation is probably due to a solar-type flare which has a pronounced effect on the luminosity of such faint stars.

II EXTRINSIC VARIABLES

A Binaries

EA	Detached eclipsing binaries, also known as Algol type stars (although the star, Algol, is actually of type EB). Each star is significantly smaller than its Roche lobe (§8A.2), and the light curve is due just to the obscuration of one star by the other with minor reflection effects in some cases. Periods range from hours to years and amplitudes up to several magnitudes.
EB, E11	Semi-detached binaries, ellipsoidal, W Ser or β Lyr stars. One of the stars fills its Roche lobe. Subdivided into types A and B according as it is the less or more massive star which fills its Roche lobe. Mass exchange is likely to be occurring. The brightness variations arise from eclipses, reflection, varying projected areas, and variations of brightness over the stars' surfaces.
EW	Contact binaries, W UMa, SV Cen stars. Both stars fill their Roche lobes, and mass loss from the whole system may occur. The brightness varies continuously through the changing projected areas of the non-spherical stars, and through surface brightness variations. Eclipses may occur but are not generally identifiable. The stars probably occupy a common envelope of material, and in some cases may eventually merge into a single rapidly rotating star.

B Rotational variables

P, PSR	Pulsars; originally radio variables, now a few are known with optical and x-ray emissions. Periods from 0.001 to 10 seconds and regularities up to one part in 10^{15}. The periodicity is that of the rotating neutron star which causes the emission (†, and §§2P.1, 8A.2).

† Stars discussed in detail in the next subsection.

Individual types of variable stars

In this section five of the above variables, which are of wider importance within astronomy, have their properties discussed in more detail. The theoretical explanations of the observed behaviour are included when understood.

Cepheids: type Cep, Cδ, CW, RR, RRab, RRc, RRs *stars*
Apart from having considerable theoretical interest, this group of stars is studied because it provides the basis of one of the better methods for determining distances out to a few megaparsecs (Appendix I). Cepheids exhibit a relationship between their mean absolute magnitudes and their periods (see below). Their absolute apparent magnitudes thus being known, we may find their distance from equation (3A.1.2) when it is rewritten as

$$D = 10^{0.2(m+5-M)} \quad \text{(pc)}. \tag{3A.5.1}$$

Since Cepheids are bright enough to be observed in nearby galaxies, the distances of the galaxies containing them are also found.

The light curve exhibits a sawtooth pattern (figure 3A.12) with a steeper rise in brightness than the subsequent fall. The RR Lyrae stars are distinguished by shorter periods and an almost constant mean absolute magnitude. Hardly any of the stars have periods between 1 and 2 days.

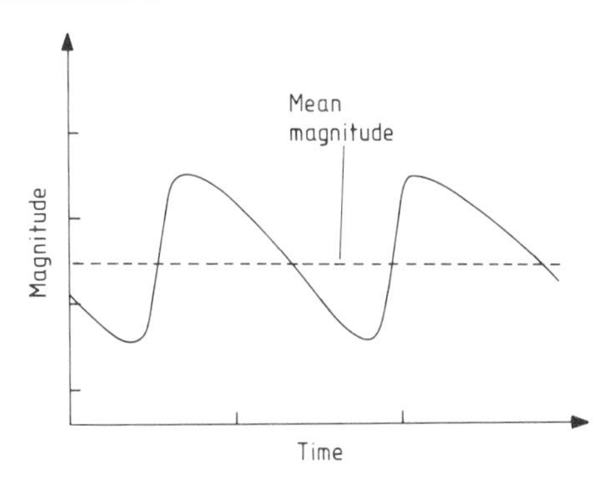

Figure 3A.12 Schematic light curve of a Cepheid or RR Lyr star (a few RR Lyr stars have curves which are much more symmetrical, and may even approximate a sine-wave variation).

RR Lyrae stars have periods from 0.25 to 1 days with a strong peak in the distribution at 0.5 days. Their mean absolute magnitude is close to +0.8, and their spectral types range from A0 to G0. They have a strong association with globular clusters, and so are sometimes known as cluster variables.

The Cepheids proper fall into two main groups associated with star population types I and II. These are often called classical Cepheids and W Virginis stars respectively. Their periods range from 2 to 100 days with a peak in the distribution at 10 days. Their mean absolute magnitudes range from -1 to -7 and are related to their periods by

$$\bar{M}_V \simeq -1.9 - 2.8 \log_{10} P \tag{3A.5.2}$$

for the classical Cepheids, and by

$$\bar{M}_V \simeq -2.8 \log_{10} P \tag{3A.5.3}$$

for the W Virginis stars (P in days). Thus the stars' distribution with period is that shown in figure 3A.13. They form a part of the 'great sequence of variable stars', which is a much more extensive and very much more approximate relationship of amplitude and period and which includes most variables. It takes the form

$$\Delta M_V \simeq 0.5 + 1.7 \log_{10} P \tag{3A.5.4}$$

where M is the amplitude of the variation. For classical Cepheids and W Virginis stars, the actual relationship is

$$\Delta M_V \simeq 0.4 + 0.4 \log_{10} P \tag{3A.5.5}$$

while the RR Lyrae stars give

$$\Delta M_V \simeq 2.0 - 1.8 \log_{10} P. \tag{3A.5.6}$$

The position of the stars on the HR diagram is complicated by the variation of effective temperature (and spectral type) throughout the cycle. The

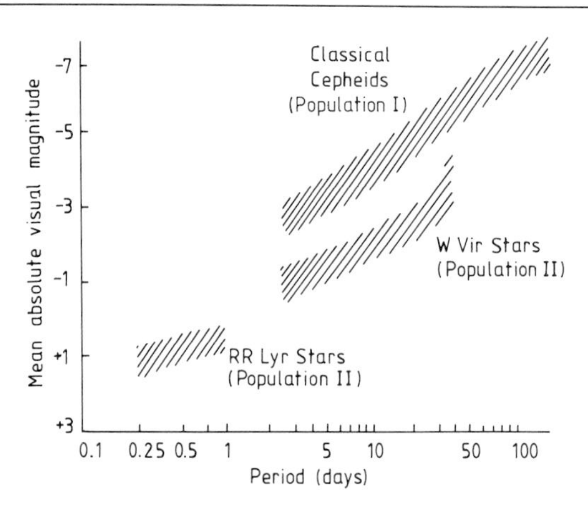

Figure 3A.13 Period/luminosity relationship for Cepheid-type stars.

stars move by about half a spectral class towards the earlier spectral types as they brighten, and then back again as they become fainter. They occupy the top end of the *instability strip* (figure 3A.14), a region of the HR diagram containing a wide variety of post-main-sequence rapidly evolving variables. A more detailed examination of the spectra reveals that the radial velocity is changing with the same period as the luminosity. Unlike binary stars, however, where the shift of the spectrum lines arises as the star as a whole moves around its orbit, here the shift is due to the change in the line profile as the outer parts of the Cepheid alternately expand and contract (figure 3A.8). The maximum *rate* of expansion coincides with the maximum luminosity, and the maximum collapse *rate* with the minimum luminosity. The maximum *size* of the star therefore occurs at the midpoint of the decline in brightness (figure 3A.15). Thus there is a phase lag amounting to about 60° (not 90° because of the skewed curves) between the luminosity and size variations. This also implies a temperature variation, and this is borne out by the observed spectral class changes.

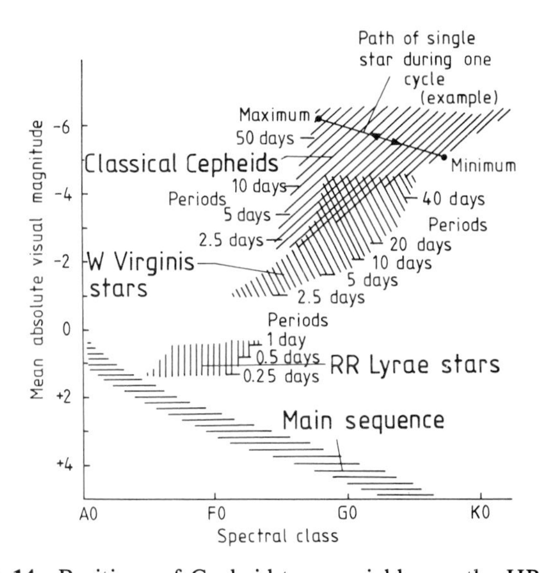

Figure 3A.14 Positions of Cepheid-type variables on the HR diagram.

These various cyclic changes and their phase relationships have provided the clue to explaining the cause of the pulsations. The fundamental process is that the ionisation of hydrogen and helium in the outer layers of the stars acts as a valve. When the star is at its smallest physical size, ionisation of the gases increases the opacity of the star's outer layers. This blocks some of the radiation coming from the interior of the star, raising the temperature and pressure within these outer layers. The increased pressure then causes the outer layers to expand. At the largest physical size of the star, recombination of the hydrogen and helium occurs, the opacity decreases, the temperature and pressure fall, and so the outer layers contract again. The total change in the star's radius is by about 10 to 15%, but since only the outer layers are involved, the change in the conditions is far greater than this. The emerging luminosity depends upon the mass overlying the hydrogen ionisation zone amongst other things, and detailed calculations show that this is a minimum soon after the minimum radius is reached. The maximum luminosity thus occurs about 20% of the period after the minimum radius is reached.

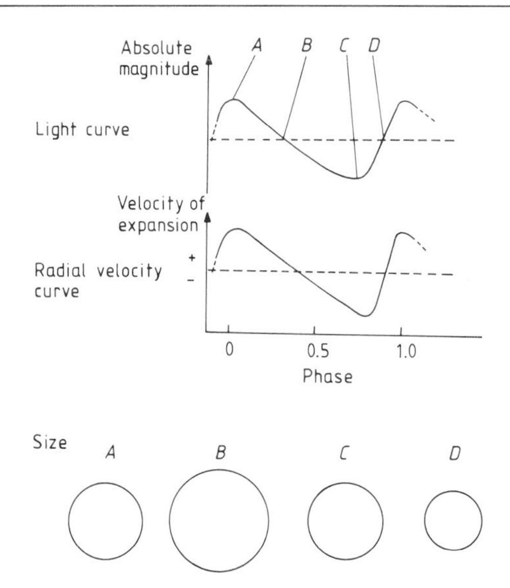

Figure 3A.15 Radial velocity and size variations of a Cepheid.

A small group of stars usually included amongst the Cepheids, although they may not necessarily belong there, are known as the double-mode or beat Cepheids. Their light curves may be resolved into two components. The longer period generally lies between 1 and 7 days, and the shorter period is about 70% as long. The light curve which results from the combination of these two, or occasionally three, modes is complex and contains the beat frequency between the basic periods. The longer period is taken to be the fundamental radial mode of oscillation of the star, and the shorter period to be its first harmonic. With both periods known, the star's mass and radius are calculable (equation 1A.1.21). The masses determined in this manner are too small by a factor or two or three compared with those expected from evolutionary considerations (§§1A.1, 8A.2). A similar but smaller mass discrepancy is found for the other Cepheid types, and at the time of writing both discrepancies remain to be explained.

Novae: types N, Na, Nb, Nc, Nr, UG, Z Cam, Nl

The term 'nova' derives from the pre-telescopic *stella nova* or 'new star'. Occasionally a new star might shine briefly out of the otherwise unchanging pattern of fixed stars, where no star had been known before. Nowadays, we may sometimes detect novae even at their faintest, so that we know that they are not genuine new stars, but pre-existing stars which have brightened into prominence for a few months or years before fading again. The amplitude of the increase in the brightness is so great, however, that even today not all novae can be found outside one of their outbursts.

Probably between 10 and 40 novae occur in the Milky Way Galaxy each year. Only two or three of these are observed from Earth; the remainder are obscured by the interstellar gas and dust in the Galaxy (Chapter 7A). Nova detection, along with comet discovery, is one of the areas of astronomy where amateur astronomers can make a real contribution. The majority of novae are still discovered by that small band of dedicated sky watchers who regularly scan the heavens armed only with a large pair of binoculars and an exhaustive knowledge of the sky.

The characteristic light curve of a nova is shown in figure 3A.16. There is a very rapid brightening by eight to ten magnitudes over one or two days, often followed after a short pause by a slightly less rapid rise of a further two or three magnitudes. The total amplitude typically ranges from eight to twelve magnitudes, or a change in luminosity for the star by a factor between 2000 and 50 000. Occasionally amplitudes may reach fifteen or sixteen magnitudes (a factor of 1 000 000 in luminosity). This rapid rise is followed by a decline of a few magnitudes in three or four weeks, and then a return to the pre-outburst brightness over the next one to ten years. This last phase may be interrupted for weeks or even months by very extensive fluctuations in brightness.

The peak brightness is typically an absolute magnitude of -8 (100 000 times brighter than the Sun). There is a useful empirical relationship between the initial rate of the decline in brightness and the absolute magnitude at maximum:

$$M_V \simeq -11 + 2 \log_{10} t_3 \qquad (3A.5.7)$$

where M_V is the peak absolute visual magnitude and t_3 is the time in days for the nova to decline by three magnitudes from its peak brightness. This relationship allows the nova to act as a standard candle for distance determination (Appendix I; see also the preceding discussion on Cepheids).

The schematic nova light curve shown in figure 3A.16 would be classified as a fairly fast nova. Slow novae may have values of t_3 of up to a year, and a decade or more for their total decline. Fast novae may have t_3 as short as ten days, and be over in a year. Apart from scaling the time axis, however, the main features of these various nova types are unchanged.

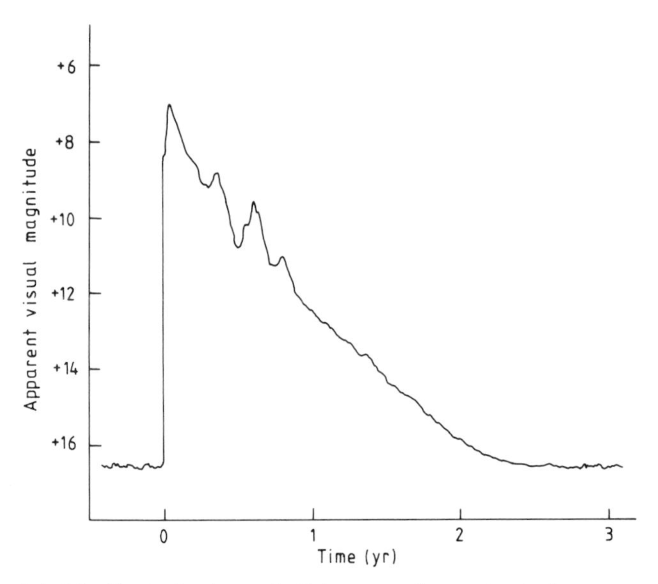

Figure 3A.16 Typical schematic light curve for a classical nova.

The spectrum of a nova alters rapidly during an outburst. Initially the optical region exhibits the absorption lines characteristic of hot stars, typically spectral type B or A. The spectral type then moves towards that of cooler stars, and emission lines start to appear. In the final stages, the spectrum is dominated by, and in some cases consists only of, emission lines. All the lines are shifted to higher frequencies by amounts which correspond to expansion velocities between 500 and 2000 km s^{-1}, and these displacements change in a complex manner as the spectrum changes. The energy emitted into the visible spectrum is shown by the light curve of the nova. The *total* energy emission of the nova, however, remains rather more constant for many weeks after the optical maximum. Initially energy output increases in the ultraviolet and roughly compensates for the optical decline, while later the infrared radiation dominates the whole of the energy emission. The total energy emitted during an outburst reaches perhaps 10^{39} J, or about 100 000 times the energy emitted by the Sun in a year.

Two other types of nova, recurrent and dwarf, show significantly different behaviour patterns from that of the classical nova just outlined.

Recurrent novae have broadly similar light curves to the classical novae, but two or more outbursts have been observed to occur. The intervals between the outbursts range from 10 to 80 years. In two other respects, however, the recurrent novae do differ from the classical novae. First, their amplitudes are smaller—six to eight magnitudes only. Secondly, their spectra are slightly anomalous during an outburst, and at other times reveal the presence of a red giant in the system. It is probable that most, if not all, novae are recurrent (see later discussion in this section) and that those observed to be so are just the ones with the shortest periods. It is possible, however, that the presence of the red giant may indicate a difference of type rather than of degree for these novae. Dwarf novae, or U Geminorum type stars, appear as quite different types of objects. They brighten rapidly, but only by six magnitudes at most, and then fade back to their pre-outburst brightness in a few days. A similar outburst is then repeated after a delay of only a few tens of days (figure 3A.17). About 10% of these stars fall into a subgroup known as the Z Camelopardalis stars. In these stars, the repetitive outbursts halt for periods of tens to hundreds of days (figure 3A.17) at a time, but then resume again. Variability on much shorter timescales is also shown by many of the dwarf novae. There is often evidence that the star is actually a close binary system, with a period of a few hours, and an amplitude due to eclipses of half a magnitude or so. In some cases the eclipse is preceded by a noticeable increase in the system's brightness. On a time scale of minutes or less, the system may flicker by up to a tenth of a magnitude. The flickering generally retains coherence for only a few tens of minutes. Its

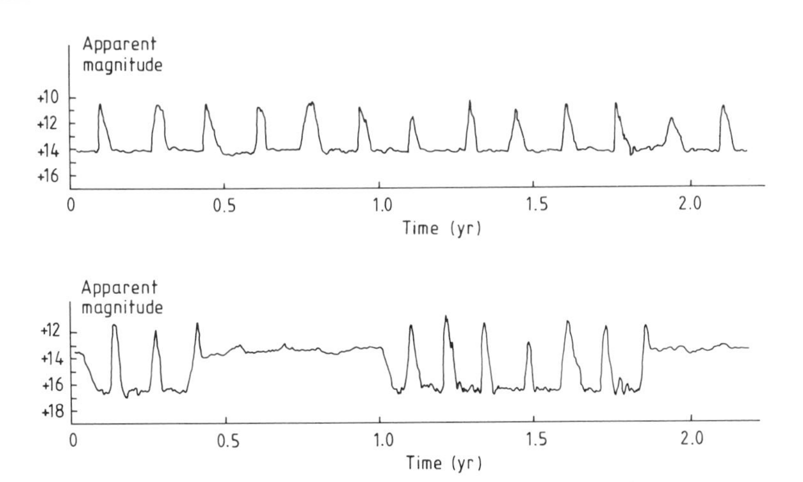

Figure 3A.17 Schematic light curves of dwarf novae. Top: U Geminorum type stars. Bottom: Z Camelopardalis type stars.

amplitude increases during the pre-eclipse 'hump' and disappears during the eclipse itself.

The explanation for novae of all types lies in a close binary system containing a white dwarf. Since more massive stars evolve more rapidly than less massive stars (§1A.2, Chapter 8A), the white dwarf must originally have been the more massive star of the system. Now it may or may not be more massive than its companion depending upon mass losses during its lifetime. Its companion is now just evolving off the main sequence (§8A.2), or may even have reached the giant region, as in the recurrent novae. It is filling its Roche lobe (§8A.2), and matter is being lost from its surface layers into the sphere of influence of the white dwarf (figure 3A.18). Viscous and turbulent energy loss then cause the material to spiral down towards the surface of the white dwarf and to accumulate there. In some cases, jets of material may be expelled by this process in directions perpendicular to the accretion disc at velocities of up to 5000 $km\,s^{-1}$. The nova's radiation outside an outburst is dominated by the emission from the interaction zone between the accreting material and that already orbiting the white dwarf, or by the emission from the gas around the white dwarf. The eclipse, when observed, is of the hot spot or gas disc around the white dwarf by the companion. The flickering must also arise within the hot spot or gas disc. Various possibilities have been suggested to explain it, including oscillations of the white dwarf or of the accretion disc, possibly linked to a magnetic field; an emission beam from the white dwarf swept around by the star's rotation, again possibly linked to a magnetic field (cf pulsars later in this section); condensations or vortices in the accretion disc; or surface waves on the white dwarf; but without any prime candidates becoming apparent.

The nova outburst itself occurs in the material accumulating on the surface of the white dwarf. Eventually the temperature and pressure therein rise to the point at which the hydrogen can undergo thermonuclear fusion (§2A.2). This occurs as a runaway reaction and is so violent that an appreciable proportion of the accreted material (50% to 90%) is expelled from the system. The explosion is the cause of the observed classical nova outburst, the large increase in brightness being due to the enormous increase in the radiating area of the system as the expelled material expands outwards. The initial decline occurs as the material cools down, the later fluctuations in brightness are due to the expanding shell becoming optically thin, and the radiating region thus changing in size and area. The basic structure of the system is unchanged by this process, violent though it may be by our standards. The exchange of material between the stars will therefore continue, and after perhaps 10 000 to 100 000 years, sufficient will have accumulated on the white dwarf for the whole process to repeat itself. Some 500 such outbursts might be expected from a classical nova system before the mass of the companion becomes too low to sustain further activity.

In recurrent novae, the companion seems to be a red giant, and this is all that may normally be detected. The amplitudes may thus be lower in these cases because the red giant's luminosity swamps the early stages of the outburst. Additionally, the possible enrichment of the material from the red giant in $^3_2$He may lower the temperature and pressure required for the thermonuclear runaway, thus leading to smaller but more frequent outbursts.

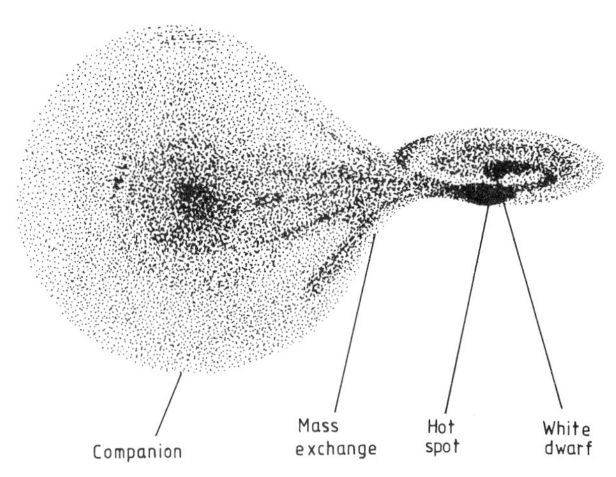

Figure 3A.18 Probable structure of a nova system.

Dwarf nova outbursts seem to have a different basis. The detailed mechanism is not known, but the gravitational energy released (§2P.5) by the accreting material probably powers the process. Possible mechanisms include magnetic fields acting to store and release the energy, positive feedback within the accretion stream, and changing opacity of the accretion disc as the hydrogen ionises and recombines (cf Cepheids). The energy of a dwarf nova outburst is insufficient to enable more than a small fraction of the accreting material to escape from the system. Thus it must eventually accumulate on the white dwarf just as in a classical nova system, and may be expected to lead eventually to a classical nova outburst for the dwarf nova system. No such phenomenon has been observed yet with certainty, but WY Sge (Nova Sge 1783) is a possible candidate, since its present day spectrum is akin to those of dwarf novae at minimum, and it exhibits occasional brightenings.

Nova-like variables are stars which appear similar to novae outside one of their outbursts, but which have not been observed undergoing an outburst themselves. Only time will tell whether or not these truly are potential nova systems.

Supernovae: type SN

In spite of the superficial commonality between the light curves of novae and supernovae which has led to their receiving similar names, there is no real physical link between the two types of object. The supernovae divide into two main types on the basis of their light curves, labelled types I and II.

Type I supernovae are the more spectacular, reaching a peak absolute magnitude near -19 (2 000 000 000 times brighter than the Sun, or comparable with the entire emission from a small galaxy). The excursions are at least 20 magnitudes, and the rise in brightness occurs at a rate of 0.25 to 0.5 magnitudes per day. After a week or so at maximum brightness, the decline occurs at an initial rate of one magnitude per week, but soon slows to one magnitude in ten weeks. The spectra are characterised by the absence of lines due to hydrogen, and by line shifts implying expansion velocities up to 20 000 km s^{-1}. No supernova (of either type) has been observed in our own Galaxy since telescopes were invented, so that all the observations are for supernovae in other galaxies. Type I supernovae are found in all types of galaxy, and occur at a rate of about one per century in the larger ones. Their presence in elliptical galaxies suggests that the masses of the objects becoming such supernovae are little more than that of the Sun, since more massive stars will have evolved to white dwarfs in such galaxies.

Type II supernovae are essentially all those which are not type I! They are thus a much more disparate group of objects. Their peak brightnesses are lower; typically absolute magnitudes of -12 to -16 may be reached. They tend to remain at their peak brightness for longer than type I supernovae, and then to decline more slowly. Their spectra contain hydrogen lines as well as those of other atoms, and indicate expansion velocities up to 10 000 km s^{-1}. They are observed to occur exclusively in the spiral arms of the more open spiral galaxies (types Sb and Sc), where perhaps one per fifty years might be expected. Recent radio detections of supernovae, however, suggest the true rate is perhaps twice this figure, and that the total of type I and II supernovae in a large spiral galaxy may reach five to ten per century.

Given the duration of a supernova near its maximum, we may expect about one galaxy in a hundred to have an observable supernova in it at any given time. Thus their study is not too difficult despite their rarity. No supernova has been observed prior to its explosion, so that their precursors are known only from theory (see below). Post-supernova objects are discussed later in this section and in §§2P.1, 3A.9, 6A.6, and 8A.2.

Type I supernovae, from their distribution in galaxies, are thought to originate from population II type stars. They are not well understood, and various suggestions for their mechanisms have been made. Among the stronger possibilities there is the complete thermonuclear explosion of the degenerate core of a highly evolved star, the cataclysmic end for a nova-type binary containing a carbon–oxygen white dwarf as accretion takes the white dwarf mass above the Chandrasekhar limit, and the result of the coalescence of two white dwarfs in a close binary system. The mass prior to the explosion is variously estimated to lie between one and eight solar masses. Some 0.2 to 1.0 $\mathcal{M}_{\odot}$ of $^{56}_{28}Ni$ is synthesised during the explosion from the carbon and oxygen in the white dwarf precursor. This is radioactive and decays to $^{56}_{27}Co$ and then to $^{56}_{26}Fe$ with half lives of 6.1 and 77 days respectively. The energy released by the decay provides the power for the later stages of the supernova's emissions, leading to the observed 'half life' of 50 to 80 days for the decline of the light curve after the first few weeks. Most of the mass is flung out into space as a hot gas cloud (see supernova remnants, §6A.6), probably leaving little or nothing behind in most cases. Sometimes a neutron star may remain and perhaps become a pulsar (see below and later in this section, and §§2P.1, 8A.2).

Type II supernovae are clearly population I objects from their close association with the spiral arms of galaxies. They are thought to be better understood than type I supernovae, although recently problems with the models have emerged, in that they fail to predict *both* an explosion leading to a supernova remnant (SNR) *and* a collapse leading to a neutron star. This probably arises from defects in our understanding of the details, however, rather than from any fundamental misunderstanding of the basic process. This is thought to involve the final stages of the life of a star whose initial mass was at least 10 solar masses and probably rather higher. As discussed in more detail in §8A.2, an unstable core of iron-group elements forms with a mass in the region of 1.5 solar masses. The temperature is probably around 8×10^{9} K, and the density some 4×10^{12} kg m^{-3}. Under these conditions the nuclei begin to fragment through photodissociation back to the lighter elements, especially to helium. Since iron has the greatest binding energy per nucleon (figure 2P.3), its break-up requires the input of vast quantities of energy. This is taken from the kinetic energy of the particles in the core, so that as the iron nuclei fragment, the pressure in the core plummets. The core therefore collapses, increasing its density until the protons and electrons combine into neutrons. The remaining pressure is further reduced, therefore, as the contribution from the degenerate electrons falls, and the core collapse accelerates. Initially the rate of this final collapse may approach the free-fall rate with an e-folding time of a hundredth of a second. The internal pressure is kept low by the loss of energy into neutrinos which originate in many different reactions:

$$e^{-} + p^{+} \rightarrow n + \nu_{e} \tag{3A.5.8}$$

$$^{14}_{7}N + e^{-} \rightarrow {}^{14}_{6}C + \nu_{e} \tag{3A.5.9}$$

$$e^{-} + \gamma \rightarrow e^{-} + \nu_{e} + \tilde{\nu}_{e} \tag{3A.5.10}$$

and then escape from the core. As the density of the material rises above 3×10^{17} kg m^{-3}, however, these neutrinos become trapped in the core. The pressure rises again, and the later stages of the collapse proceed on a time scale dictated by the diffusion of the neutrinos out from the core. For the probable conditions in the core at this stage, this gives an e-folding time of about an hour. During this stage, the neutrino luminosity may reach 10^{42} W, or ten million times the supernova's peak optical brightness! The neutrino emission of course is for a much shorter time, and so the total energies emitted in the forms of neutrinos and photons throughout the outburst are about 10^{45} to 10^{46} J, and 10^{42} J respectively. The detection of such a neutrino pulse is one of the aims for the neutrino 'telescopes' currently in use or under construction, but has yet to be achieved. The final fate of the collapsing core is to become a neutron star, or possibly a black hole or white dwarf at the upper and lower limits of the process (see later in this section and §§2P.1, 3A.9, 8A.2).

While the core is collapsing as outlined above, the outer parts of the star are undergoing an explosion. The precise mechanism causing this is not yet clear. One possibility is called the bounce-shock process. This suggests that as the inner parts of the collapsing core start to slow down, through the developing baryon degeneracy pressure or neutrino pressure or both, their collapse may halt and even rebound (the 'bounce') slightly, reverberating with a period of about a millisecond. The outer parts of the core are still collapsing at near free-fall rate however. The resulting collision between the infalling and stationary portions of the core produces an intense shock wave which propagates outwards at perhaps 20% of the speed of light, with the wave's energy reaching perhaps 10^{44} to 10^{45} J. The deposition of a significant fraction of this energy into the overlying outer layers of the star by some (unknown) process would suffice to disrupt them and to cause the observed explosive phase of the supernova. Other suggested possibilities for initiating the explosion involve thermonuclear runaways, magnetic fields, and Rayleigh–Taylor instabilities.

Whatever the precise process, the outer parts of the star, with a total mass five or six times that of the Sun, are exploded outwards at velocities in the region of 10 000 km s^{-1}. The increase in the luminosity occurs as the surface area of the emitting region increases, and provides the first detectable evidence (at present) of the occurrence of the supernova. The material continues to expand outwards at high velocities; for example, velocities over 5000 km s^{-1} are still observed in the Crab nebula, over 1000 years after the supernova which produced it. It eventually becomes an emission nebula and is called a supernova remnant (SNR). These objects are of great contemporary interest and are dealt with as a separate topic (§6A.6). Since the material in the SNR is likely to have undergone significant processing inside the star, it is a major contributor to the increasing abundance of the heavier elements, and via the *r*, *s* and *p* processes explains many of the observed relative abundances of the elements (§§1A.2, 2A.2, 3A.6, Chapter 7A). The collapsing core probably ends as a neutron star and then may be observable as a pulsar in some cases (see below). Alternative end points, as with the type I supernovae, may be as white dwarfs and as black holes (§3A.9).

Pulsars: Type P, PSR, MXB

At the time of writing (1985) it is just fifty years since Baade and Zwicky speculated on the possible existence of stars formed almost exclusively from neutrons, and not quite twenty years since such unlikely objects were found to exist in reality. These stars are known as neutron stars and are formed mostly from baryon-degenerate matter (§2P.1). Their masses are similar to that of the Sun, but their radii only about 10 km. Their mean densities are thus around 10^{16} to 10^{17} kg m^{-3} (10^{13} to 10^{14} times the density of water). Their primary observational presence is as pulsars and as binary x-ray sources.

The term pulsar derives from *pulsa*ting *r*adio source, and this accurately describes their main *observational* characteristic. Well over 300 have now been catalogued. They exhibit a regular series of flashes in their radio brightnesses (figure 3A.19). Their periods average just under a second, but can range from 1.5 milliseconds to nearly 5 seconds. The individual pulses are highly variable (figure 3A.19), but the average of several hundreds of pulses gives a remarkably stable profile. Even more remarkable is the stability of the period: typically it is constant to one part in 10^{13}. The ubiquitous quartz wrist watches, for comparison, have a stability of only one part in 10^5 and the world's timing standards, based upon atomic clocks, are stable to one part in 10^{14}.

The pulse profiles change very little with wavelength over the radio region, but their arrival times show an increasing delay as the wavelength increases. This is due to the reduction in the velocity of the radio waves by the interstellar medium (§7A.3). Since the delay depends upon the column electron number density along the line of sight, if the actual number density is known, the distance of the pulsar may be found. Often a value of 3×10^4 m^{-3} may be used for the number density, and the pulsars are then found to lie at distances of 100 to 18 000 pc. Only in one case (PSR 1929 + 10) is the object close enough for the trigonometrical parallax (Appendix I) also to be found. That gives a distance for the pulsar of 150 pc, while the dispersion distance is 300 pc. The normal uncertainty in the dispersion distance is probably rather less than this factor of two, however. In four cases, PSR 0531 +21, PSR 0833 −45, PSR 0540 −69.3, and PSR 1937 +21.4 (the first two of which are associated with the Crab nebula and Vela SNRs, the third is in the large Magellanic Cloud, and the last is the 'millisecond pulsar'), optical pulses have also been found. The periods and pulse shapes agree with the radio data. The Crab nebula pulsar also pulses at x- and gamma-ray wavelengths. Several other x-ray pulsars are known (see x-ray binaries below), but the Vela pulsar appears to be without pulses in this region.

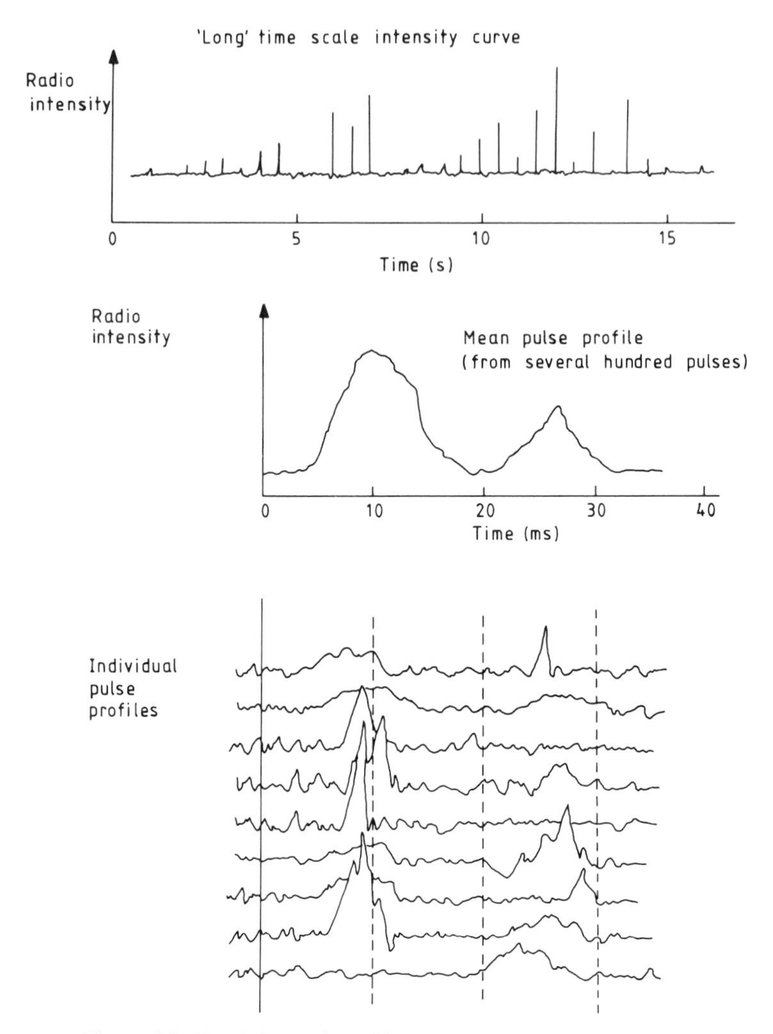

Figure 3A.19 Schematic radio observations of a typical pulsar.

The emission is usually strongly linearly polarised; up to 100% in some cases. Weaker circular polarisation has also been observed. The direction of the linear polarisation can vary, and in some cases sweeps smoothly through up to 180° during each pulse. The polarisation indicates the presence of an intense magnetic field; some estimates place the strength at 10^7 to 10^9 T ($\sim 10^{12}$ times the strength of the Earth's magnetic field).

One of the most important clues to the nature of pulsars has come from the observation that in all cases, though constant to a very high degree of precision, each pulsar's period in fact increases slowly with time. The rate of increase is measured by the rate of change of period as a proportion of the period,

$$(1/P)\,\mathrm{d}P/\mathrm{d}t \qquad (3A.5.11)$$

and this has values from 10^{-16} to 10^{-11}. It is the reciprocal of the pulsar's characteristic time—that is, the time required for the period to increase from zero to its present value, assuming a constant rate of increase of period. Pulsars' characteristic times, often loosely called the age of the pulsar, range from a few thousand to a few million years. Generally the shorter period pulsars are slowing most rapidly. One exception to this trend towards increased periods is the existence of 'glitches'. These are very sudden decreases in the period by up to 0.0001%, at intervals of a few years. The rate of increase of period speeds up temporarily after a glitch, eventually reverting to its earlier value after some tens of days. Much smaller amplitude glitches are more frequent, and are sometimes called the pulsar's 'timing noise'.

The radio output from pulsars varies with the period and the characteristic time. Values generally lie between 10^{18} and 10^{23} W. If, however, the increasing period represents the slowing down of the rotation of a neutron star (see below), then the total energy loss must be nearer to 10^{23} to 10^{31} W. The excess energy loss may go into energetic particle production. Several pulsars are embedded in SNRs, and this particle emission may provide their energy source, although since not all SNRs have associated pulsars, this cannot always be the case (§6A.6). The energy radiated by the Crab nebula, however, (10^{31} W) is roughly equal to the estimated rotational energy loss of its associated pulsar (PSR 0531 +21).

Two other individual pulsars are worthy of mention, the binary pulsar (PSR 1913 +16) and the millisecond pulsar (PSR 1937 +214). The first of these is a binary system probably containing two 1.4 solar mass neutron stars. The orbital period is about 7.75 hours and the pulsar period 0.059 s. The orbit is highly elliptical with an eccentricity of about 0.6, and the semi-major axis is about 2×10^6 km. The interest in it lies largely in the fact that it forms a very high precision clock in an intense gravitational field. It therefore enables many checks upon the predictions of general relativity to be made. For example, general relativity predicts that the periastron point should be advancing at about 4° per year (cf 43″ per century for Mercury), and that the orbital period should shorten at a rate of

$$(1/P)\,\mathrm{d}P/\mathrm{d}t \simeq -9 \times 10^{-17} \qquad (3A.5.12)$$

through the loss of angular momentum by gravitational radiation. The observations give 4.2° per year, and -8×10^{-17} respectively. The latter figure is particularly significant in that the predictions of other gravitational theories (scalar–tensor theory, most cases of Brans–Dicke theory etc) have much poorer agreements with the observations than does Einstein's general relativity theory. The millisecond pulsar has a period of only 1.557 ms, and a characteristic time of several hundred million years, compared with an expected one of a few years only, if this object continued the trend of other pulsars for shorter characteristic times as their periods decrease. The interest in this pulsar and the more recently discovered PSR 1953 + 29, which has a period of 6.133 ms, stems from the difficulty of accounting for their properties by the standard pulsar model (see below). Possible models to explain these two systems include the coalescence of a binary pair of neutron stars, mass exchange from a companion to the neutron star, and weak magnetic fields reducing the rate of loss of angular momentum. The absence of glitches in the millisecond pulsar and the extreme regularity of its pulses offer the possibility of its providing a more accurate time standard than an atomic clock, and of allowing astronomy to reclaim its traditional role as time-keeper to the world!

The rapidity and stability of the pulses from pulsars provide the main clues to their natures. For such stability, the basic mechanism must be based upon either rotation or vibration, while the speed of the changes limits the size of the emitting region (*1) to a few thousand kilometres at most. Thus, given that the stability requires an object of solar mass or thereabouts, only white dwarfs and neutron stars are left as possible candidates. Theory can then give a guide to the periodicities to be expected from such objects:

Vibrating white dwarf: period $\simeq$ 2 s
Rotating white dwarf: period > 1 s
Vibrating neutron star: period $\simeq$ 0.001 s
Rotating neutron star: period > 0.0005 s.

Thus with the observed periods lying in the region of 0.0015 to 4 seconds, the only possibility is that of a rotating neutron star. Although the reason for the rapidity and stability of the pulse period is thus well established, the actual mechanism for the emission remains in dispute. Most models base the emission mechanism upon the intense magnetic fields expected on neutron stars. The simplest models then have the magnetic axis at an angle to the rotational axis and the emission occurring continuously into two conical beams around the magnetic poles (figure 3A.20). A pulse is observed in this case in the same manner that the continuous emission from a lighthouse at sea is pulsed: the radiated beams of radiation can only be observed while they point down the line of sight to the observer, and they are invisible when the rotation sweeps them around to point elsewhere. The double structure of many pulses (figure 3A.19) may be explained if the emission intensity peaks towards the edges of the beams. Charged particles will be trapped by the pulsar's magnetic field, and since the inner portions will be corotating with the pulsar, the particles will be accelerated along the field lines away from the neutron star. Eventually they may reach velocities close to the speed of light, and when they break away from the magnetic field, they may provide the energy input to the surrounding SNR (if present). At least some of the cosmic rays (§7A.4) may originate from these particles. They may also be the cause of the increasing period of the pulsar through the angular momentum that they carry away. With this scenario, the very high stability of the millisecond pulsar serves to confirm its possession of a comparatively weak magnetic field. Synchrotron emission (Chapter 6P) from such accelerating particles provides an alternative but less satisfactory mechanism for the pulses. In order to obtain pulses by this method, some arbitrary 'hot spot' on the magnetosphere must be postulated. Furthermore, the separation of the emission region from the neutron star itself would reduce the stability of the emitted pulses to the point where it becomes difficult to explain the observed high degree of constancy.

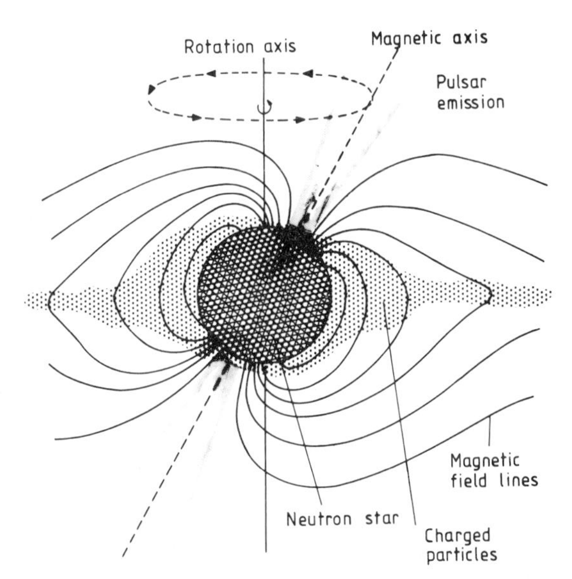

Figure 3A.20 Schematic cross section of the oblique rotator model for a pulsar.

The glitches are thought to occur following a sudden contraction and spin-up of the neutron star. The precise process involved is unclear, but since the centrifugal force will decline as the period lengthens, a solid crust to the neutron star, initially in hydrostatic equilibrium, would experience an increasing strain with time. If this crust should suddenly fracture (a 'starquake'), allowing contraction, then we might get the observed sudden spin-up of the pulsar. The initially higher slow-down rate following a glitch may then arise as the angular momentum change is slowly transmitted from the charged particles in the interior of the neutron star (which rapidly follow the change in the rotation velocity through the action of the magnetic field) to the neutrons via a weak frictional coupling.

Although there is an association between SNRs and pulsars, there are many SNRs without observable pulsars, and some pulsars which might be expected to have an SNR but which do not. On the oblique rotator model, a pulsar would only be observed when the line of sight could intersect one of the emission beams. Thus there must be many pulsars in existence which are not observed. Nonetheless there are only two observed pulsars in over 120 SNRs, and this is too large a discrepancy to be explained by this picture. Either many supernovae do not result in neutron star formation, or the neutron star does not always go into a pulsar mode directly after its formation. A possible, though rather speculative, mechanism whereby the pulsar mode might be delayed is if it requires an external trigger such as high energy cosmic rays for its initiation. The magnetic field immediately after its formation might then be too strong for such cosmic rays to penetrate through to the neutron star. In most cases, therefore, the SNRs would have dispersed before the pulsar 'switched on', and the neutron stars in most observed SNRs will not yet have a sufficiently weak magnetic field for the trigger to have operated. An alternative or possibly complementary explanation is that not all supernovae produce neutron stars. The type I supernovae in particular (see earlier discussion) seem likely to leave little behind.

Binary x-ray stars. Some twenty or so of the observed sources of x-radiation in the sky are found to be pulsating rapidly. The first such object to be discovered, Her X-1, is a reasonable representative of the group. It has a pulse period of 1.24 s, and in addition the pulses disappear for about six hours at a time once every 1.7 days. Its optical identification is with the variable HZ Her, this being an eclipsing binary with a period of 1.700 17 days. It is perhaps a hundred times as bright as the Sun in the optical and the spectral class is that of an early main sequence star. Its x-ray luminosity, however, is some ten thousand times the total solar emission, and the variability places an upper limit of 100 000 km on the x-ray emitting region. An A0 V star is about two million kilometres in diameter. The x-ray emission must thus be associated with the companion in the binary system, and this must be a compact object. The other x-ray pulsars have periods ranging from 0.07 to 835 seconds, and almost all of them show signs of belonging to binary systems. Unlike 'normal' pulsars, however, the pulse periods in these stars are decreasing with time. One additional observation, which applies only to Her X-1, is that in addition to its 1.24 s and 1.7 d periods, its x-ray emission exhibits a 35 day on–off cycle: the x-rays are emitted strongly for 12 days, disappear for 7 days, reappear for 6 days, and then disappear again for 10 days. The optical 1.7 d variation persists at all times. A 1.24 s variation has also been found in the gamma-ray region for this object.

The generally accepted model for these systems has a neutron star as a member of a binary system. Its companion, which is not necessarily of lower mass, is losing material to the neutron star, either through an intense stellar wind, or through transfer at the inner Lagrangian point, having expanded to fill its Roche lobe (§8A.2). The accreting material is preferentially guided towards the poles of the neutron star's magnetic field, where its impact with the surface results in the observed x-ray emission. Rotation and an inclined magnetic field then produce pulses in the same fashion as in a 'normal' pulsar. Since some 10^{16} J of gravitational energy are released as one kilogram of material falls on a neutron star (*2), an exchange rate of only $10^{-8} \mathcal{M}_{\odot}$ per year would suffice to power the observed x-ray emissions. The increasing spin of the neutron star, which leads to the observed decrease in the pulse period, arises via the tidal effects of the accreting matter. The longer period variations result from the binary motion. The third period, of 35 days, which is found in Her X-1, may arise as the precession of the accretion disc around the neutron star causes it to mask the x-ray emitting regions. The 1.24 s periodicity can be detected in the optical counterpart, HZ Her. This may be due to the absorption of the x-ray beam by the companion as it sweeps across its surface, the energy being thermalised and re-emitted into the visual spectrum.

Since the neutron star can be observed in the x-ray region, and the companion in the visual, the orbits of both stars may often be determined. The masses of the two components of the system can thereby be found (§1A.1). The results give the neutron star's masses lying between 0.4 and 2.5 $\mathcal{M}_{\odot}$, but with uncertainties by up to a factor of two. The most probable average mass lies between 1.2 and 1.6 $\mathcal{M}_{\odot}$, and compares well with the Chandrasekhar limiting mass for a white dwarf (equation 2A.1.29).

X-ray bursters. A brief mention of what appears to be a third observational manifestation of neutron stars is worth giving, even though at the time of writing the evidence for it is less than totally conclusive. About half the Galactic x-ray sources emit hard x-rays in sharp bursts, and similar objects are found in several globular clusters. The bursts occur at intervals of a few hours, except in the case of the object called the rapid burster, where the interval may be only a few seconds. The burst rise time is less than a second, and it reaches a peak x-ray luminosity of 10^{30} to 10^{32} W. The bursts decay in a few tens of seconds after reaching their maxima. Some of the sources also have a steady background emission of x-rays at about 10% of the burst intensity.

The most widely accepted explanation for the bursters postulates a binary system with matter accreting on to a neutron star from its companion. The difference between these objects and the x-ray binary pulsars (above) may be due to the presence of a much weaker magnetic field in the burster system. The accretion stream is not therefore channelled towards the magnetic poles to any significant degree. The material, which is mostly hydrogen and helium, thus accumulates over the whole surface of the neutron star. The continuous x-ray emission, when seen, comes from the accretion stream. The bursts are due to thermonuclear flashes in the material on the surface of the neutron star. This is a very similar process to that postulated for novae (see above). The much greater total energy emission from nova outbursts (10^{38} J) and their much longer repetition periods arise because far more material must accumulate on the surface of a white dwarf than on a neutron star before the thermonuclear runaway begins.

The strong gamma-ray source known as 'Geminga', and the occasional isolated gamma-ray bursts from other parts of the sky, which are observed about once a day on average, are also generally attributed to neutron stars. Beyond this, however, their nature is not understood (although theories abound at the time of writing).

T Tauri stars: type In T

The importance of T Tauri stars lies not in their numbers (only about 1000 are known) but in their key position in our evolutionary picture of stars. As discussed further in §8A.1, these stars, with the probably related Herbig Ae and Be stars and FU Orionis objects, are almost certainly stars in which nuclear reactions have only recently commenced. They are thus in the process of evolving towards the start of their main sequence lives.

Their main observational characteristics are:

(i) A spectrum which includes emission lines at 406.3 and 413.2 nm, due to neutral iron (the group-defining criterion). Other emission lines due to hydrogen and ionised calcium, and forbidden lines of ionised sulphur may be seen, as well as strong lithium absorption. The general spectral type lies between F5 and G5, but can be as late as M0.

(ii) A close association with interstellar gas and dust. Often the stars occur in groups of five or six, known as a T-association, within an emission nebula such as the Orion nebula.

(iii) An absolute magnitude similar to or a few times that of the Sun. Irregular pseudo-sinusoidal variations with an amplitude of up to three magnitudes and a characteristic time of a few months. Variable polarisation is also often observed.

(iv) P Cygni line profiles for some of the lines. A small subgroup called the YY Orionis stars have inverse P Cygni profiles, with the absorption component on the long wavelength side of the emission component (§3A.2). P Cygni line profiles are indicative of an expanding atmosphere (contracting or infalling in the case of the inverse profiles). The velocities for this motion found from the lines are mostly around 100 to 200 km s^{-1}, and lead to estimates of the mass loss rate of over 10^{-9} $\mathcal{M}_{\odot}$ per year. The spectrum lines, and hence also the velocities and mass losses, can be very variable with time. In a few cases, the profiles can switch from normal to inverse P Cygni profiles or vice versa, indicating a switch from expansion to contraction for the outer regions of the star.

(v) There is a large infrared excess, sometimes with spectral features at 3.1, 10 and 20 microns, the latter possibly being attributable to water ice and silicate particles.

(vi) The stars appear on the HR diagram about two magnitudes above the main sequence, running parallel to it between spectral types F5 and M0.

Various empirical models have been suggested to account for these observations. None, however, is totally convincing in all its details. The least improbable scenario involves a one to two solar mass star just forming a solar-type outer atmosphere. The atmosphere starts to be driven outwards by the deposition into it of Alfvén wave energy from the deep convection zone of the star, transmitted to it by the strong magnetic field which is still present. A shock zone then develops as this encounters the still infalling material from the surrounding nebula. The zone would be at a high temperature and unstable. The emission lines form in this region, changing rapidly as conditions change. High velocity expulsion of material could occur as jets, alongside continuing and more diffuse accretion. Dust grains at some distance from the star provide the source of the infrared excess. Additional irregular variations may occur through the operation of solar-type flares and chromospheric activity. Much, however, remains to be clarified before a satisfactory model for these stars is achieved.

3A.6 ABUNDANCES

The composition of stars has already been reviewed in Chapter 1A. Here we take a brief look at the means whereby the abundance of an element in a star's atmosphere may be determined, and then look at some of the exceptions to the general rule of solar-type compositions for stars.

Determination

The strength of a spectrum line is usually measured by its equivalent width. This is the width of a section of the nearby continuum which has the same area as that contained between the envelope of the spectrum line and the continuum (figure 3A.21). Such a definition is particularly convenient for use when working on tracings of spectra. If the data are stored on a computer, or are presented in some other form, then the determination of the equivalent width will be similar in intent, but the details may vary. The symbol for the equivalent width is normally W or W_λ. It is measured in terms of wavelength, and in the optical region varies from 0.0001 nm for the faintest lines to 10 nm for the strongest. Since the determination of equivalent width is a long and involved process, with many opportunities for inaccuracies to be introduced, the errors in the finally determined value are usually large. Standard deviations are 5% for the highest quality data, 20% for typical data, and very much worse if the data are poor or the lines weak.

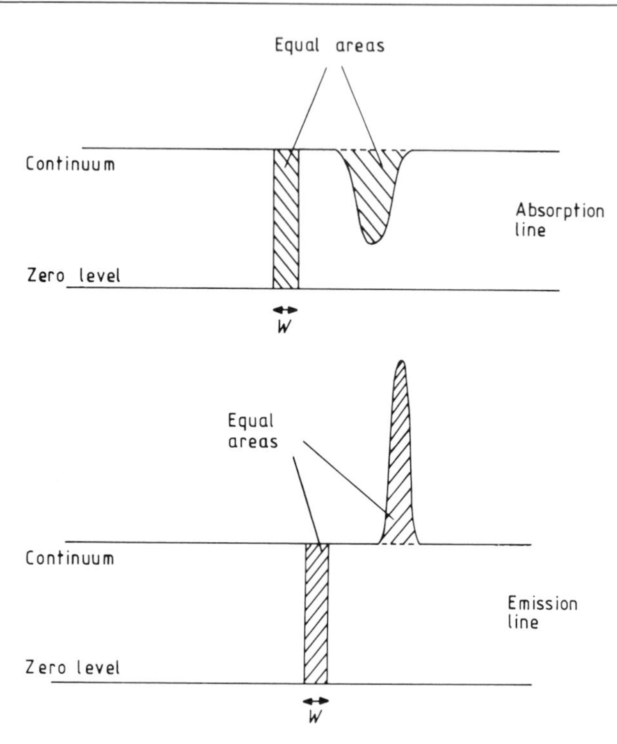

Figure 3A.21 Equivalent widths of emission and absorption lines.

The equivalent widths of the lines in a spectrum are capable of providing information on many aspects of the object. Their main use, however, is in the determination of the relative abundances of the elements by the method known as coarse analysis. The more recently developed fine analysis (see below) is gradually superseding coarse analysis for the final determinations of abundances, but it is not a complete replacement because it usually uses the results of coarse analysis as its starting point. Furthermore, fine analysis is a very much lengthier process than coarse analysis and is much more restricted in its range of application at the moment. Thus coarse analysis is discussed here since it is likely to remain a useful technique for the foreseeable future.

The foundation of coarse analysis is the curve of growth. This is the relationship between the equivalent width of the line and the number of atoms or ions producing it (figure 3A.22). The details of the curve of growth depend upon many factors, such as temperature, pressure, surface gravity, etc and it varies between different lines. The overall pattern of behaviour, however, is generally similar to that shown in figure 3A.22. There are three main phases to the development of a spectral line, and these produce the three quasi-linear portions of the curve of growth. The first of these (*A* on figure 3A.22) is known as the linear section. The line is weak and the number of absorbing atoms is much smaller than the number of appropriate photons. Over this section, therefore, there is a direct proportionality between the line strength and the number of absorbing particles,

$$W \propto N. \tag{3A.6.1}$$

As the number of absorbing particles increases, the centre of the line becomes almost black (saturates), but the wings of the line as yet are unaffected by pressure broadening (§3A.2). The line strength therefore becomes almost independent of the number of absorbing particles, and we reach the 'constant' portion of the curve of growth (*B*):

$$W \simeq \text{constant}. \tag{3A.6.2}$$

Finally, further increases in the number of absorbing particles cause broadening of the line profile. Absorption can then occur far from the line centre. This is section *C* of the curve of growth, and the equivalent width varies as the square root of the number of absorbing particles:

$$W \propto \surd N. \tag{3A.6.3}$$

In a star, many other factors such as the presence of other species of atoms and ions etc affect the line development, but a generally similar behaviour pattern may be found.

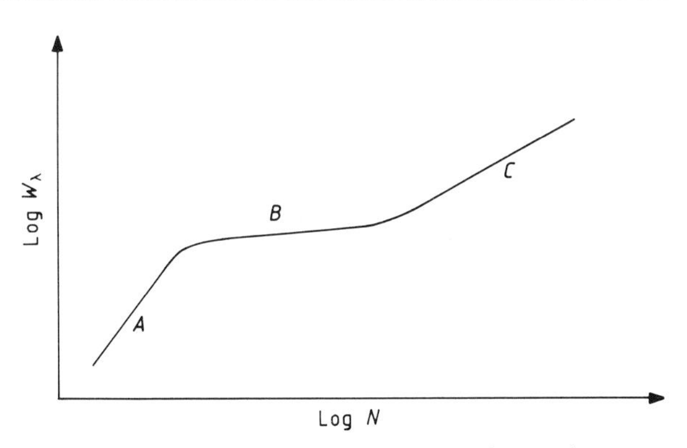

Figure 3A.22 Schematic curve of growth.

In order to determine the abundance of an element, a generalised curve of growth is needed which applies to many or all of the atom's lines. This may be accomplished in several ways, but probably the most widespread method is to normalise the equivalent width and the number of atoms. The equivalent width is normalised by dividing it by the wavelength of the line. Thus $\log_{10}(W_\lambda/\lambda)$ is plotted as the ordinate of the curve of growth. The abscissa is then $\log_{10}(Nf)$, where f is the oscillator strength (equation 3P.1.6), and Nf is thus the effective number of atoms contributing to the formation of the line. The observations are plotted on a graph of $\log_{10}(W_\lambda/\lambda)$ against $\log_{10}(f)$. Since the oscillator strengths can vary by large factors from one line to another, the resulting curve will also be a curve of growth. The shift along the abscissa required to superimpose the theoretical and observed curves of growth then gives the value of $\log_{10}(N)$ directly. The relationship of N to the abundance of the atom depends upon the lines used in the determination. If they originate from an excited level then N will be the number of atoms excited to that level etc. If TE or LTE can be assumed then the Boltzmann equation (equation 3P.4.10) can be used to determine the populations of the other levels, and so the total number of atoms or ions. Saha's equation (equation 3P.4.1) can then be used to relate this to the numbers of atoms in other states of ionisation, and so the total abundance of the element can be found. Alternatively, the populations of the other levels and ionisation stages may be determinable directly from the lines produced by them.

Fine analysis requires much higher quality data, and the corrections for instrumental degradation etc must be carefully applied. The technique uses a detailed model atmosphere, together with assumed abundances, temperature, pressure, turbulence and surface gravity to predict line strengths and line profiles. These are then compared with the observations, and the assumed parameters altered until a good fit is achieved. The initial values would normally be obtained from a coarse analysis (see above). Fine analysis, as it becomes more detailed, gradually merges into spectrum synthesis. The spectrum is then calculated point by point over a wide range of wavelengths for comparison with the observations. This is a very complex procedure since the usual simplifying assumptions (LTE etc) have to be abandoned. The number of reliable models of stars based upon spectral synthesis is relatively few at the moment, but this is undoubtedly the direction in which spectroscopic analysis will proceed in the future.

Results

The normal composition of stars has already been reviewed in Chapter 1A (figure 1A.11). The main variations from that concern the abundances of the heavier elements relative to those of hydrogen and helium (figure 1A.12) and depend mostly upon the age of the star. In this section we are concerned with significant deviations from this normal pattern, usually called abundance anomalies. They are determined spectroscopically and thus refer to the surface layers of the star. They do not necessarily apply to the interiors of the stars, nor are variations in the internal composition due to nucleosynthesis (figure 1A.13 for example) revealed, except in the one or two really exceptional cases when such material is brought to the surface by some process.

White dwarfs

By far the largest group of stars to differ from the standard 'solar' type of abundance are the white dwarfs. Some of their anomalies have already been mentioned in §§1A.2 and 3A.2. The interpretation of the spectra is very difficult, and it therefore remains to be proven that the observed differences do arise through altered abundances. The main apparent variations from the norm, however, lead to the following groups:

(i) Hydrogen-rich; these white dwarfs probably have near-solar compositions. Since often only the hydrogen lines are visible, however, this is not certain.

(ii) Helium-rich; no sign of any hydrogen at all in these stars.

(iii) Metal-rich; the hydrogen lines are very weak or completely absent. The main spectrum lines are due to ionised calcium. Sometimes iron, magnesium and other elements also appear strongly.

Because of the problems in inerpreting these spectra, no estimates of the degree of over- or underabundance can be made at the time of writing. There are also white dwarfs with an almost pure continuum spectrum; any lines are less than 5% of the continuum intensity and thus lost in the noise of the data. For such stars no indication of their composition at all is possible.

On evolutionary grounds (§8A.2), we might expect white dwarfs to be the collapsed end products of stars. Their bulk compositions may then range from almost pure carbon through to almost pure iron. The theoretical models of relativistic electron-degenerate spheres (§§2P.1, 2A.1) have the best agreement with the observations for compositions of magnesium or iron.

Wolf–Rayet stars

These are very hot stars whose spectra are formed almost entirely from intense broad emission lines superimposed upon a continuum similar to that of an O- or B-type star. Hydrogen lines are generally absent, but strong helium lines are always present. The stars subdivide into two groups: the first with strong lines of nitrogen in addition to those of helium and called the WN stars, the second with strong lines of carbon and oxygen in addition to those of helium and called the WC stars. There are also stars which appear to be intermediate between the Wolf–Rayet stars and very hot main sequence stars, and these are called the Of stars.

Amongst the WN stars, there appear to be roughly equal amounts of hydrogen and helium by number, compared with the cosmic ratio of ten to one. Though very variable, the abundance of hydrogen appears to increase towards the cooler end of the sequence. Nitrogen is overabundant by mass in comparison with the helium by about a factor of 40, while carbon is underabundant by about a factor of 150.

The absence of hydrogen in the WC stars is less certain since the carbon lines blend with any possible hydrogen lines. Indeed some workers suggest a near normal abundance for hydrogen in both types of Wolf–Rayet star, attributing the anomalies to our lack of understanding of the photospheric processes in these stars. A low hydrogen abundance, however, still remains as the more probable interpretation of the spectra. With respect to helium, carbon is enriched by a factor of six by mass, nitrogen also being overabundant to about the same degree.

It seems a possibility that Wolf–Rayet stars are the remnant cores of large stars. The original mass would have been over 25 $\mathcal{M}_{\odot}$, but over 50% of this has been lost in very intense stellar winds (Chapter 5A). The highly processed core of the star is thus revealed. The WN stars may then be showing the products of the CNO cycles. The slowest reaction (figure 2A.5) is

$$^{14}_{7}\mathrm{N} + ^{1}_{1}\mathrm{H} \rightarrow ^{15}_{8}\mathrm{O} + \gamma \qquad (3A.6.4)$$

and this bottleneck leaves most of the material in the form of nitrogen, depleting the carbon. The WC stars may be more evolved and exhibiting the reaction product (carbon) of the triple-alpha process (§2A.2). Some planetary nebulae (§6A.5) have central stars with Wolf–Rayet spectra. Their masses are much lower than 'normal' Wolf–Rayet stars, however, and their mode of origin is unclear.

Lithium and T Tauri stars

Most of the properties of T Tauri stars have already been reviewed (§3A.5). One peculiarity not mentioned in detail there, however, is that lithium is often found to be overabundant, up to 100 times the solar abundance being possible. An even more extreme overabundance is found in the lithium stars. These are late-type giants, usually of spectral classes S and C (§3A.2), but sometimes hotter, and with generally peculiar spectra.

Now the solar abundance of lithium and its neighbours boron and beryllium is remarkably low (figure 1A.11) compared with other nearby elements. This is due to the ease with which lithium, boron and beryllium take part in nuclear reactions. For example, lithium would be completely consumed in the reactions

$$^{6}_{3}\mathrm{Li} + ^{1}_{1}\mathrm{H} \rightarrow ^{4}_{2}\mathrm{He} + ^{3}_{2}\mathrm{He} \qquad (3A.6.5)$$

and

$$^{7}_{3}\mathrm{Li} + ^{1}_{1}\mathrm{H} \rightarrow ^{8}_{4}\mathrm{Be} + \gamma \tag{3A.6.6}$$

$$^{8}_{4}\mathrm{Be} \rightarrow ^{4}_{2}\mathrm{He} + ^{4}_{2}\mathrm{He} \tag{3A.6.7}$$

in under 10 000 years for the conditions at the centre of the Sun. The reactions occur more slowly, but still significantly, at the lower temperatures and pressures to be found well away from the centre of the star. Mixing via convection will thus lead to the depletion of these elements throughout the whole star in only a few million years.

The $^{6}_{3}\mathrm{Li}$ isotope is not produced by any known nucleosynthetic reaction. Its origin, and also that of some of the $^{7}_{3}\mathrm{Li}$ isotope, seems to be in the spallation reactions of cosmic rays and the interstellar medium (§7A.4). The heavier cosmic ray nuclei fragment into light nuclei upon collision with interstellar protons.

The overabundance of lithium in T Tauri stars may thus be seen to be due to their very recent formation from the interstellar medium. The latter has become enriched in lithium and the other light elements through the operation of cosmic-ray spallation over aeons, and insufficient time has elapsed since the onset of nuclear burning in T Tauri stars for it all to have been destroyed. The origin of the lithium in the lithium giants is much less clear. The ratio of the $^{7}_{3}\mathrm{Li}$ to the $^{6}_{3}\mathrm{Li}$ abundance is often as high as ten. Now $^{7}_{3}\mathrm{Li}$ can be produced by electron capture on $^{7}_{4}\mathrm{Be}$:

$$^{7}_{4}\mathrm{Be} + \mathrm{e}^{-} \rightarrow ^{7}_{3}\mathrm{Li} + \nu_{\mathrm{e}} \tag{3A.6.8}$$

as well as in the spallation reactions. So it may be that in these stars beryllium is being transported from the core, where it is produced in a sideshoot of the proton–proton chain (figure 2A.4), into the higher layers of the star. Conditions may there be hot enough to convert it to lithium, but not hot enough then to destroy the lithium.

Ap stars

The name for these stars derives from their spectral class and the peculiarities then found in their spectra. They are generally to be found a little above the main sequence on the HR diagram between spectral types B5 and F5. Their main peculiarity compared with other A-type stars is the unusually high line strengths of some elements, and this is interpreted as indicating overabundances of those elements. The stars at the hot end of the sequence tend to have an excess of manganese compared with iron. Additionally mercury, phosphorus, gallium, strontium, platinum, and some other elements may also appear overabundant. In the mid-range stars silicon and some of the rare-earth elements appear to be in excess, while at the cool end europium, chromium and strontium are the overabundant elements.

The other major characteristic of Ap stars is their strong magnetic fields. The field strength lies between 10 and 500 mT, and is often highly variable, sometimes with a period of a few days. Probably all Ap stars have such magnetic fields, since those where one has not been detected have very broad spectrum lines which render observations very difficult.

There are one or two extremely odd stars in this group such as Przybylski's star (HD 101065) and 3 Cen A. The former possesses only the lines due to the rare earths, especially holmium, and varies rapidly with several periodicities, while the latter seems to have helium predominantly in the form of ${}^{3}_{2}He$, and very strong lines of gallium, krypton and phosphorus.

The explanation for the properties of the Ap stars is unclear. Almost certainly the magnetic field must play a part in the process. It may lead to unusual diffusion effects transporting processed material from the centre to the surface of the star, or even to unusual patterns of nucleosynthesis. The magnetic fields may also provide a mechanism for the periodic spectral variations via a pulsar-like oblique rotator model, with the rotation and magnetic axes misaligned. Inadequate laboratory data and/or inadequate modelling may also account for some of the apparent peculiarities, particularly amongst the mid-range Si and rare-earth element Ap stars.

A group of stars which are similarly placed on the HR diagram and which also may have their peculiarities explained by a diffusion process are the Am stars. They occupy spectral classes A4 to F0 and appear overabundant in the rare earths and heavy elements. There is also a slight overabundance of the iron-group elements, and a possible underabundance of calcium and scandium. They are slow rotators with sharp spectral lines, and do not have the magnetic fields associated with the Ap stars. Most are also members of binary systems, sometimes both of the components being Am stars.

Carbon stars

Spectral classes R and N (§3A.2) are often combined into a single class, C, standing for carbon stars. They are cool giants with strong absorption bands in their spectra due to C_2, CN, and CH. Their carbon/oxygen ratio may be up to five times the normal value. They also exhibit an overabundance of lithium, and ${}^{13}_{6}C$ and technetium are sometimes observed. Since there are no stable isotopes of technetium, and the longest half life is only a few million years, nucleosynthetic products must be being brought to the surface. They are generally believed to be highly evolved stars with masses over three times that of the Sun, but little else about them is understood.

S stars

These stars have similar temperatures to the K and M spectral classes (§3A.2), and like C-type stars their spectra contain strong CN bands and lines due to lithium and technetium. They differ from these other stars, however, in having absorption bands due to zirconium and yttrium oxides instead of the more usual titanium and scandium oxides. They are almost all long-period variables (§3A.5). Again, the underlying processes behind the observations are not known.

Other anomalies

There are a great many spectral and other abundance peculiarities: symbiotic stars, Ba II stars, RS CVn stars, R CrB stars, to mention only a few. The more detail that is known about any star, the greater the likelihood of finding some anomaly, so that in the final analysis, every star must be non-standard in some way or other. The interested reader is referred to more specialised literature and to the research journals for information on such stars, since a plethora of very brief descriptions would be of little use, and more detailed coverage would be inappropriate in a fairly general book such as this.

3A.7 ROTATION

The spectroscopic effects of rotation have been covered in §§3P.3 and 3A.2. Here we consider the results which have been obtained.

The major problem in determining the rotation of an individual star is that only the projected equatorial rotational velocity may be found. Unless the inclination of the rotational axis can be ascertained, a very rare occurrence, then the observed velocity is just a lower limit to the actual velocity. We are therefore normally reduced to making statistical studies of large numbers of stars. Two basic approaches are useful: to look at the mean velocities, and to look at the maximum velocities.

The maximum rotational velocity which is possible for a star is that at which gravitational and centrifugal accelerations balance each other at the equator, and this is given by

$$\omega = (G\mathcal{M}/R_e^3)^{1/2} \text{ (rad s}^{-1}\text{)} \tag{3A.7.1}$$

or

$$V_e = (G\mathcal{M}/R_e)^{1/2} \text{ (m s}^{-1}\text{)} \tag{3A.7.2}$$

where ω is the angular velocity, G is the gravitational constant, $\mathcal{M}$ is the mass of the star, R_e is the equatorial radius of the star and V_e is the equatorial velocity of the star; and this is shown in figure 3A.23. At higher rotational velocities, the star would fission into a close binary system. The largest observed rotational velocities occur amongst the Be stars (*3). Their maximum values generally lie in the region of 300 to 400 km s^{-1}, well below the break-up velocity of 500 to 600 km s^{-1}. The effective gravity at the equator for such stars is still reduced by 40% to 50%, however, and this may account for their extensive outlying gaseous envelopes.

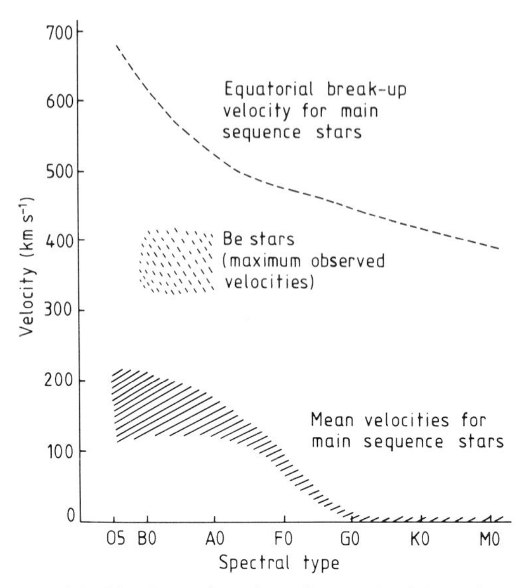

Figure 3A.23 Rotational surface velocities of stars.

The mean values of $V_e \sin i$ for main sequence stars are also shown on figure 3A.23. If the distribution arose from a random orientation of the rotational axes combined with a single rotational velocity for each spectral class, then the true equatorial rotational velocities would be about 25% higher than these observed mean velocities. The remarkable feature of the mean velocity curve is the reduction in the observed velocities in the F stars, and the very low values in the G, K and M stars. Since all stars are thought to condense from interstellar gas clouds (Chapter 7A, §8A.1), there is no intrinsic reason why there should be more angular momentum per unit mass in the more massive stars than the less massive stars. Nor is it a tenable hypothesis that all smaller stars are viewed pole-on so that their projected velocities are small. Clearly, some additional process must be invoked to slow down the less massive stars. The generally accepted hypothesis is that angular momentum is transferred from the protostar during the last stages of its collapse from the interstellar gas cloud to any remnants of the cloud still surrounding it (§8A.1). The transfer may be via magnetic or viscous connection of the protostar to the enveloping gas. The transfer of angular momentum drives the gas away from the protostar, and aided by thermal, radiation and stellar wind pressures, it is eventually lost completely. During this process, condensation in the expanding and cooling nebula may lead to the formation of a planetary system. The protostar is left rotating very slowly, and its original angular momentum is lost to the planetary system (if any) and escaping gas. Additional support for this scenario comes from the imbalance of angular momentum in the solar system, the Sun having 99.8% of the mass of the solar system, but only 0.5% of the angular momentum. The reason why such a process does not slow down the hotter stars may be that they become nuclear-burning at an earlier stage of the collapse, and so evaporate their surrounding gaseous remnants before significant angular momentum exchange has had a chance to occur.

3A.8 MAGNETIC FIELDS

The effect of a stellar magnetic field upon the star's spectrum is to cause the lines in it to split into two or more components via the Zeeman effect (§3P.1). In practice, the variation of the field's strength and direction over the surface of the star and the various other line-broadening effects likely to be acting (§3A.2) mean that the Zeeman components blend into each other. Fortunately for the observer wishing to study stellar magnetism, the Zeeman effect not only splits the lines, but causes the resulting components to be polarised. In particular, the longitudinal component of the magnetic field leads to opposite circular polarisations for the side components. The wings of spectrum lines from a magnetic star will thus retain a net polarisation, even though the individual components are not separated. Hence polarimetic measurements in the wings of spectrum lines enable magnetic field strengths to be found.

The range of magnetic fields found on stars ranges from a fraction of a millitesla to 10^9 T or more. Normally, however, stellar magnetic fields weaker than about 20 mT are undetectable; only on the Sun (Chapter 4A) can very weak fields be studied. In the radio region fields as weak as 10^{-9} T can be detected from a study of the radio emission lines from the interstellar medium (§7A.2).

Almost the only non-degenerate stars found to have high intensity magnetic fields are the Ap stars (§3A.6). With the exception of those stars whose lines are too broad for detection of any magnetic effect, all Ap stars studied have been found to have magnetic fields. The field strengths average about 50 mT, but can be as high as 3.5 T. Generally the fields are strongest in the stars with the sharpest lines; i.e. in the slowest rotating stars of the group. The magnetic field is usually highly variable and may even reverse its polarity on occasion. When the star's other variations are periodic, then the magnetic field variations follow the same period. The model of the stars is as an oblique rotator (§3A.6), so that the period is just the star's rotation period. Additionally, however, the field itself must be varying. The source of the field and the reasons for its variability remain unknown at the time of writing.

One or two other hot stars are found to have magnetic fields, but they are probably related to the Ap stars. Thus the helium-rich B9 giant, HD 175362, has a field which varies from -500 mT to $+700$ mT over a period of 3.670 days. (By convention, magnetic fields are positive when directed towards the observer.) Amongst the cooler stars, a few peculiar M and S stars have variable fields averaging a few tens of mT, but sometimes reaching several hundred mT.

The strongest magnetic fields are to be found amongst the collapsed objects, white dwarfs and pulsars. Conservation of the magnetic field during the collapse to a compact object would lead to magnetic field strengths of several tens of tesla on white dwarfs and up to 10^{10} T on neutron stars, even if the initial field were only as strong as that on the Sun (0.1 mT). So this is only to be expected. The observations are difficult, however, and based upon the effect on the total spectrum since the lines are too broad to be useful. Circular polarisations of 0.5% or so are thus found in some white dwarfs, indicating actual magnetic fields of 10^2 to 10^3 T. The polarisation is usually variable, implying a similar behaviour for the magnetic field. In G 195–19 it is cyclic, with a period of 32 hours.

The magnetic fields on neutron stars are much more uncertain, with no clear-cut observational confirmation of their existence. The models generally accepted for pulsars (§3A.5) assume the presence of very strong magnetic fields. Field strengths of a few times 10^8 T are required to provide the observed spin-down rates and to power supernova remnants by these models. They also rely upon a magnetic connection to explain the increased spin-down rate after a glitch. Two possible electron synchrotron lines (Chapter 6P) at 20 and 48 keV have been detected in x-ray binary pulsars. If this interpretation is correct, then field strengths of several times 10^8 T are implied. This is in line with the expectations, but remains to be proven.

3A.9 BLACK HOLES

At the time of writing, black holes remain a theoretician's dream. Several strong candidates and a vast number of poor candidates have been proposed as examples of actual black holes (see later discussion), but none have yet been unequivocally demonstrated to exist. Nonetheless, it would seem inevitable that sooner or later they must be found. Neutron stars took thirty years between their prediction and discovery. It is now over two hundred years since the concept of a black hole was formulated, but most of the serious work upon them dates fom the 1960s, so their actual discovery can be expected any decade now!

Black holes are a prediction of general relativity. The space–time continuum near any mass is warped, and so a beam of light passing the mass has its path deviated from a straight line. This effect forms one of the tests of general relativity; light skimming the surface of the Sun should be deflected by 1.75 seconds of arc, and this is almost exactly the observed value. Increasing gravitational field strengths are due to increasingly warped space–time continua and passing light beams will be deflected to increasingly greater degrees. Eventually the curvature of space–time may become so sharp that all light paths lead back to the object's surface. Light (and other electromagnetic radiations) originating at such an object's surface will thus by unable to escape and be detected by an external observer. We then have a black hole.

We may obtain a quantitative description of a black hole from general relativity. However, a combined classical and relativistic approach, though less rigorous, gives the same result for electrically neutral, non-rotating black holes far more simply. We just set the escape velocity (equation 2P.5.7) to the velocity of light:

$$\text{Escape velocity} = (2G\mathcal{M}/R)^{1/2} \tag{3A.9.1}$$

where $\mathcal{M}$ is the mass of the object and R its radius. Setting the escape velocity equal to the velocity of light, we therefore find

$$R_S = (2G/c^2)\mathcal{M}. \tag{3A.9.2}$$

R_S is known as the Schwarzschild radius, and gives the size required for an object to become a black hole. Thus

$$R_S = 1.484 \times 10^{-27}\mathcal{M}\ (\mathrm{m}) \tag{3A.9.3}$$

for $\mathcal{M}$ in kg, or

$$R_S = 2.59\mathcal{M}\ (\mathrm{km}) \tag{3A.9.4}$$

for $\mathcal{M}$ in units of the solar mass. A solar-mass black hole would thus have a radius a little less than three kilometres. Solar-mass neutron stars (§§2P.1, 2A.1, 3A.5) have radii of ten kilometres or so, and thus are not very far away from being black holes. The surface separating the rest of the Universe from that part of the black hole from which nothing can escape is called the event horizon. The uncharged non-rotating black hole (often called a Schwarzschild black hole) has a spherical event horizon with a radius equal to the Schwarzschild radius. However, rotation and electric charge will affect the black hole, leading to lenticular event horizons. Other than the mass, charge and angular momentum, however, no other property of the material prior to its forming a black hole remains detectable afterwards. Any material captured by the black hole after its formation can similarly only affect its mass, charge and angular momentum. The material inside the event horizon is causally disconnected from the rest of the Universe. Its ultimate fate is unknown, although Einstein's equations predict a central singularity with infinite densities being reached. Various speculations to eliminate the physically difficult situation of a singularity have been put forward. These include the influence of quantisation of gravity, tunnelling through space–time to other parts of the Universe, or even to other universes entirely, and the idea that singularities exist but that the rest of the Universe is always 'protected' from them by event horizons, an hypothesis often called 'cosmic censorship'. These ideas are not dealt with further here; instead, we concentrate on those aspects of black holes which might be of influence in astrophysics.

One of the more remarkable aspects is that a black hole cannot form in the first place! An external observer watching a clock which was falling with material collapsing towards a black hole, would see it running more and more slowly as it encountered stronger and stronger gravitational fields. This effect is known as relativistic gravitational time dilation, and a clock in an intense gravitational field is slowed compared with one in a weak or zero field by a factor of

$$\left(1 - \frac{2G\mathcal{M}}{c^2R}\right)^{1/2} \tag{3A.9.5}$$

or, from equation (3A.9.2),

$$(1 - R_S/R)^{1/2}. \tag{3A.9.6}$$

Thus when the clock reached the Schwarzschild radius, it would appear to the external observer to have stopped altogether. In a similar way, the collapse to a black hole would appear to slow down, and it would tend asymptotically to the event horizon, taking an infinite length of time actually to form the black hole. In practice, the collapse would approach very close to the event horizon very rapidly, and the still collapsing object would then possess most of the properties of a black hole as seen by an external observer, even though strictly it was not one.

The second problem associated with the formation of a black hole is that radiation always has a local velocity equal to c, and this is equal to the escape velocity from the event horizon. The effect of the gravitational field upon radiation, however, is to shift its wavelength, not to change its velocity. The radiation escaping from a collapsing object therefore does so at ever increasing wavelengths, the redshift becoming infinite for radiation originating at the event horizon. Since the energy of a photon is inversely proportional to the wavelength, such radiation would not carry any energy away from the black hole.

If a black hole were to be observed by a distant observer in the process of formation, say by the collapse of a neutron star, then there would be initially a very rapid collapse. The object would go from a radius of about 10 km to within a metre or so of the event horizon in about a thousandth of a second. Its luminosity would decay exponentially, and the photons would have exponentially increasing redshifts, the time constants in both cases being about 10^{-5} s. The observed collapse would slow down, however, as the event horizon was approached, and it would reach zero luminosity only after an infinite length of time, when its size would be that of the Schwarzschild radius (for an uncharged, non-rotating object).

The object will become *practically* unobservable, however, within a fraction of a second of the initiation of the final collapse; its total luminosity would be less than 10^{-60} W one second into the collapse, for example. The detection of such an object, which from now on we will call a black hole even though this is not strictly true, must thus rely upon the only remaining observable quantities: mass, charge and angular momentum. So far the various black hole candidates (see later discussion) are based upon observations of the companion of a black hole in a binary system (i.e. mass), or upon the energy released by material falling into a rotating black hole (i.e. angular momentum; see below). Charge is not generally of significance in astrophysics since the high conductivities of plasmas allow it to dissipate rapidly, and thus electrically charged black holes are unlikely to be found.

Rotation of a black hole causes more than just a change in the shape of the event horizon. A second boundary, the static limit, is formed. This actually exists for stationary black holes as well, but is then coincident with the event horizon. Rotation causes the two boundaries to separate. The static limit marks the point at which the observer cannot remain stationary with respect to the rest of the Universe, for to do so would require him to move faster than the speed of light with respect to the black hole. This is a consequence of Mach's principle which states that inertia is determined by the effect of all the matter in the Universe. Normally distant matter predominates in this respect, but nearby masses do have a contribution. In particular, if they are rotating, then they will tend to drag the local inertial frames of reference (*4) around with them. Thus an inertial observer near a rotating object will see the distant stars rotating. If he is to be stationary with respect to these distant stars, then he will have to apply a counter-rotation, and render his local frame of reference non-inertial. Normally the effect is very small: one rotation per ten million years, for example, for inertial observers near the Earth. Near a black hole the effect is much greater: 10^6 rpm for an inertial observer near a black hole formed by the collapse of our present Sun, for example. The dragging of the frames of reference will also lead to an inertial observer being dragged around the black hole, and an orbiting observer will find his orbit precessing rapidly. At the static limit, the velocity with respect to the black hole which is required to counteract this effect becomes the speed of light. Thus any observer within the static limit must be carried around the black hole.

Until the event horizon has been reached, it is possible for an object to move radially with respect to the black hole. Thus objects may cross the static limit in both directions. This has led to a suggestion by Penrose of a process whereby the rotational energy of a black hole might be extracted. A particle crossing the static limit might split into two, one portion entering the black hole and the other re-emerging with a greater total energy than that of the original complete particle, the extra energy coming from the rotational energy of the black hole. Because of this means of extracting energy from a black hole, the region between the static limit and the event horizon has become known as the ergosphere. The relative velocities required for the two fragments for the process to be effective seem likely to be more than half the speed of light. It may therefore not provide a realistic astrophysical energy source.

A more straightforward and probable energy-generating mechanism is that of the infall of matter to a black hole. Such material would generally form an accretion disc in orbit around the black hole, because of the initial angular momentum it may have possessed with respect to the black hole. Eventually it would fall into the black hole as viscous drag causes the orbit to decay. But prior to that the potential energy released during the infall would go into heating the accretion disc, and thus become available to power astrophysical processes. Since up to 42% of the rest-mass energy of the accreting material may be released in this way, it may provide a mechanism to explain the energy emissions of very luminous compact sources.

Two of the more esoteric aspects of black holes remain to be mentioned: their areas and temperatures. The surface area of the event horizon of a Schwarzschild black hole is

$$A_{\mathrm{S}} = (16\pi G^2/c^4)\mathcal{M}^2. \qquad (3A.9.7)$$

With rotating black holes, this becomes

$$A_{\mathrm{R}} = \frac{8\pi G^2}{c^4}\left[1 + \left(1 - \frac{J^2c^2}{G^2\mathcal{M}^4}\right)^{1/2}\right]\mathcal{M}^2 \qquad (3A.9.8)$$

where J is the angular momentum.

A fundamental result, sometimes called the second law of black-hole dynamics, for reasons which will become apparent in a moment, is that in *no* interaction of a black hole with other matter can the area of the event horizon *decrease*. Thus from equation (3A.9.7), for example, we see that if two similar Schwarzschild black holes combine, the resulting single black hole has an area twice the combined areas of the precursors. The reason for the name given to this law derives from the very close relationship between thermodynamics and black-hole physics. Identifying entropy with the area of the event horizon, the second law of black-hole physics is identical with the second law of thermodynamics:

$$\mathrm{d}S \geqslant 0. \qquad (3A.9.9)$$

A qualitative justification for this analogy comes from equating the information content of a body with its negative entropy: the more information or structure involved in an object, the lower its entropy. Now, the only structures determinable for a black hole are its mass, charge and angular momentum, and these together determine the event horizon area. With entropy as the number of indistinguishable states of an object, we have for a Schwarzschild black hole

$$S \propto (kG^2/hc^4)\mathcal{M}^2 \quad (3A.9.10)$$

where k is Boltzmann's constant and h is Planck's constant. So, from equation (3A.9.7),

$$S \propto (k/16\pi h)A_S \quad (3A.9.11)$$

and there is a direct relationship between the area of a black hole and its entropy.

There are also analogies for the zeroth, first and third laws of thermodynamics. The zeroth law of black-hole physics then leads to the unexpected result that the temperature of a black hole is *not* zero; i.e. it is not truly black, and we must expect it to be a source of energy in some manner. The nature of this energy was first deduced by Hawking, and is now generally known as Hawking radiation. The process depends upon virtual particle pairs (*5), one member falling sufficiently far towards the black hole to separate it from its counterpart and so prevent the normal annihilation. The second particle may then escape from the black hole's neighbourhood and appear in the external Universe. To an outside observer, it would appear that the black hole was emitting particles and antiparticles. The energy to form these particles must now be found since they are no longer destroyed within the Heisenberg time (*5). This comes from the black hole; one way of thinking of the process is as a particle (or antiparticle) travelling backwards in time and coming out of the black hole, then being scattered by the gravitational field into an antiparticle (or particle) travelling forward in time. The event horizon area decreases in this process, but the second law of black-hole physics is not contravened since an inflow of *negative* energy is involved.

The Hawking process leads to an energy emission from a black hole, and this may be used to assign a characteristic temperature to the event horizon. Since the gravitational stress increases as the black hole becomes smaller and less massive, the efficiency of the Hawking process increases as the size of the hole decreases. The temperature is thus inversely dependent upon the mass of the black hole, and for Schwarzschild holes we find

$$T \simeq 6 \times 10^{-8}\mathcal{M}^{-1}\ \text{(K)} \quad (3A.9.12)$$

where $\mathcal{M}$ is the mass of the black hole in units of the solar mass. The total luminosity of a solar-mass black hole due to the Hawking process is thus about 10^{-28} W, and it would take about 10^{67} years to lose half its mass in this way! It may be much more significant for smaller black holes, however (see below).

As mentioned at the beginning of this section, actual examples of black holes have yet to be found unequivocally. There are, however, several objects wherein the presence of a black hole is strongly suspected. The main line of reasoning is based upon the mass limits for compact objects. Electron degeneracy pressure is unable to support a white dwarf against gravity once its mass exceeds the Chandrasekhar limit of about 1.4 $\mathcal{M}_{\odot}$ (§§2P.1, 2A.1). Neutron stars similarly have an upper limit to their masses, and this is variously estimated to lie between two and three solar masses. A neutron star of over 3 $\mathcal{M}_{\odot}$ must therefore inevitably collapse to a black hole, and any compact object with a mass greater than this must be a strong black-hole candidate. The most promising object of this type is the binary star HDE 226868, also known as Cyg X-1 from its x-ray emission. The optical primary is an O9 supergiant with a mass of about 20 $\mathcal{M}_{\odot}$. The most probable mass for the companion is about 12 $\mathcal{M}_{\odot}$, with an almost absolute lower limit of 3.3 $\mathcal{M}_{\odot}$. The companion must also be a compact object since its variations on a time scale of milliseconds place an upper limit on the size of the emission region of 10^3 km. Although alternative explanations for the observations have been suggested, including the presence of a third body in the system, these require very severe constraints upon the parameters of the system, and the companion of HDE 226868 is probably best interpreted as a black hole. V861 Sco is a very similar system. It has a B0 supergiant in a binary system with a period of about 8 days. The companion again seems to be a compact object, this time with a mass of about 8 $\mathcal{M}_{\odot}$. Other similar, but less clear-cut, cases include Cir X-1 and LMC X-3, and some of the models for the intriguing emission-line object SS433 have a 0.4 $\mathcal{M}_{\odot}$ white dwarf in orbit around a 5 $\mathcal{M}_{\odot}$ black hole.

Rather less clearly, there may be evidence for black holes with masses millions of times that of the Sun. Jets of material are found being emitted from many galaxies, and may originate from massive black holes. The centre of the Seyfert galaxy NGC 4151 seems to have a mass of 10^9 $\mathcal{M}_{\odot}$ in a region about 3×10^9 km across, and this is about the Schwarzschild radius for such a mass. Many of the theories purporting to explain the energy emissions from quasars, Seyfert galaxies and even from the centre of the Milky Way Galaxy require massive black holes in order to produce the observed energy emissions from the observed volumes.

Speculation has also suggested the existence of very small black holes. Conditions no longer exist in which these mini, or quantum black holes, might form, but they might have occurred in the conditions of the early stages of the Big Bang origin of the Universe. The interest in them lies in the Hawking radiation coming from them. Since the rate of such radiation is inversely proportional to the mass (equation 3A.9.12), the last stages will occur explosively. A black hole with a mass of 10^5 kg would have a half life due to Hawking radiation of less than a second, and the final energy release would occur as a burst of particles and energy, theoretically reaching an infinite luminosity. Black holes formed during the Big Bang with masses around 10^{11} kg might be expected to be reaching this stage now. They might be expected to provide observable gamma-ray, cosmic-ray or neutrino bursts. Such bursts observed to date, however, do not seem likely to have originated in this way; but the possibility remains as an inducement to observers!

PROBLEMS

3A.1 Derive equation (3A.1.2), starting from equation (3A.1.1). State clearly any assumptions made in the derivation.

3A.2 Show that the absorption line profile produced by a thick, spherically symmetric, constantly expanding gaseous envelope around a star is given by

$$I(\Delta\lambda) = I(0)\left[1 - \left(\frac{c}{V\lambda}\right)^2 \Delta\lambda^2\right]$$

$$\text{for } \Delta\lambda = 0 \text{ to } \frac{V\lambda}{c}\left(1 - \frac{R_1^2}{R_2^2}\right)^{1/2}$$

$$= I(0)\left\{1 - \frac{R_2^2 - R_1^2}{R_1^2}\left[1 - \left(\frac{c}{V\lambda}\right)^2 \Delta\lambda^2\right]\right\}$$

$$\text{for } \Delta\lambda = \frac{V\lambda}{c}\left(1 - \frac{R_1^2}{R_2^2}\right)^{1/2} \text{ to } \frac{V\lambda}{c}$$

where V is the velocity of expansion, R_1 is the radius of the photosphere and the inner radius of the line-producing region, R_2 is the outer radius of the line-producing region, and $I(\Delta\lambda)$ is the line intensity (energy removed from the continuum level) at $\Delta\lambda$ from the undisplaced wavelength λ. Hence plot out the absorption line profile to be expected from a gaseous envelope whose thickness is a tenth of the photospheric radius.

3A.3 How far away would a (human) space traveller be when the Sun had become only just visible to the naked eye?

These space travellers meet some large green aliens from a planet around Betelgeuse. Their eyes are similar to human eyes, except that their pupils are 0.5 m across (compared with a maximum of 0.007 m for human eyes). How far away from Betelgeuse would the aliens have travelled when their star was only just visible to them?

3A.4 By combining the information in figures 3A.3 and 3A.10, and by assuming that energy emission per unit area varies as the fourth power of the temperature, estimate the sizes of stars (*a*) at the top of the main sequence, (*b*) at the bottom of the main sequence; (*c*) the coolest supergiants and (*d*) the hottest white dwarfs, all relative to the size of the Sun.

3A.5 Using the mass/radius relationship for neutron stars of constant density found in problem 2A.3,

$$\mathcal{M} = 3.73 \times 10^{42} R^{-3} \text{ (kg)}$$

estimate the mass of the Crab nebula pulsar (PSR 0531 + 21). You may need some of the following information:

Moment of inertia of a uniform sphere	$= 0.4\,\mathcal{M} R^2$
Energy emitted by the Crab nebula	$= 5 \times 10^{30}$ W
Period of the Crab pulsar	$= 0.0331$ s
Characteristic time of the Crab pulsar	$= 25\,000$ yr.

Comment on the likely relationship of your answer to the true mass.

3A.6 Given that the area of a rotating uncharged black hole (a Kerr black hole) is given by

$$A_K = \frac{8\pi G^2}{c^4} \left[\mathcal{M}^2 + \left(\mathcal{M}^4 - \frac{c^2}{G^2} J^2 \right)^{1/2} \right]$$

where $\mathcal{M}$ is the mass of the black hole and J is the angular momentum of the black hole, and remembering Hawking's theorem that the area of a black hole never decreases during interactions, show that the coalescence of two contra-rotating Kerr black holes of equal masses, which results in the formation of a single Schwarzschild black hole, can lead to the release of up to 50% of the original combined masses of the two black holes into radiation.

4A The Sun

SUMMARY

Basic properties of the Sun, limb darkening, small-scale suface features, activity cycles, solar neutrino problem, properties of the corona.

See also stellar compositions (§§1A.2, 3A.6), energy transfer (§§2P.4, 2A.3), spectrum line formation (§3P.1), Zeeman effect (§3P.1), radiative transfer (§3P.2), atomic and ionic populations (§3P.4), photometry (§3A.1), spectral class (§3A.2), spectrophotometry (§3A.2), Fraunhofer lines (§3A.2), rotation (§3A.7), magnetic fields (§3A.8), solar wind (§5A.1), synchrotron radiation (§6P.2), and the evolution of medium mass stars (§8A.2).

INTRODUCTION

The Sun is a reasonably typical, though rather larger than average, main sequence star. This has been assumed as well as being explicitly stated many times in the foregoing text, and the Sun was often used as an example to illustrate the properties of stars in general. Thus much of the data on the Sun as a star has already been covered. Further data on its origin and evolution may be found in Chapter 8A, while the solar wind is discussed in §5A.1, and solar cosmic rays in §7A.4. In this chapter we are concerned with phenomena which are not determinable for stars in general. That is not to say that such phenomena do not occur on other stars, but simply that they may be studied only on the Sun by reason of the far greater flux from the Sun than from other stars, or because the disc of the Sun and features on it are angularly resolved.

4A.1 BASIC DATA

The astronomical unit

The mean distance of the Earth from the Sun is used as a unit of distance by astronomers. It is called the astronomical unit, and abbreviated as AU (Appendix I). It is one of the most fundamental measurements in astronomy since almost all other distance measurements are based upon it. Modern estimates of its value are based upon the determination of the distance of the Earth from a nearby planet or asteroid, and then the application of Kepler's third law,

$$a^3 = (G/4\pi^2)(\mathcal{M}_1 + \mathcal{M}_2)P^2 \quad (4A.1.1)$$

where $\mathcal{M}_1$ and $\mathcal{M}_2$ are the masses of the Sun and the planet, a is the planet's orbital semi-major axis and P is the planet's orbital period, with appropriate corrections for the perturbations caused by the other planets. The determination of the distance of the planet or asteroid is via radar or triangulation. The results are usually quoted in terms of the solar parallax: the angular shift of the position of the Sun for observers separated by a distance equal to the Earth's radius perpendicular to the Earth–Sun line. It may be more simply regarded as the angular radius of the Earth seen from the centre of the Sun. The two methods lead to incompatible values:

$$\begin{aligned}\text{Solar parallax} &= 8.7984'' \pm 0.0004'' \quad &\text{(triangulation)}\\ &= 8.794\,179'' \pm 0.000\,001'' \quad &\text{(parallax).}\end{aligned}$$

The discrepancy probably arises from errors in the triangulation method due to inadequate orbital coverage of the asteroid (Eros) which has provided the best data, although there are also problems with the corrections for the effects of atmospheres with the radar method. The best of the radar results, however, give

$$1\ \text{AU} = 1.495\,978\,923 \times 10^{11} \pm 1.5 \times 10^3\ \text{m}. \quad (4A.1.2)$$

Much of the time, a useful approximation is

$$1\ \text{AU} = 150\,000\,000\ \text{km}. \quad (4A.1.3)$$

Radius

Given the distance of the Sun, it ought to be a simple matter to determine its size. In practice the measurement is complicated by the diffuse edge to the Sun, and by possible periodic and other changes in the size.

The generally accepted value for the angular radius of the solar photosphere is 959.63″, leading to a diameter of 6.9600×10^5 km. The diffuse edge of the Sun, caused by its being a gaseous and not a solid object, extends over about 0.2″ (150 km). The various apparent changes in the size of the Sun are comparable with or smaller than the 'fuzziness' of the edge, and they have large uncertainties. They occur on several time scales: short-period, 5 to 50 minutes, and long-period, 11 and 76 years. There is also a possible secular contraction amounting to 0.1″ per century. This, if real, must be a part of a very much longer cycle, since otherwise it would imply that the Sun will disappear in just about a million years! The 76-year cycle has an amplitude of about 0.2″, and taking this into account, the average radius of the Sun becomes 959.8″ (6.961×10^5 km). A cycle of a similar length appears in the sunspot data (§4A.4); so that there seem to be fewer sunspots when the Sun is larger and vice versa.

Solar constant

The average amount of energy per unit area received by the Earth, above the atmosphere, is called the solar constant. The highly variable effect of the atmosphere is difficult to compensate for, so that terrestrial measurements of the solar constant are of low accuracy. Only the use of spacecraft has brought sufficient precision to its measurement to determine that it is not, in fact, a constant. The solar maximum mission found a value of 1368.2 W m^{-2} for the value of the solar constant in the middle of 1980. Departures from this figure by up to 0.15% were also measured, the cause being the presence of sunspots. In the early 1980s, the value has consistently decreased at a rate of 0.04% per year. Rather less certainly, there may have been an increase by up to 0.5% in the two years following the solar minimum in 1976.

Solar luminosity and effective temperature

The solar constant and the value for the AU lead rapidly to an estimate of the total amount of energy emitted by the Sun (its luminosity), and this is

$$L_{\odot} = 3.8478 \times 10^{26}\ \mathrm{W}. \tag{4A.1.4}$$

The energy, outside the Earth's atmosphere, from a star with a bolometric magnitude of zero (§3A.1) is 2.48×10^{-8} W m^{-2}; thus we find the apparent bolometric magnitude of the Sun to be

$$(m_{\mathrm{Bol}})_{\odot} = -26.85. \tag{4A.1.5}$$

The effective temperature of an object (§3A.3) is the temperature of a black body (§2P.3) which emits the same total amount of energy per unit area as the object in question. The solar radius and luminosity give the emission per unit area for the Sun as

$$E_\odot = 6.319 \times 10^7 \text{ W m}^{-2} \tag{4A.1.6}$$

and so the solar effective temperature is

$$(T_{\text{eff}})_\odot = 5778 \text{ K}. \tag{4A.1.7}$$

This is the most generally useful temperature criterion. However, as the surface layers of the Sun are only approximately in LTE (§2A.3), other methods of measuring the temperature lead to slightly different values (§3A.3). Furthermore, limb darkening (§4A.2) causes even the effective temperature to vary across the visible disc of the Sun. Equation (4A.1.7) thus gives only the average temperature for the whole Sun.

Mass

The mass of the Sun may be found from Kepler's third law (equation 4A.1.1), once the astronomical unit and the mass of a planet are known. The best determination of a planetary mass is naturally that of the Earth. However, it is the centre of mass of the Earth–Moon system that describes a Keplerian orbit around the Sun, rather than the Earth itself, and so we must use the combined mass of the Earth and the Moon, rather than that just of the Earth. This is estimated to be 6.050×10^{24} kg; hence we find the solar mass to be

$$\mathcal{M}_\odot = 328\,901.4(\mathcal{M}_\oplus + \mathcal{M}_☾) \tag{4A.1.8}$$

$$= 1.990 \pm 0.001 \times 10^{30} \text{ kg}. \tag{4A.1.9}$$

The limitation on the accuracy of the determination of the solar mass lies almost entirely in the accuracy with which the gravitational constant is known.

Composition

The composition of the surface layers of the Sun has been discussed in §1A.2 (figure 1A.11). The probable internal composition is also discussed there and shown in figure 1A.13. Nothing further is included here, but the reader should note the possible implications of the 'solar neutrino problem' (§4A.6) and some of the proposed solutions to it, for both the internal and surface compositions of the Sun.

Rotation

The Sun rotates only very slowly (§§3A.7, 8A.1), and the rotational period increases with latitude. The two methods of determining the rotation are through the Doppler shifts of spectrum lines from the approaching and receding limbs of the Sun, and following markers such as sunspots as the rotation carries them across the visible disc. The two methods apparently led to differing results, and this was interpreted as implying intrinsic motions for the sunspots. Recently, however, it has become apparent that the discrepancy probably arises through an instrumental effect in the spectroscopic data. The results of the two methods are now very close, and sunspot motions (§4A.3), if they exist at all, must be small. Groups of spots, however, may have some intrinsic motion since they consistently give slower rotational velocities than the single spots.

At the equator of the Sun, the rotation rate is 14.45 degrees per day: a rotational period of 24.91 days, and an equatorial velocity of 2.03 $\mathrm{km\,s^{-1}}$. The spin rate has recently increased over a 30° band centred on the equator. Earlier investigations found values of

14.372 degrees per day	(1883–1885)
14.38 degrees per day	(1934–1944)
14.349 degrees per day	(1956–1958)

corresponding to rotational periods of 25.05, 25.03, and 25.09 days. Just before the Maunder minimum (§4A.3) the rate rose to 14.9 degrees per day (a period of 24.2 days). This may be indicative of some association of the rotational variations with the longer term solar cycles, higher velocities occurring with the lower sunspot numbers. Thus the Gleissberg 80-year cycle was near a minimum in 1884, and a maximum in 1957, the 11-year average sunspot numbers being then 60 and 150, respectively.

The differential rotation of the Sun, based upon the 1934–1944 data, is given by

$$\omega(\varphi) = 14.38(\pm 0.01) - 2.96(\pm 0.09)\sin^2\varphi \ (^\circ\mathrm{d}^{-1}) \qquad (4\mathrm{A}.1.10)$$

or

$$P(\varphi) = 25.03\left(1 + \frac{1}{4.858\,\mathrm{cosec}^2\varphi - 1}\right)\ (\mathrm{d}) \qquad (4\mathrm{A}.1.11)$$

where $\omega(\varphi)$ is the solar angular rotational velocity at latitude φ and $P(\varphi)$ is the period at solar latitude φ. The rotational period is thus 27.9 days at a latitude of 45°, and, by extrapolation, 31.5 days at the poles. The differential rotation varies to a much greater extent than the equatorial velocity. The $\sin^2\varphi$ coefficient, for example, was -2.6 in 1884 and -3.4 in 1957. It was enhanced by about a factor of three just before the Maunder minimum, so that there is no clear association of changes in the differential rotation with the longer solar cycles. One further recently discovered peculiarity is that the northern hemisphere rotates, as a whole, slightly more slowly than the southern hemisphere.

Since angular momentum must be conserved, changes in the rotational rate at the surface cannot simply extend all the way to the Sun's centre. Furthermore, the vibration spectrum of the Sun suggests that the solar interior is rotating much faster than the surface; core rotation periods of 3 to 12 days are variously estimated. Thus the observed changes in the rotation probably arise in a relatively thin outer skin, with conservation of angular momentum by concomitant changes over portions of the interior. A possible mechanism might be changes in the linkage between this outer layer and the more rapidly rotating core, arising through changes in the solar convection zone, but this or any other process still remain to be proven.

4A.2 LIMB DARKENING

The visible disc of the Sun appears darker towards the edge than in the centre. This effect is called limb darkening and it arises because unit optical depth (§3P.2) is reached at a higher point in the Sun's outer layers along the slanting angle of view at the limb than along the perpendicular viewpoint at the centre of the disc (figure 4A.1). Since the effective temperature is decreasing outwards for this region, cooler and therefore darker material is seen at the edge of the disc than in the centre.

Limb darkening occurs in almost all stars. But since it requires the star to be angularly resolved if it is to be seen directly it is not normally measurable. In a very few stars, which have been resolved by techniques such as speckle interferometry, etc (*Astrophysical Techniques*, Chapter 2),

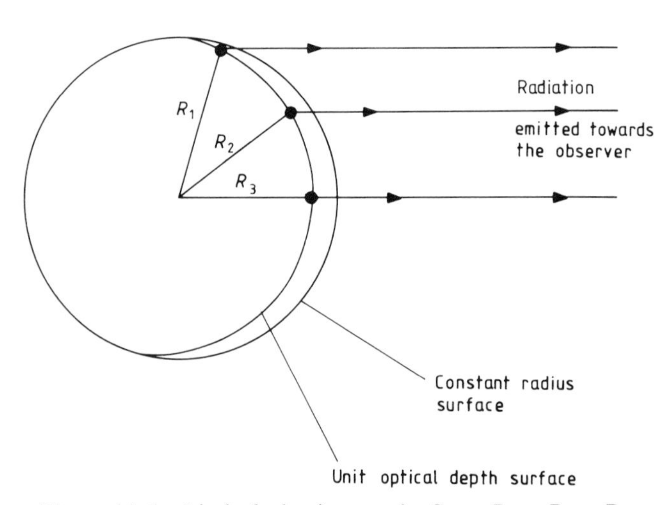

Figure 4A.1 Limb darkening on the Sun: $R_1 > R_2 > R_3$.

limb darkening has been seen directly. In other cases such as the analysis of eclipsing binary light curves, or spectral synthesis, it is one parameter in the modelling of the data, and so an estimate of its magnitude may be obtained. It is found that the limb darkening in some stars can be very different from that on the Sun. Red giants more or less fade away completely at their edges, without any clearly defined limb at all. White dwarfs, by contrast, although not resolved by any technique, must be expected to have as clearly defined a surface as an atmosphere-less planet, and no limb darkening at all.

In other portions of the spectrum, different layers of the Sun are seen. In particular, in the radio region, unit optical depth occurs at heights of hundreds of thousands of kilometres above the visible photosphere. This is well into the corona where the temperature rises with increasing height (§4A.3, Chapter 5A). The limb is thus observed to be brighter than the centre of the disc, and we have limb brightening rather than limb darkening.

We may investigate limb darkening quantitatively by solving the equation of transfer (§3P.2). A remarkably simple approximation enables this to be accomplished analytically in the case of the Sun, and to give realistic results. We treat the surface layers as thin, and assume the absorption and emission coefficients to be independent of frequency. From equation (3P.2.13) we then have

$$\cos\theta\,\frac{\mathrm{d}I}{\mathrm{d}\tau}=I-\frac{j}{\varkappa} \tag{4A.2.1}$$

since by our assumption there is no frequency dependence. Integrating over a sphere we obtain

$$\int_{\text{sphere}}\cos\theta\,\frac{\mathrm{d}I}{\mathrm{d}\tau}\,\mathrm{d}\omega=\int_{\text{sphere}}I\,\mathrm{d}\omega-\int_{\text{sphere}}\frac{j}{\varkappa}\,\mathrm{d}\omega \tag{4A.2.2}$$

or, since the optical depth is independent of angle,

$$\frac{\mathrm{d}}{\mathrm{d}\tau}\int_{\text{sphere}}I\cos\theta\,\mathrm{d}\omega=\int_{\text{sphere}}I\,\mathrm{d}\omega-\int_{\text{sphere}}\frac{j}{\varkappa}\,\mathrm{d}\omega. \tag{4A.2.3}$$

Now the integral on the left-hand side is just the flux, $\mathscr{F}$, or the net energy emitted by a surface, so that we have

$$\mathrm{d}\mathscr{F}/\mathrm{d}\tau=4\pi J-4\pi j/\varkappa \tag{4A.2.4}$$

where J is the mean intensity. Since we are dealing with the solar atmosphere, there are no sources or sinks of energy. Thus

$$\mathrm{d}\mathscr{F}/\mathrm{d}\tau=0 \tag{4A.2.5}$$

and so

$$J=j/\varkappa. \tag{4A.2.6}$$

Now considering a thin layer of the atmosphere (figure 4A.2), it will emit energy in direction θ amounting to

$$j\rho \, ds = j\rho \sec\theta \, dx. \tag{4A.2.7}$$

This will be absorbed by the overlying layers, so that if the layer under consideration is at optical depth τ, then the amount of radiation which eventually emerges will be

$$dI = j\rho \sec\theta \, dx \, e^{-\tau\sec\theta}. \tag{4A.2.8}$$

Now from equation (3P.2.6) we have

$$d\tau = -\varkappa\rho \, dx \tag{4A.2.9}$$

so that

$$dI = -(j/\varkappa)\sec\theta \, e^{-\tau\sec\theta} \, d\tau \tag{4A.2.10}$$

and from equation (4A.2.6)

$$dI = -J(\tau)\sec\theta \, e^{-\tau\sec\theta} \, d\tau \tag{4A.2.11}$$

where the dependence of the mean intensity upon optical depth has been explicitly shown, for clarity. To find the total emergent intensity we therefore simply integrate over optical depth

$$\int_0^{I(0,\theta)} dI = -\int_\infty^0 J(\tau) \, e^{-\tau\sec\theta} \, d(\tau\sec\theta). \tag{4A.2.12}$$

To find J as a function of τ we make a further assumption, often known as Eddington's first approximation, that the radiation field consists of a constant outward intensity, and a smaller constant inward intensity, i.e.

$$I(\tau,\theta) = I_1(\tau) \qquad 0 \leqslant \theta \leqslant \pi/2 \tag{4A.2.13}$$

$$= I_2(\tau) \qquad \pi/2 < \theta \leqslant \pi. \tag{4A.2.14}$$

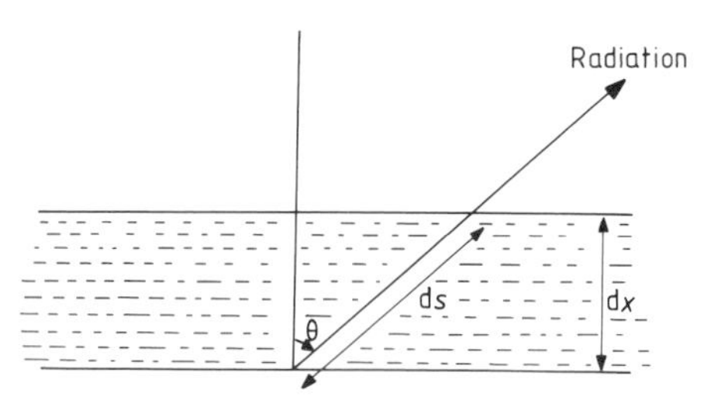

Figure 4A.2 Notation for the derivation of the limb-darkening formula.

Since by definition

$$J = \frac{1}{4\pi} \int_{\text{sphere}} I \, d\omega \tag{4A.2.15}$$

we quickly find, as would be expected,

$$J = \tfrac{1}{2}(I_1 + I_2). \tag{4A.2.16}$$

Similarly, with

$$\mathscr{F} = \int_{\text{sphere}} I \cos\theta \, d\omega \tag{4A.2.17}$$

we obtain

$$\mathscr{F} = \pi(I_1 - I_2). \tag{4A.2.18}$$

Now, multiplying the equation of transfer by $\cos\theta$ and integrating over a sphere gives

$$\mathscr{F} = \frac{d}{d\tau} \int_{\text{sphere}} I \cos^2\theta \, d\omega = \int_{\text{sphere}} I \cos\theta \, d\omega - \frac{j}{\varkappa} \int_{\text{sphere}} \cos\theta \, d\omega \tag{4A.2.19}$$

and so, integrating with respect to τ,

$$\mathscr{F}\tau + \text{constant} = \int_{\text{sphere}} I \cos^2\theta \, d\omega \tag{4A.2.20}$$

$$= \int_0^{2\pi} \int_0^{\pi} I \cos^2\theta \sin\theta \, d\theta \, d\varphi \tag{4A.2.21}$$

$$= (2\pi/3)(I_1 + I_2). \tag{4A.2.22}$$

Hence, from equation (4A.2.16),

$$J(\tau) = (3\tau/4\pi)\mathscr{F}(\tau) + \text{constant}. \tag{4A.2.23}$$

Now as we have seen (equation 4A.2.5) the net flux is constant in the outer layers of the Sun, and so

$$\mathscr{F}(\tau) = \mathscr{F}. \tag{4A.2.24}$$

Thus, at the surface, with $\tau = 0$,

$$J(0) = \text{constant}. \tag{4A.2.25}$$

But since

$$I_2(0) = 0 \tag{4A.2.26}$$

because there is effectively no energy incident on the surface of the Sun from space, equations (4A.2.16) and (4A.2.18) give

$$J(0) = (1/2\pi)\mathscr{F}. \tag{4A.2.27}$$

Thus, finally, equations (4A.2.23), (4A.2.25) and (4A.2.27) give

$$J(\tau)=\frac{3}{4\pi}\mathscr{F}\tau+\frac{1}{2\pi}\mathscr{F} \tag{4A.2.28}$$

$$=\frac{1}{4\pi}\mathscr{F}(2+3\tau). \tag{4A.2.29}$$

So now, returning to equation (4A.2.12), we find

$$I(0,\theta)=\frac{1}{2\pi}\mathscr{F}\int_0^\infty e^{-\tau\sec\theta}\,d(\tau\sec\theta)+\frac{3}{4\pi}(\cos\theta)\mathscr{F}\int_0^\infty \tau\sec\theta e^{-\tau\sec\theta}\,d(\tau\sec\theta) \tag{4A.2.30}$$

$$=\frac{1}{2\pi}\mathscr{F}+\frac{3}{4\pi}(\cos\theta)\mathscr{F} \tag{4A.2.31}$$

and we have an expression for the emergent intensity at an angle θ to the surface. Normalising to the vertical (central) intensity we find

$$I(0,\theta)/I(0,0)=0.4+0.6\cos\theta. \tag{4A.2.32}$$

The observed limb darkening on the Sun, as a wavelength of 500 nm, is actually given by

$$I(0,\theta)/I(0,0)=0.35+0.65\cos\theta \tag{4A.2.33}$$

—an excellent agreement given the very simple assumptions underlying the theoretical equation.

From equation (4A.2.33), we see that the limb of the Sun has about 35% of the intensity of the centre of the disc. The observed temperatures correspondingly fall by about 25% from the centre of the disc of the Sun to its limb.

4A.3 SOLAR SURFACE FEATURES

The surface layers of the Sun are conventionally divided into three regions: the photosphere, the chromosphere and the corona (figure 4A.3), although in fact the boundaries are not well defined. The photosphere is the normally visible surface of the Sun. Its temperature, as we have seen (§4A.1), is just below 6 000 K, and the density around 10^{-3} kg m^{-3}, or about the density of the Earth's atmosphere at a height of 50 km. Conditions change very rapidly in the 5000 km above the photosphere; the (kinetic) temperature drops to 4500 K before returning towards 6000 K, and then shooting up to 10^5 to 10^6 K in a very thin transition zone at the top of the chromosphere. The density drops rather more smoothly towards 10^{-14} kg m^{-3}. This region is called the chromosphere, and short-lived, needle-like extensions of it, called spicules, extend upward for several thousands of kilometres above its normal upper surface. The outermost layer is called the corona and it extends out to the region of the planets. Its density is around 10^{-15} kg m^{-3}, and its kinetic, excitation, ionisation etc temperatures are around 10^6 K. It is dealt with further in §4A.5. The surface features of the Sun mostly occur within the photosphere, with a few extending up into the chromosphere or occasionally into the corona.

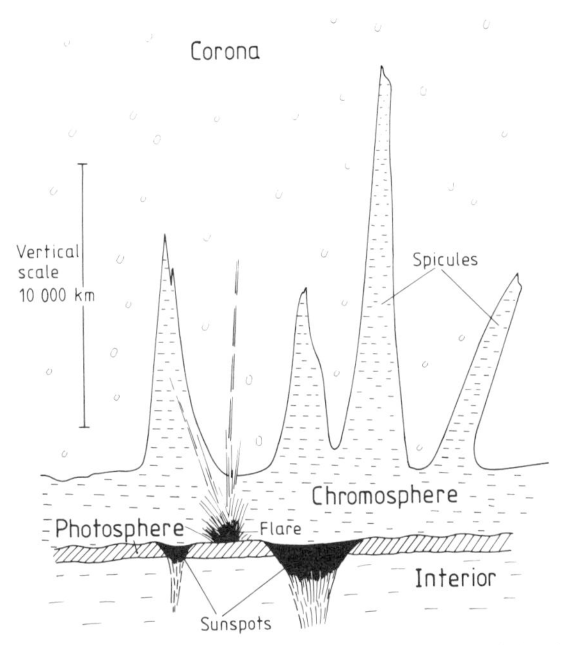

Figure 4A.3 Schematic view of the outer layers of the Sun.

Granulation

The most widespread feature of the solar surface is the granular appearance of its undisturbed regions (figure 4A.4). The surface is divided into small regions, 500 to 3000 km across, separated from each other by slightly darker channels, perhaps 500 km wide. Each granule appears and then disappears over about a 10 minute interval. There is a temperature difference of a few hundreds of degrees between the centre and edge of a granule. The brighter granules, at least, are moving outwards at velocities up to 0.5 km s^{-1}, and there are rather higher downward velocities in the darker inter-granular lanes. The granulation results from the effects of the solar convection zone (§2A.3). This nominally stops a few hundreds of kilometres below the visible surface, but its effects continue to propagate out to the photosphere, the granule corresponding to the centre of a rising convection cell.

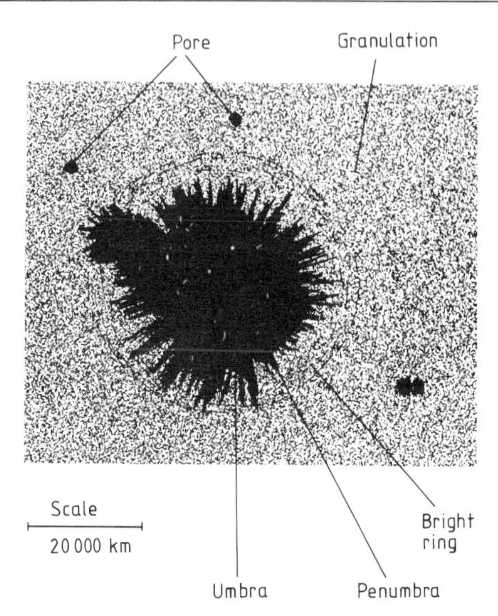

Figure 4A.4 Schematic view of a sunspot.

If an image of the Sun is obtained through a very narrow-band filter centred on a strong absorption line, then the resulting picture is of a layer 2000 to 3000 km above the normal surface. Such an image is known as a spectroheliogram, since it can also be obtained via a scanning spectroscope called a spectrohelioscope (*1). The observed layer corresponds to unit optical depth in the line, which is reached at a smaller physical depth (greater height) because of the increased absorption coefficient across the spectrum line. The Fraunhofer lines, especially Hα and Ca II H and K, are commonly used for the purpose. Spectroheliograms thus enable parts of the chromosphere to be seen directly. Granulation, but on a much larger scale than in the photosphere, is thereby found to occur in the chromosphere. Individual chromospheric granules are outlined by the spicules. These, as the name suggests, are 'little spikes' which typically penetrate 5000 km into the corona. They are most easily seen at the edge of the Sun, where they appear as an almost continuous 'hedge'. In fact, only about 2% of the solar surface is covered by spicules; their apparent contiguity occurs simply as many sets are projected on top of each other at the solar limb. The chromospheric granules are more or less circular and about 30 000 km across. Their lifetimes are around 20 hours. The spicules, being concentrated at their edges, form a filigree or fishnet pattern when seen on the disc of the Sun rather than at its edge. Material flows from the centre to the edge of each cell. They are thought to be related to the deep, underlying convection pattern, perhaps mirroring the convection cells at depths of thousands of kilometres. The movement of the material drags the solar magnetic field with it so that it becomes concentrated at the granule edges. The spicules may then result from the greater efficiency with which energy is deposited into the chromosphere by the intensified magnetic field.

Sunspots

By far the best known surface features of the Sun are the small ephemeral dark regions called sunspots. Almost as well known is that their average numbers fluctuate on an eleven-year cycle, whose influence may be traced in terrestrial phenomena. The sunspot cycle is discussed along with other periodicities later in this section. Here we are just concerned with the individual morphology and behaviour of the spots.

Although we have just called sunspots 'small', 'ephemeral' and 'dark', this is only true within a solar context. An individual spot can be from a few thousand to a hundred thousand kilometres in diameter, its lifetime can be between 1 and 200 days, and its temperature at the centre about 4300 K. Thus even a medium-sized sunspot would be a hundred times brighter than the full Moon were it possible to see it in isolation; it is dark only in comparison with the much brighter photosphere.

The main features of a sunspot in white light are shown in figure 4A.4. In the centre is a very dark, more or less circular region, generally without any visible structure, called the umbra. When seen near the limb of the Sun, it is apparent that the umbra is depressed by some 500 km with respect to the normal photosphere. Surrounding the umbra is a brighter region with a radial structure, which is called the penumbra. It is usually just under two and a half times the size of the umbra. This in turn is sometimes surrounded by a ring just brighter than the photosphere, but this is normally very difficult to observe. The smallest spots are called pores, and are just umbrae without associated penumbrae. The temperature of the material in the umbra is about 4300 K, and that of the penumbra about 5700 K. The photospheric temperature at the centre of the visible disc, for comparison, is about 6050 K. Most sunspots are small; over half are less than 10 000 km in diameter at their maximum development. Many of them appear as pairs, separated by up to ten degrees in solar longitude (bi-polar configuration), the leading spot in terms of the solar rotation usually being the larger. Very large individual spots are possible, but at the larger sizes it is more common for a sunspot group to form. This can become a very complex array of many individual large and small spots, with many mutual interactions, covering an area several hundred thousand kilometres across.

The size of a sunspot or sunspot group is generally measured in units of millionths of the area of the solar hemisphere. The frequency of occurrence of spots decreases with their size, and is given by

Area ($\times 10^{-6}$ solar hemispheres)	1–250	250–500	500–1000	>1000
Percentage occurrence	86	9	4	1

Spots larger than about 500 millionths of a hemisphere are potentially visible to the unaided eye.

CAUTION

Looking at the Sun directly, with the naked eye, or through a telescope, binoculars etc is likely to cause permanent eye damage or blindness. Commercially produced solar filters should always be used during solar observing, or the image of the Sun should be projected onto a screen. Never look at it directly; even the often recommended smoked glass or exposed film is not safe.

An individual spot has a relatively short life. A useful rule of thumb is that the lifetime, in days, is one tenth of the spot's maximum area in millionths of a hemisphere. A morphological classification of spots and groups, known as the Zurich system, is a good guide to their development throughout their lifetimes. The Zurich system has nine classes, as follows:

A A single pore, or group of pores, without any bi-polar configuration.

B A group with a bi-polar configuration.

C A bi-polar group; one spot possessing a penumbra.

D A bi-polar group whose main spots have penumbrae; one or more of the spots having some simple structure. The total length of the group less than 10° of solar longitude.

E A large bi-polar group. Two main spots possessing penumbrae and complex structures. Numerous small spots. Total length generally in excess of 10° of solar longitude.

F A very large complex or bi-polar group. Length over 15° of solar longitude (150 000 km).

G A large bi-polar group without small spots between the main components. Length over 10° of solar longitude.

H A uni-polar spot with a penumbra, and a size greater than 2.5° of solar longitude.

J A uni-polar spot without a penumbra, and a size less than 2.5° of solar longitude.

These classes are illustrated schematically in figure 4A.5. The Zurich system is based upon the development of sunspots, as well as upon their intrinsic properties. The development patterns range from

A or A–B–A for small spots

through

A–B–C–B–A for a small group

and

A–B–C–D–E–G–C–H–J–A for a large group

to

A–B–C–D–E–F–G–D–C–H–J–A for the very largest groups.

The initial phases of the development generally occur much more rapidly than the decline; the rise to maximum occupies typically 20% to 30% of the total lifetime.

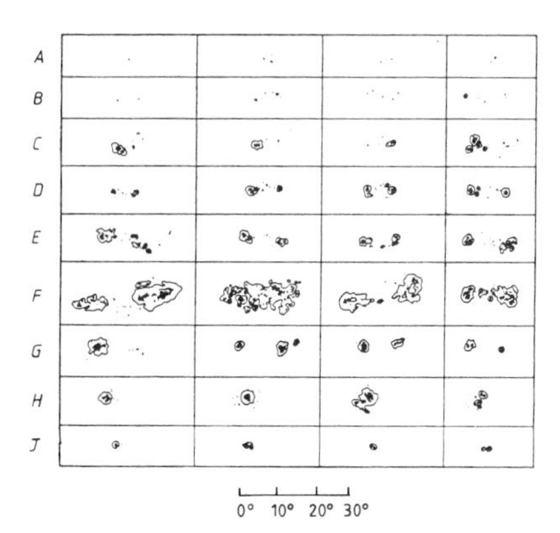

Figure 4A.5 The Zürich classification of sunspot groups. Four examples of each class are shown. The scale at the bottom indicates heliographic longitude. (Reproduced from *Sunspots* by R J Bray and R E Loughhead (1964), London: Chapman and Hall, by permission of the publishers.)

Within a group, the preceding spot is generally the larger, better developed and longer lasting. The preceding spot is also generally two or three degrees of heliographic latitude closer to the solar equator during the initial stages. Over the lifetime of the group, the preceding and following spots may move apart by several degrees of longitude, and move away from the equator by up to five degrees of latitude. Any general motion of the whole group in solar longitude is much more difficult to ascertain. As previously mentioned, systematic motions of the spots were suspected at one time from the apparent difference between the rotational period derived from the spots and that derived from Doppler shifts. Correction for the effect of scattering in the optics of the spectroscopes used to find the Doppler shifts has now eliminated the disagreement. Any residual motions of the spots in longitude must therefore be very small.

Within sunspots systematic motions of the material may be observed. The results which are obtained, however, depend upon the strength of the spectrum line which is used to obtain them, a phenomenon known as the Evershed effect. Since, as we have seen, the stronger absorption lines originate from a point higher in the atmosphere than the weaker lines, this actually implies a differential velocity structure with height above the photosphere. At low levels (i.e. the level of the visible photosphere), there is an almost radial outflow of material within the sunspot penumbra at velocities up to $2\ \mathrm{km\,s^{-1}}$. Inside and outside the penumbra, the velocity falls towards zero. At higher levels, by contrast, an inflow of material is found over a wide area centred on the spot.

One of the most fundamental properties of sunspots is not evident from direct visual observations. This is the presence of intense magnetic fields within and close to the spots. The Zeeman splitting of the spectrum lines (§3A.8) may be used to detect the fields. But in contrast to the more general stellar case, the fields are so intense, and the resolution so good, that the individual Zeeman components may be seen, rather than a more general line-broadening effect. The magnetic fields are found to be perpendicular to the surface of the Sun at the centre of the umbra. Further out within the spot, they diverge outwards at angles to the normal of about

$$\theta = 90^\circ \rho / p \tag{4A.3.1}$$

where ρ is the distance from the centre of the spot and p is the penumbral radius. The field strength is typically 300 mT within medium to large sunspots at maximum, compared with the average solar magnetic field of 0.1 mT, and it is given for an individual spot by the empirical expression

$$H = 370A/(A + 60)\ (\mathrm{mT}) \tag{4A.3.2}$$

where A is the spot's area in millionths of a hemisphere. Within a spot the field strength varies according to

$$H(\rho) = H(0)(1 - \rho^2/p^2). \tag{4A.3.3}$$

The magnetic field builds up rapidly as or even before the spot itself appears, and it remains near to its maximum value throughout most of the spot's existence. It starts to decline in strength only as the spot itself is disappearing.

The polarity of an individual spot can be either north or south. Most spots, however, appear as pairs separated in solar longitude, and these will then be of opposite polarities. Even single spots are usually found to have an associated disturbed region of the photosphere where a companion spot might be expected to occur, and the polarities of the spot and the disturbed region again oppose each other. A more fundamental entity than the sunspot is thus the bi-polar magnetic region (BMR) which underlies it. BMRs may exist without any spots appearing at all. Very large spots and groups usually have very complex polarities with many fields intertwined. The evidence for the BMR is then often obscured. Throughout one solar cycle (§4A.4), the preceding spot or preceding region of a BMR will be of one polarity in one hemisphere, and of the opposing polarity in the other hemisphere. In the following cycle the polarities reverse (figure 4A.6).

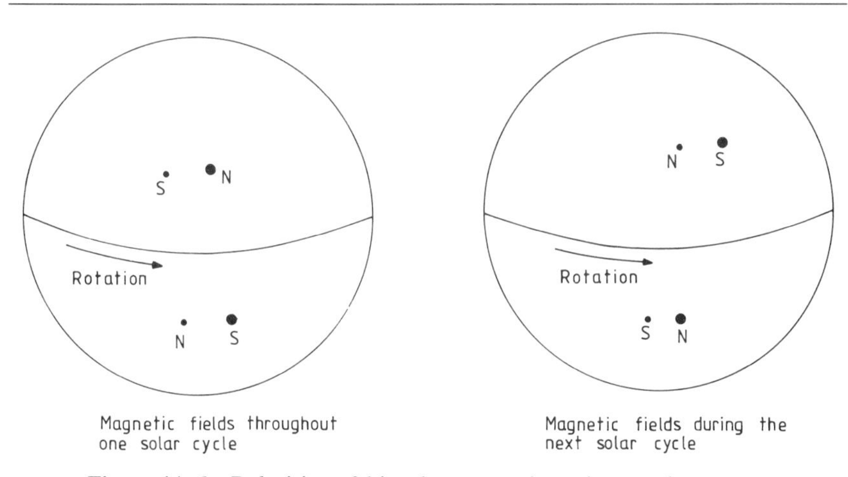

Figure 4A.6 Polarities of bi-polar magnetic regions and sunspots.

The immediate cause of a sunspot is the magnetic field itself. The material of the Sun's surface layers is hot enough for many atoms to be ionised. Charged particles can move freely along magnetic lines of force, but can only diffuse slowly across them due to collisions with other particles. Thus, through viscosity, the material as a whole is constrained from moving freely across lines of the magnetic field. The energy of the Sun, however, is transported to near the visible surface by convection, and this requires the rising material to spread outwards as it reaches the top of a convection column before it descends again. Hence, within the strong radial magnetic fields of sunspots, the efficiency of convective energy transfer is reduced as the horizontal motion of the material is hindered. The material in the region of the strong magnetic field therefore cools down and appears dark in comparison with the nearby undisturbed solar surface. Though the immediate cause of the spots is thus clear, the cause of the intense magnetic fields is by no means understood. One suggested mechanism is discussed later in relation to the solar cycle (§4A.4).

Flares

A solar flare is one of the many associated phenomena of sunspots. It is a sudden and short-lived brightening of the chromosphere. Flares are best seen on Hα spectroheliograms, only very occasionally being bright enough for detection in white light. They are commonest near the larger, more complex multipole spots. Several flares per day can then occur within the more active such regions. Flares are classified according to their areas and brightness on a scale of 1 to 4. The lifetime increases with the importance of the flare: from 20 minutes for a class 1 flare to over 3 hours for classes 3 and 4. There are also very small brief subflares labelled 1 – . The physical size of a flare is typically 20 000 to 40 000 km across, and the total energy emission ranges from 10^{23} to 10^{25} J. In addition to the optical emission, a flare produces complex radio, ultraviolet, x- and gamma-ray emissions, and also high-energy subatomic particles. These latter are predominantly electrons, protons and neutrons with velocities of 10^3 to 10^5 km s^{-1}. They are the main sources of the solar cosmic rays (§7A.4), and are thought to induce the radio bursts associated with the flare. There are five types of solar radio burst: I to V, with all except the first associated with flares. They comprise intense bursts of radio emission at metre wavelengths, with each individual component of a burst declining in frequency rapidly with time. The type III bursts occur within a minute or two of the optical appearance of the flare and are thought to be due to synchrotron emission (Chapter 6P) from the high-energy particles from the flare, the electron component probably being the most important. The type V burst follows shortly thereafter. It may be due to plasma oscillations (equation 7A.3.4) in the disturbed material left behind as the high-energy particles stream through the outer layers of the solar atmosphere. The type II bursts are delayed by 10 to 20 minutes from the start of the flare; they are also probably plasma oscillations, but due to lower energy particles, shock fronts, and the turbulence left by these. Type IV bursts are similarly delayed. They may arise as synchrotron radiation from the more energetic electrons in the turbulent regions. Radio emission at centimetre wavelengths originates from the same level of the solar atmosphere as the flare itself. This radiation shows that complex changes are occurring in both the temperature and the magnetic field structure prior to the flare taking place. The pre-flare build-up may reach brightness temperatures of 10^7 K just before the occurrence of the flare. Circular polarisation of up to 100% can be found, and it can reverse completely over distances of only a few thousand kilometres and times of a few minutes, indicating similar complexities for the magnetic fields.

Flares are important from the human standpoint for several reasons. First the particle emissions can damage spacecraft, and could be dangerous to an astronaut outside, 'space walking'. Fortunately the particles are emitted as jets, and so only occasionally does such a stream of particles intersect the Earth's locality. When this does occur, it may induce aurorae in the polar regions. Whether this is by the direct interaction of the flare's particles, or whether these act as a trigger to dump particles from the Van Allen belts is not yet clear. The short-wave emissions from the flare can be sufficiently intense to lower the Earth's ionosphere, leading to long-wave radio communication fade-outs. Some rather fanciful speculations have linked 'superflares', if such exist, to the mass extinctions found in the fossil record. This would require the Earth's magnetic field to be very weak at the time of the superflare due to being in the midst of one of its periodic reversals. Radiation damage would then be the cause of the extinctions.

The cause of flares is not at all clear. Almost certainly some form of magnetic storage and dumping of energy, perhaps with explosive reconnection of the fields, is involved. So far, however, the magnetic field configurations and mechanisms for realising this in practice have run into problems explaining the extreme rapidity of the energy release. The particle emissions from flares show evidence of excess deuterium and tritium, but whether the thermonuclear reactions implied by this are the cause or consequence of the flare remains to be seen. Admitting the existence of some process capable of releasing some 10^{25} J into a small volume of the solar atmosphere in about 100 seconds, then the subsequent development of the various flare phenomena can be followed reasonably successfully via computer models of the expanding gas and the associated shock fronts.

UV Ceti stars are very faint M-type dwarfs. They are also known as flare stars. Their normal luminosity is perhaps only a ten-thousandth of that of the Sun, and so when a flare occurs on their surfaces, it enhances their brightnesses sufficiently for the change to be detected from Earth. The flares appear to be similar to those occurring on the Sun, but the observations are currently incapable of providing details of their behaviour. See §3A.5 and figure 3A.11 for further information.

Plages and faculae

Plages, or flocculi as they used to be known, are bright regions near a sunspot or other centre of activity in the chromosphere. They are therefore observed on spectroheliograms, particularly those in the light of the Hα or Ca II *H* and *K* lines.

Faculae are bright regions in the photosphere. They are seen in white light, usually only when the centre of activity is close to the edge of the Sun, and limb darkening has increased the contrast.

There is a good deal of confusion over the terminology of these objects. Plages are also known as chromospheric faculae, faculae then being called photospheric faculae. Flocculus, the old term for plage, is now sometimes used to refer to small light or dark condensations within a plage. Thus care must be taken when reading the literature to determine the precise definitions of the terms actually being used. The various phenomena are closely related to each other, however, and we may ascribe them all to a single cause. For that reason, here we shall refer only to plages as encompassing the whole event, unless more detailed specification is actually needed.

The plages are associated with magnetic fields. Thus they appear around sunspots, but also anywhere that the magnetic field has become disturbed. In particular, they appear before sunspots, and remain visible after sunspots have disappeared. With uni-polar spots, the plages around the second disturbed region reveal the true bi-polar nature of the magnetic field. The plages on Ca II *K* spectroheliograms coincide with, and can be used to delineate, the regions of enhanced vertical magnetic field. They start to appear at field strengths of around 2 mT. The material forming a plage is hotter by some 100 K and denser than its surroundings.

The cause of plages lies in the magnetic field. Unlike sunspots, however, here it serves to enhance rather than inhibit the energy flow. The source of the plage's energy is probably the deposition into it of energy from the convection zone by magnetohydrodynamic or acoustic waves; the same energy source probably as that of the corona (Chapter 5A). The small increase in the magnetic field strength is sufficient to enhance the deposition of this mechanical energy without suppressing the convection in the underlying layers, and so to cause the increased radiation that we see as a plage.

Prominences and filaments

Prominences and filaments are in fact the same phenomenon; the former is seen at the edge of the Sun and is brighter than its background, while the latter is the same object but seen as a dark silhouette when projected against the bright disc of the Sun. They are not strictly surface phenomena since they extend well above the chromosphere, and on occasion can extend 10^6 km or more into space. Nonetheless, as extensions of the chromosphere, they are dealt with here rather than as a part of the corona.

The more spectacular apparition of the objects is as prominences. They may be seen during solar eclipses or through the use of a coronagraph (*2).

However, as their radiation is largely in the hydrogen, helium and ionised calcium lines, narrow-band filters or a spectrohelioscope centred on these lines will show them more easily. They often give the appearance of huge flames leaping upwards from the solar surface (figure 4A.7). In fact, their resembance is closer to rain clouds on the Earth; the material within a prominence is cooler and denser than its surroundings, and is often moving downwards towards the photosphere. Their shapes can be very complex, and extensive morphological classification schemes have been devised. Very frequently, however, their shapes are obviously related to those of magnetic fields, with loops and arches common occurrences. Overall, the species known as quiescent prominences are a vertical, blade-like structure, typically 50 000 km high, 8000 km thick, and 10^5 to 3×10^5 km long. The 'blade' need not be precisely vertical, and the longitudinal shape, though tending to linearity, may be somewhat curved or sinuous. The lifetime of a typical prominence lies between 2 and 6 months, over which time its large-scale form will change only slowly. The other main type of prominence, the active prominence, is a rather different structure, and is discussed later in this section.

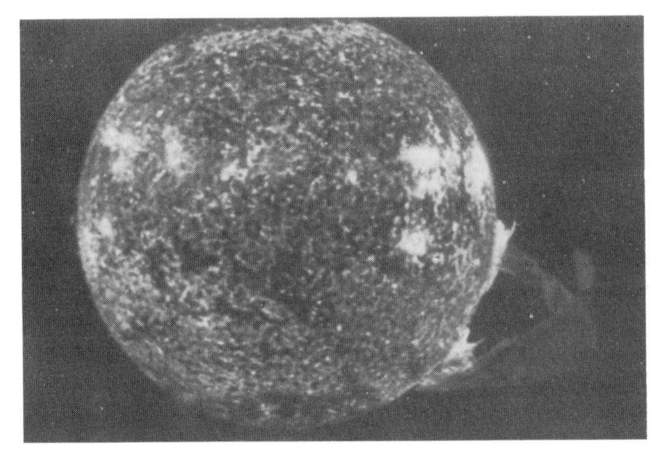

Figure 4A.7 Skylab spectroheliogram in the light of ionised helium at 304 nm (reproduced by permission of the US Naval Research Laboratory).

Prominences generally form close to sunspot groups a few weeks after the spots themselves appear. They are usually inclined at a few tens of degrees to the solar meridian initially, and on the polar side of the group. They can also form near plage regions from which no sunspots develop. After two or three rotations of the Sun, the differential rotation (§4A.1) acts to pull the prominence into more of an east–west alignment, and also to lengthen it by up to 100 000 km per rotation.

After the disappearance of the sunspots or the plage region, the prominence remains in existence. It is no longer anchored to the spot's position, however. It slowly drifts polewards until it gradually merges with other prominences to form a 'crown' near the pole. It then loses its integrity, and it becomes impossible to identify it as a separate phenomenon or to say when it finally disappears.

The density of the material in the prominence is about 10^{-9} kg m^{-3} or about 100 times that of the surrounding corona. Since its temperature of about 10 000 K is about 1% that of the corona, there is approximate pressure equilibrium between the prominence and the corona. But since the density is 100 times that of the corona, there cannot be vertical hydrostatic equilibrium through the effects of gas pressure alone. An additional force is required to support the weight of the prominence, and this is almost certainly provided by a magnetic field. Prominences tend to occur at the boundaries of two extensive regions of opposing polarities. Thus, in its region the magnetic field lines will be horizontal, or nearly so. The charged particles in the plasma which comprises the prominence are therefore inhibited from vertical motion in the same way that the magnetic field of sunspots inhibits horizontal motions. The weight of the prominence then compresses the magnetic field until it becomes sufficiently strong to support the material against the solar gravitational field.

Since the prominence is embedded in the very high temperature corona, it is constantly being heated. The coronal electrons are the main source of this energy input. Their mean free paths inside a prominence would be about 100 km, so that their energy of some 400 W m^{-2} is deposited primarily around the edges of the prominence. If the prominence is not to heat up and dissipate, its energy losses must balance this energy input. The losses are primarily by Lyman continuum radiation. The rate for this under typical prominence conditions is about 5×10^{-5} W m^{-3}. Thus a minimum thickness for the prominence of about 8000 km is required for the energy losses to be sufficient to balance the coronal heating. This is very close to the thicknesses actually observed for prominences, providing strong support for the validity of this hypothesised energy regime. That prominences significantly thicker than the minimum are not observed may be attributable to their becoming optically thick in the Lyman continuum at only slightly greater thicknesses. Self-absorption will then reduce the efficiency of the energy loss mechanism and cause the structure to evaporate back towards its minimum thickness.

Active prominences

In addition to the large, long-lived, relatively unchanging quiescent prominences just discussed, smaller, shorter lived and much more variable prominences may be seen. They are variously called active, surge, or sunspot prominences. Their lengths are typically only 50 000 km, but their heights and thicknesses are similar to those of the quiescent prominences. Their shapes predominantly comprise loops extending up from the chromosphere, or complex condensations above the chromosphere. Both types show obvious influences of magnetic fields. Very rapid changes in their appearances can occur as the magnetic fields change, often as a flare occurs in a nearby sunspot group. The material within active prominences is flowing downwards to the solar surface. Like the quiescent prominences, there is pressure and energy equilibrium between the prominence and the surrounding corona; unlike the quiescent prominences, however, the magnetic field strengths seem insufficient to provide hydrostatic equilibrium. The material is therefore constantly being lost to the prominence, and it can only continue to exist through the constant inflow into it of material from the corona.

Centres of activity

The various solar surface phenomena can usefully be linked into the concept of a centre of activity. This gives a summary of the mutual development of the phenomena, and emphasises their interrelationships and common cause. It is very much a summary, however, and wide deviations from the generalised development pattern outlined below often occur.

The centre of activity is based around a local intensification of the magnetic field at a point on the solar surface. As we shall see later, the magnetic fields in bi-polar magnetic regions are thought to be due to a flux tube emerging through the surface layers; thus the magnetic disturbance almost always develops into a BMR. The total lifetime of a centre of activity may approach a year, but the most of the 'excitement' occurs during the first month.

The centre of activity starts with a disturbance in the appearance of the chromosphere, soon followed by the development of plages. Within a couple of days a sunspot develops at the preceding (western) edge of the centre of activity, and the plages intensify. After about five days, the trailing spot may form towards the eastern edge of the centre of activity. Simultaneously, flares may occur between the spots, and active prominences appear nearby. The maximum level of activity is reached after about ten to fifteen days. The spots may then have formed a complex group, flare activity is at a maximum, and the plages and chromospheric structures may cover a region 100 000 to 200 000 km across. After the maximum, most of the phenomena start to decline, although the plages continue to extend their area. After a month or so, only the western spot remains of the main group, and a quiescent prominence has formed and stabililised on the polewards side of the centre of activity. Thereafter the quiescent prominence continues to develop, while all other activity subsides. Four to six months after commencement, the original site of the centre of activity will have returned to normal. Only the quiescent prominence then remains, and this drifts towards the poles. There it merges with other prominences, and becomes indistinguishable as a separate entity, perhaps a year after the initiation of the centre of activity.

Cyclic activity

The properties of the Sun vary on many time scales, even though by normal astronomical standards the Sun is not a variable star. By far the most widely known variation is the one called the sunspot cycle. This is considered first, before a rather briefer look is taken at the shorter and longer term changes.

Sunspot cycle

Individual spots come and go on a time scale of a few weeks. If the statistics of many spots are studied, however, then a pattern starts to emerge, even though the occurrence of an individual spot remains unpredictable. There are three main properties of sunspots which exhibit regularities on the same time scale: the total area of the Sun covered by spots, the mean latitudes of the spots at a given time, and their magnetic field polarities. The latter variation has already been mentioned (figure 4A.6), but the areal variation is the most obvious to an observer, and we therefore start by looking at that in detail.

The area of an individual spot or group is measured in units of millionths of a solar hemisphere (see earlier discussion). Such areas could then be added to give the total area of the visible disc of the Sun covered by spots at any given time. Generally, however, such a procedure would be time-consuming and would provide data with redundantly high levels of accuracy for this purpose. A more rapidly produced, estimated measure is thus normally used to gauge the levels of sunspot activity. This is called the sunspot number, or, more precisely, the Wolf or Zurich relative sunspot number. It is obtained by the rule:

$$R = k(10g + s) \qquad (4A.3.4)$$

and 27-day averages are taken to smooth out the effects of solar rotation. In this equation, g is the number of sunspot groups, and s is the number of single spots, observed on the solar disc at a given time. The factor k is a personal correction to match all observers' and observatories' estimates to a common scale. It takes account of different instrumentation, observing conditions and personal quirks in assigning spots to groups or as several single spots etc. The correction factor must be determined for each observer and observatory by extensive calibration to the internationally accepted figures.

The value of the sunspot number can range between 0 and 200 on any individual day. If its 27-day mean is recorded over a number of years, however, then a more-or-less regular pattern of variation emerges (figure 4A.8). Since the beginning of the eighteenth century, the period of the variation has averaged almost exactly eleven years; figure 4A.8 shows the best-fit positions for the maxima using a 10.96 year period. Individual cycles, however, have varied in length from 8.8 to 13.8 years. The cycles are asymmetrical, with the rise to maximum averaging two years less in duration than the fall to minimum. The sunspot number at a maximum can range from 45 to 190 (figure 4A.8), corresponding to spot areas of about 750 and 3100 millionths of a hemisphere.

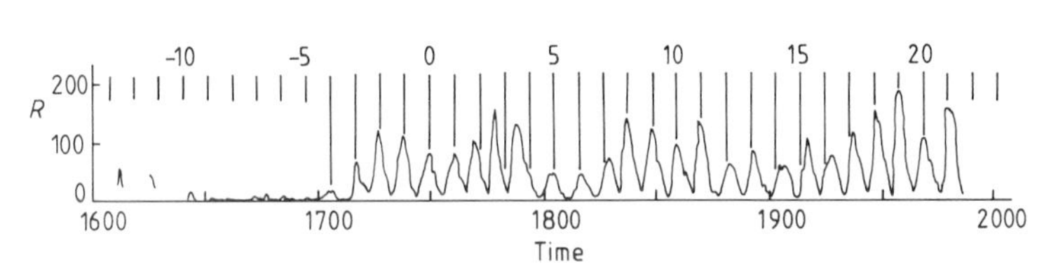

Figure 4A.8 Sunspot numbers.

The next maximum should be towards the end of 1990. It will be cycle 22, sunspot cycles by convention being numbered from the 1750 maximum. Even discounting the individual variations of the cycles, this prediction is by no means certain; figure 4A.8 shows a dearth of spots between 1645 and 1700 with hints of cyclic patterns re-emerging prior to 1645. Although the early observations are less reliable than modern ones, this total lack of solar activity in the latter half of the seventeenth century is confirmed by associated indicators such as the disappearance of aurorae and the solar corona over the same interval. High levels of terrestrial ${}^{14}_{6}C$ are found in tree rings produced at this time, a phenomenon associated with low solar activity (see below). The period is known as the 'Maunder minimum', after the astronomer who first pointed out its significance from studies of the historical data. It also coincides with the 'mini-ice age' in Europe in the late seventeenth century, when the average temperature of the Earth decreased by 1 K. It is not certain that the two events are causally connected, but a possible linkage might lie through a reduction in the ultraviolet transparency of the Earth's atmosphere at times of low solar activity. This could arise because Galactic cosmic rays (§7A.4) would penetrate the inner solar system more easily when the extended solar magnetic field was reduced through the lack of a contribution from solar activity. Their interaction with the Earth's atmosphere produces various oxides of nitrogen, and these absorb the solar ultraviolet radiation, reducing the total energy available at the surface of the Earth by perhaps 1 to 2%. (Very high levels of solar activity can also have the same effect through the solar cosmic rays.)

Slightly better established is the relationship between the $^{14}_{6}C$ levels in the Earth's atmosphere and the level of solar activity. The relationship is an anticorrelation; the higher the level of solar activity, the lower the terrestrial abundance of $^{14}_{6}C$. The cause is again the shielding of the Earth from cosmic rays by the extended solar magnetic field. The $^{14}_{6}C$ is produced by the interaction of cosmic ray particles with the $^{14}_{7}N$ in the atmosphere, so that less is produced when solar activity is high. The $^{14}_{6}C$ isotope is radioactive, with a half life of 5730 years. It is incorporated into living material along with the much more abundant $^{12}_{6}C$ isotope, and thus provides the basis of the carbon-14 dating method used by archaeologists. In a solar context, its importance lies in the clues that it gives to solar activity far further into the past than provided by the data from direct observations. Over five thousand years of solar activity may be traced in this way from the $^{14}_{6}C$ in tree rings, the direct observations in the last three hundred years enabling the data to be calibrated. It is thus found that the Maunder minimum is not an isolated instance of the cessation of solar activity. Earlier similar events may have occurred in 1500 BC, 700 BC, 400 BC, 700 AD, and 1500 AD, the last being known as the Spörer minimum. The duration is typically 50 to 100 years. The $^{14}_{6}C$ data also show periods of more intense activity than at the moment, with the slight possibility that we may be at the start of such a period just now. More certainly, intensified activity occurred for a few decades around 2700 BC, 2200 BC, 1800 BC, 0 AD, and 1300 AD.

Another, and strong, correlation occurs between sunspot activity and the deuterium/hydrogen ratio in tree rings. This ratio is temperature-dependent, since molecules containing heavy hydrogen are less mobile than those containing normal hydrogen. The data from the tree rings show a marked 22-year periodicity. Now as we have seen (figure 4A.6), the magnetic field reversals of bi-polar magnetic regions means that two of the 11-year solar cycles must elapse for the solar activity to repeat itself truly. Thus the solar cycle has also a fundamental 22-year period, providing that the magnetic observations which cover only the last five decades are not a recent anomaly. The best correlation of the two cycles requires a 13-year phase delay, so it may not be the case that the sunspots are the cause of the 22-year terrestrial temperature variations, but that there is a more basic change in the Sun which 13 years later also causes the emergence of the sunspots.

The other cyclic variation on the Sun associated with the sunspots is that of their mean positions on the Sun. Changes occur in the sense that the spots appear closer to the solar equator as the cycle progresses. The sunspots are confined generally between solar latitudes of 10° and 35° north and south of the equator. Occasionally spots may appear down to 0° and up to 60°, however. No preferential longitudes occur, although groups may reappear on the same meridian as a previous group, more often than to be expected on a truly random basis. Close to a sunspot minimum, spots of the new cycle start to appear at ± 35° of heliographic latitude. Around the maximum of the cycle, they are appearing within 10° of ± 15° latitude, while as the end of the cycle approaches they appear near ± 8°. When each spot is plotted on a latitude/time diagram, we get the 'butterfly diagram' (figure 4A.9), whose name derives from its appearance. The migration of the point of emergence of a spot throughout a solar cycle is known as Spörer's law. Sometimes spots at the start of one cycle may overlap the existence of spots of the previous cycle. Any doubt over their attributions, however, may be resolved by their positions and polarities.

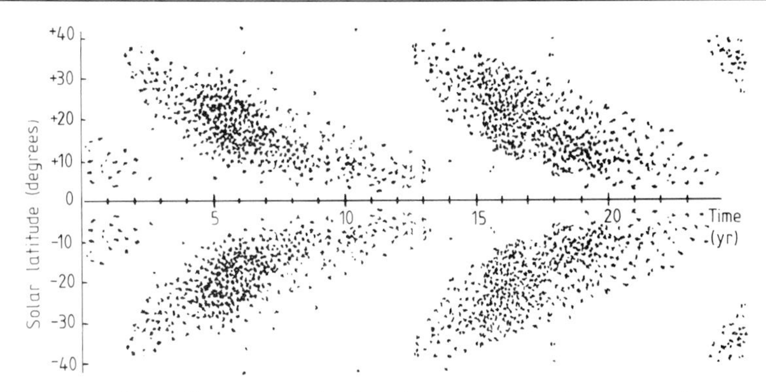

Figure 4A.9 Heliographic latitudes of sunspots (the 'butterfly' diagram).

The immediate cause of sunspots, or more fundamentally of a BMR and centre of activity, is a kink in a subsurface tube of high magnetic flux which penetrates the surface of the Sun. The origin of the flux tubes, and of the various cyclic changes, however, is not understood. None of the numerous theories even begin to provide a satisfactory theoretical basis for understanding solar activity. One of the more successful hypotheses, though, is due to Babcock. This envisages the general magnetic field of the Sun being wound up and intensified by differential solar rotation (see §8A.2 for a more complete discussion of the interaction of magnetic fields and plasmas). As the field strength rises towards 25 mT, magnetic pressure becomes significant. Since the material as a whole is in hydrostatic equilibrium, this additional source of pressure leads to a reduction in the gas pressure within the high-flux region. This in turn leads to a reduction in the density of the material, and so the flux tube experiences a bouyancy force lifting it towards the surface. Twisting and kinks in the flux tube further intensify the field strength towards the several hundred millitesla observed in actual sunspots. When one of the kinks penetrates the surface, it produces the effects we call a centre of activity. The flux tube will continue to expand outside the Sun, eventually reconnecting to leave a subsurface flux tube and a 'bubble' of magnetic flux floating off to join the interplanetary field. The critical field intensity is reached at high latitudes early in the cycle: only later does the differential rotation intensify the equatorial fields, and so the migration of sunspot appearances (Spörer's law) results. On this model, the magnetic reversal from one cycle to the next results from the reversal of the general magnetic field of the Sun through some (unknown) process. Unfortunately for this model, the general solar magnetic field is thought to originate in the 'bubbles' of magnetic flux coming from the BMRs, so that it becomes a 'bootstrap' process *par excellence*!

A recent discovery of large-scale streaming motions in the photosphere may be relevant to the whole sunspot formation process and their cycles. The currents of material have relative velocities of a few metres per second, and run parallel to the equator. Usually there are two fast and two slow streams in each hemisphere. The positions of the currents start near the poles and migrate towards the equator over a 22-year interval. The sunspots tend to appear at the boundaries between the currents. The significance of these observations, and whether they are a cause or an effect of solar activity remains, however, to be demonstrated at the time of writing.

Other cycles

The behaviour of the Sun varies on time scales other than the basic 11 (or 22) year sunspot cycle. The longer term variations are apparent from figure 4A.8: the heights of the maxima are quite obviously modulated. A cycle length of 76 to 80 years, the Gleissberg cycle, is probable, with a possible additional 179 year cycle superimposed upon it. Since the basic cycles have only been observed reliably for 250 years, and these exhibit considerable variability, the validity of these longer term variations is likely to remain unproven for some time to come. In support of their reality, however, a variation in the solar radius by about 0.5″ with a period of 80 years may also exist (§4A.1).

On very short time scales, the size and shape of the Sun is oscillating in a variety of complex modes. The periods of these changes range from 5 minutes to an hour, and they represent various forms of whole-body vibrations of the Sun.

Other stars must exhibit similar behaviour to these various small-scale solar features. However, the integrated, whole-disc observations which are all that are possible for most stars, are not capable of revealing it. In one or two cases surface activity can be detected or inferred. Most notably there are the faint UV Ceti stars (§3A.5) which occasionally brighten through flare activity; T Tauri stars may also have had flares detected occurring on them. Starspots are possible causes of the variations on time scales of a few days for some RS CVn stars. Radio emission from flares on these stars has also been found. Some evidence for variations in the levels of such activity has been discovered, but the observations are insufficient at present to detect any cyclic components in the changes. Some Be stars have cyclic variations which may be attributable to surface activity cycles. Amongst main sequence stars, a few have been found to exhibit changes in their Ca II *H* and *K* emission-line strengths which can be interpreted via a combination of the star's rotation and variable chromospheric activity due to photospheric centres of activity. If this interpretation is correct then such activity appears to be limited to the more slowly rotating stars with periods of a couple of days or more.

4A.4 SOLAR NEUTRINOS

Neutrinos are expected to be emitted by the Sun since they originate at various stages of the fusion reactions supplying its energy (§2A.2). However, only one neutrino in some 10^{10} from the centre of the Sun will be lost by interaction with the overlying layers of the Sun, as it travels outwards. Neutrinos thus potentially provide a direct and near-instantaneous picture of the conditions at the core of the Sun. The information is delayed only by the neutrino's flight time to the Earth: some eight to ten minutes. The electromagnetic radiation, by contrast, is based upon the emission by the core over about the last 10^7 years. The nuclear fusion reactions expected to be possible for conditions at the centre of the Sun are the proton–proton chain and the CNO cycle, and their variants (figures 2A.4, 2A.5). The neutrinos produced and their energies are shown on those diagrams. Combining such information with a specific solar model enables a neutrino spectrum for the Sun to be predicted. Figure 4A.10 shows such a spectrum for the standard solar model.

Although the interaction cross section of neutrinos is very low (§2P.4) so that observing them is very difficult, their potential for providing direct observational data on the interior of the Sun is so important that considerable efforts have been expended on their detection. The only successful experiment to date is that due to R Davis Jr and his group. This experiment started work in the 1960s, and has now accumulated considerable quantities of data. The detection of neutrinos in this case is based upon their interaction with the ${}^{37}_{17}Cl$ isotope:

$$\nu_e = {}^{37}_{17}Cl \rightarrow {}^{37}_{18}Ar + e^- . \qquad (4A.4.1)$$

Only the higher end of the neutrino spectrum (figure 4A.10) is detectable by this reaction, since the threshold for its occurrence is 814 keV. The interaction cross section ranges from 10^{-50} m^2 at 814 keV to 10^{-46} m^2 at 10 MeV. Thus, in practice, the experiment will detect only the neutrinos from the side reaction of the proton–proton chain which leads to the decay of $^{8}_{5}B$. The detector is a tank of some 600 tonnes of tetrachloroethene (C_2Cl_4), in which one or two atoms of chlorine are converted to argon by the neutrinos each day. The argon atoms are detected via their subsequent radioactive decay back to chlorine.

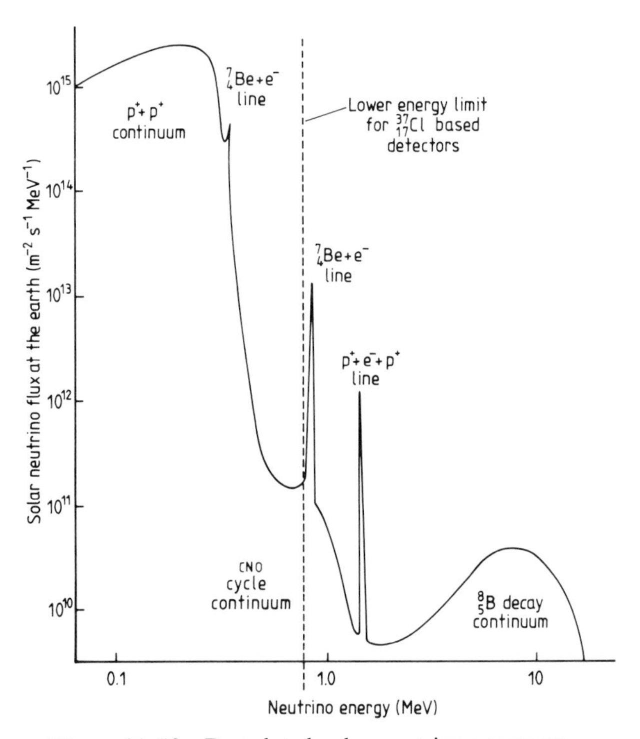

Figure 4A.10 Postulated solar neutrino spectrum.

The expected rate of neutrino detection from the predicted spectrum (figure 4A.10) is about three times the observed rate. Expressed in terms of the solar neutrino unit (1 SNU = 10^{-36} captures per second per target atom), the expected rate is 7.3 ± 1.5 SNU, and the observed rate is 2.2 ± 0.4 SNU. This discrepancy is known as the solar neutrino problem, and it has yet to be satisfactorily explained. It has achieved prominence for its apparent implication that a fundamental error has been made either in our understanding of the physics of stellar interiors, or in our understanding of subatomic physics (or both). Possible explanations have been suggested in §2A.2. The preferred explanation from an astrophysical viewpoint would be the oscillation in neutrino type (figure 4A.11) implied if the particle has a non-zero rest mass, since this would alter the generally accepted stellar and solar models to the least extent. At the time of writing, any such non-zero rest mass or oscillation of type remains to be confirmed. Various detectors for the lower energy solar neutrinos are due to come on stream in the next few years, and these may finally determine whether the solar neutrino problem is real, and, if so, whether it is the astrophysicist's or nuclear physicist's worry!

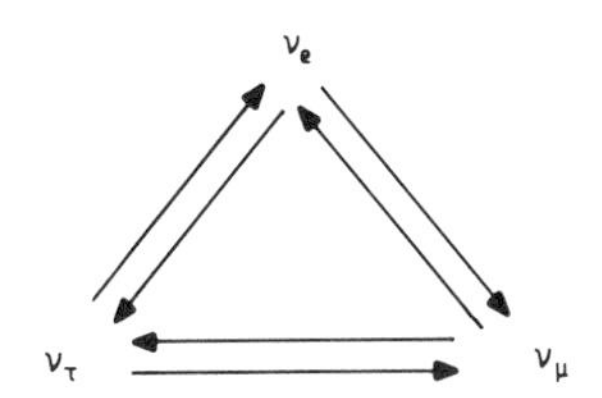

Figure 4A.11 Neutrino oscillation.

4A.5 SOLAR CORONA

The outermost layer of the Sun is called the corona. It is a very rarefied, very hot plasma, which eventually merges into the interstellar medium (Chapter 7A) at 40 to 70 AU out from the Sun (i.e. from just beyond Pluto's orbit to nearly twice that distance). The interaction zone forms a shock front known as the heliopause. The whole of the currently known planetary system is thus embedded in the outer layer of the Sun. Material from the corona is steadily being lost to the interstellar medium by a mass outflow called the solar wind. This latter aspect of the corona is considered along with stellar winds in general in Chapter 5A. Here the properties of the corona itself are discussed.

Morphology

For many years a certain romance and mystique has been associated with the corona and its investigation. This has arisen through its all too brief moments of visibility during solar eclipses, and the travel to remote parts of the world which eclipse observation often necessitates. Until coronagraphs (*2) could be lifted into space, only the inner corona could be studied except during such eclipses.

Nonetheless, many eclipse expeditions in the latter half of the nineteenth century led to the basic appearance of the corona being well understood by the start of this century. It is a whitish halo surrounding the solar disc, and extending outwards to between one and five solar radii from the photosphere. Its brightness fades rapidly with distance from the photosphere, and overall it is rather fainter than the full Moon. The total luminosity varies with the solar cycle, changing from 30% to 60% of the intensity of the full Moon between minimum and maximum levels of solar activity.

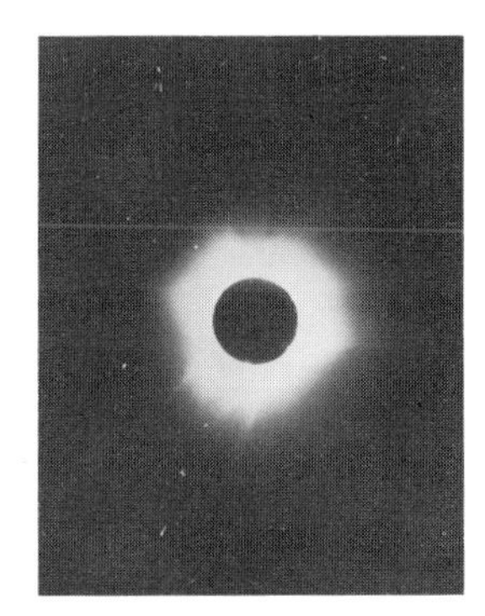

Figure 4A.12 The solar corona. Left: at sunspot minimum. Centre: at intermediate sunspot numbers. Right: at sunspot maximum. (Reproduced by permission of the Royal Astronomical Society.)

The shape of the corona is much more variable than its luminosity (figure 4A.12). At solar maximum it is more or less circular, with a radially striated appearance and a naked-eye radius three to four times that of the Sun. At solar minimum, the radius reduces to twice that of the Sun, and its shape is elliptical. The polar regions are distinguished by radial striations known as the polar rays. Near the equator the structure is more amorphous. At intermediate levels of solar activity, the appearance is that of the minimum corona with huge radial projections, the coronal fans, superimposed. The fans are related to centres of activity and increase in number as the sunspot number increases. At maximum, the corona is largely formed of such fans, leading to its greater size and circularity. During the Maunder minimum (§4A.3) historical references to the corona during eclipses are non-existent. Thus it may be that during such prolonged disappearances, the corona is reduced below the normal minimum, perhaps even to the point of invisibility to the unaided eye.

Structures within the corona are obviously related to magnetic fields. The polar rays, for example, are curved in a manner reminiscent of a dipolar field. They may arise through the inhibition of motion of the material across the field lines, so that denser streamers retain their integrity and do not diffuse into their surroundings, while at the same time being constrained to follow the direction of the magnetic field. Other loops and arch systems also occur within the corona, and are often associated with prominences and active regions on the photosphere.

In other wavelength regions, the corona may have a very different appearance from this optical structure. At metre radio wavelengths, the corona is optically thick. Thus, in the absence of solar flares and other non-thermal sources, the radio observations reveal the lower parts of the corona directly. They generally show a corona about twice the size of the Sun, and elongated along the equator. Wavelengths up to one metre show limb brightening due to the increasing temperature of the corona with radius. Over two metres wavelength, the appearance is just that of a simple disc with a central maximum in intensity. At short wavelengths (extreme ultraviolet, x- and gamma rays) the corona is again seen directly. Its appearance is then very much more variable, consisting mostly of small-scale bright regions associated with photospheric centres of activity, and coronal fans. Its brightness links closely with the local magnetic field strength, leading to 'coronal holes' in regions of the lowest field strengths.

Coronal optical spectrum

The optical corona contains light from three different sources, labelled the *E*, *F*, and *K* coronae. The *F* corona is not a genuine corona in the sense that it originates in the region within a few solar radii of the Sun. It is, in fact, solar photospheric radiation scattered by interplanetary dust near the Earth. It is the forward-scattering component of the zodiacal light (*3). At more than two solar radii, it is by far the strongest component of the coronal light.

The *K* and *E* coronae are due to the immediate outer layers of the Sun. There is about a hundred times more energy in the *K* corona than in the *E* corona, and it forms the major part of the corona out to about two solar radii. The spectrum of the *K* corona is that of a pure continuum with the spectral distribution of a 5700 K black body. Like the *F* corona it is scattered photospheric radiation. This time the scattering is Compton scattering by the free electrons in the solar atmosphere. This is independent of wavelength, hence the *K* corona mimics the spectrum of the photosphere, except that the absorption lines are not seen. These have disappeared because they are broadened to widths of 20 nm or more by the Doppler shifts induced by scattering from electrons moving at upwards of 5000 km s^{-1}. The lines thus merge into each other and are no longer individually observable. This phenomenon allows separation of the contributions from the *K* and *F* coronae. The *F* corona contains photospheric absorption lines of normal intensities, but they are apparently reduced by the superimposition of the *F* corona continuum on the *K* corona continuum. The proportion of the observed spectral intensity which must be subtracted in order to render the absorption line intensities normal is thus the contribution from the *K* corona, while the remainder is that from the *F* corona.

The *E* corona is the only component which originates from the coronal material itself. It consists generally of weak emission lines superimposed on the spectra from the *K* and *F* coronae. The strongest such lines are due to highly ionised metals, particularly Fe X, XI, XIII, XIV, XV, Ni XIII, XV, and Ca XV. The lines are not those normally associated with these ions; indeed, the problem of their identification led at one time to the postulation of a new element, coronium, in order to explain them. They are now known to be forbidden transitions (§3P.1) from metastable levels, which are normally swamped by the allowed transitions. Only under the very rarefied conditions in the corona do the excited states' lifetimes against collisions become long enough for the lines to become sufficiently intense to be seen. The typical lifetimes of the excited levels against the forbidden transitions are about 10 ms, compared with about one second between collisions for coronal conditions.

Coronal x-rays

Solar x-ray images reveal the corona directly. Such emission originates in the hotter, more active portions of the corona, hard x-rays coming from regions with kinetic temperatures of 3 to 5×10^6 K, and soft x-rays from regions at 2 to 3×10^6 K. They thus provide flags marking the most active portions of the corona. These are usually found to be closely associated with the most active photospheric regions. A very patchy structure for the corona is revealed by the observations. Some regions, the coronal holes, are almost totally without emission. The rest of the corona comprises a loose interlacing of loops and arches following the local magnetic fields. The coronal rays and large-scale loops are usually associated with the most active portions of the x-ray corona. The observations, however, are very recent, and much undoubtedly remains to be discovered about the short-wave emissions from the corona.

Physical conditions

The coronal electron density may be found from the intensity of its *K* component, and varies from about 10^{15} m^{-3} at 1.003 $R_\odot$, through 10^{13} m^{-3} at

$1.5\ R_{\odot}$, and $10^{11}\ m^{-3}$ at $3\ R_{\odot}$, to less than $10^{6}\ m^{-3}$ at the distance of the Earth from the Sun. Assuming the composition to be largely ionised hydrogen, the corresponding total densities are therefore 10^{-12}, 10^{-14}, 10^{-16} and $10^{-21}\ kg\,m^{-3}$ respectively.

As will already have become apparent, the temperature of the corona is very high. The value obtained, however, depends upon the method used to measure the temperature. The low density of the corona leads to low effective temperatures: from a few thousand degrees at its base to a few tens of degrees Kelvin at the distance of the Earth from the Sun. Most other measures of temperature give very much larger figures. Thus the lack of Fraunhofer lines in the *K* corona, ionisation levels, forbidden line intensities, metre wavelength radio observations of the quiet Sun, and the x-ray emission all suggest temperatures near 8×10^{5} K. Several other methods, however: density gradient, line widths, etc, suggest temperatures of about 1.5×10^{6} K. The difference may not be significant, though, since the errors of the measurements are large. The corona is isothermal to at least three solar radii, and possibly much further; the electron and ion temperatures in the solar wind near the Earth (Chapter 5A) from spacecraft observations are still between 10^{5} and 10^{6} K. There is a moderate global increase in the coronal temperatures between solar minimum and maximum.

If particle densities and kinetic temperatures are known, then the pressure may be estimated. This ranges from $10^{-2}\ N\,m^{-2}$ at the base of the corona to $10^{-11}\ N\,m^{-2}$ at 1 AU (10^{-7} to 10^{-16} atm). This simple picture is complicated by the directed motions of the particles (the solar wind; Chapter 5A). This can lead to directed pressures one or two orders of magnitude larger than the normal gas pressure. Thus the Earth's magnetosphere (*4) is highly compressed on the side towards the Sun and dragged out into a long tail in the anti-solar direction, through the effect of this imbalance in pressure.

The composition of the corona is identical with that of the solar surface layers (§1A.2), with the exception of the much higher levels of ionisation which are present.

Coronal energy

A major problem in understanding the corona is to explain its enormously high temperature. The second law of thermodynamics prohibits heat from flowing from a cooler object to a hotter one, so the 6000 K photosphere cannot heat the corona to the observed 10^{6} K temperature.

A possible alternative energy source long favoured by astronomers is that of mechanical or acoustic energy. Acoustic and magnetohydrodynamic (Alfvén) waves will be generated in profusion in the turmoil of the convecting outer layers of the Sun. Such waves which propagate outwards will encounter regions of progressively lower densities. The wave energies will thus become shared amongst fewer and fewer particles. Eventually the particles will be driven to velocities higher than the local speed of sound, a phenomenon known as the whiplash effect (*5). The waves become shock waves, and the wave energy is rapidly deposited into the medium as random motions of the particles. In the Sun, such deposition of energy occurs predominantly near the base of the corona, and the particle density is then so low that the kinetic temperatures are driven to the observed extreme values. The reality of this mechanism, however, now seems less well established. In the x-ray region the solar and other stellar coronae may be observed directly. The various spacecraft-borne detectors of recent years have revealed that x-ray luminosities amongst stars of similar spectral types can vary by a factor of 100 or even 1000. Since the convection zones of such stars should be similar, their coronal x-ray emissions should also be similar, if they are powered by the acoustic or magnetohydrodynamic waves just outlined. It is possible, however, for magnetic fields to vary to a much greater extent amongst stars of similar spectral types. More recently, therefore, alternative hypotheses based upon the dissipation of magnetic energy by field line reconnection, or by magnetohydrodynamic waves, and heating by electric currents have been advanced. Such processes are supported by the observed association of regions of strong x-ray emission with significant levels of magnetic activity. But at the time of writing the situation remains to be clarified. Possibly acoustic heating may dominate at solar minimum and in 'normal' parts of the corona, with additional, perhaps large, contributions from other sources for disturbed portions of the corona and at large distances from the Sun.

PROBLEMS

4A.1 What fraction of the total pressure at the centre of the present Sun is due to degenerate electrons?

4A.2 By assuming circular orbits for the Earth and Venus, show that

$$D^2 = V^2 - E^2 + 2ED \cos \theta$$

where D is the instantaneous Earth–Venus distance, V is the radius of Venus' orbit, E is the radius of Earth's orbit and θ is Venus' elongation. Hence estimate a value for the astronomical unit, given the following data:

Date		Elongation of Venus (decimal degrees)	Radar pulse travel time from Arecibo to Venus and back (s)
Jan	1	24.5347 W	1498.9865
Feb	1	17.3673 W	1610.0682
Mar	1	10.5386 W	1679.6078
Apr	1	0.8563 W	1719.6345
May	1	3.6092 E	1715.1824
Jun	1	12.9588 E	1658.9150
Jul	1	21.4015 E	1552.4947
Aug	1	29.6082 E	1394.9481
Sep	1	36.8827 E	1201.1949
Oct	1	42.5571 E	991.1310
Nov	1	46.0635 E	760.5433
Dec	1	44.5406 E	534.9735

5A Stellar Winds

SUMMARY

Properties of the solar wind, properties of stellar winds, theory of coronal and radiative stellar winds.

See also stellar compositions (§§1A.2, 3A.6), radiation laws (§2P.3), spectrum line broadening (§3A.2), solar corona (§4A.5), formation of H II regions (§6A.3), planetary nebulae (§6A.5), interstellar medium (§§7A.1, 7A.2, 7A.3), cosmic rays (§7A.4), and the formation and evolution of stars (§§8A.1, 8A.2).

INTRODUCTION

The existence of relatively steady outflows of material from some stars has been apparent for over half a century. P Cygni line profiles (§3A.2) and lines skewed to shorter wavelengths have long been accepted as due to regions of outwardly expanding material. The velocities derived for the expansion from such line profiles exceed the star's escape velocity in many cases. The material must therefore be lost to the star and the observed rates of mass loss often reach $10^{-5}\ \mathcal{M}_{\odot}$ per year, and more rarely reach $0.1\ \mathcal{M}_{\odot}$ per year. More recently much weaker outflows (stellar winds, as they are generally known) have been found to occur in many stars, and in particular the solar wind, predicted by Parker in 1958, was discovered by the long-distance space probes of the early 1960s.

5A.1 THE SOLAR WIND

The mean properties of the solar wind throughout much of the solar system have been established by many spacecraft missions, and by terrestrial observations of the behaviour of comet tails (which are influenced by the wind). Near the Earth, but sufficiently far away to be beyond the influence of the magnetosphere, it is found to be a proton–electron gas with an ion or electron number density ranging from 5×10^6 m^{-3} at solar minimum to 5×10^7 m^{-3} at solar maximum. The directed velocity of the particles ranges from 400 to 800 km s^{-1} between solar minimum and maximum, with the direction generally radial to the Sun. The random components of the velocities lead to kinetic temperatures for the material of 10^5 to 10^6 K for the ions, and 2×10^5 to 10^6 K for the electrons, at the extremes of the solar cycle.

These mean figures, however, are only very rough guides to the actual values at any given moment. The wind is very 'gusty', and changes rapidly with time and position. A pattern to the variations is often to be found though; the wind speed rises abruptly by perhaps 200 km s^{-1} over 24 hours, followed by a decline back to the previous value over 5 to 10 days. The density may triple during the velocity rise, but returns to its earlier value in only one day; there are also concomitant rises and falls in the kinetic temperature and the pressure. The explanation for this behaviour lies in the supersonic nature of the wind; the velocity of sound is about 50 km s^{-1} for solar wind conditions near the Earth. A high velocity region of particle emission, perhaps associated with an active region and/or a coronal fan, may have a velocity of 500 to 600 km s^{-1} compared with the normal expansion velocity of the surrounding material of 300 to 400 km s^{-1}. The rotation of the Sun causes the stream lines (*1) to be spirals, except within about 10 $R_\odot$ of the Sun, where there is corotation with the Sun. In a radial direction, therefore, the high velocity material will be overtaking the slower material at a closing velocity three or four times the local speed of sound (figure 5A.1). The compression region formed at the leading edge of the high velocity stream becomes very narrow because of the supersonic closing velocity. Shock waves propagate forward from it into the lower density material, and backwards into the high velocity stream. This double shock structure does not develop fully until two or three AU from the Sun, but has been observed by several of the outer solar system space probes. The gust in the wind observed at the Earth is just the passage of the developing compression and depletion zones past the Earth. At distances of 20 or more AU, the spiral pattern will begin to wind up. The forward and trailing shocks from successive revolutions of the spiral will start to overlap and interact leading to very chaotic and variable conditions.

The solar wind also carries a weak magnetic field with it. Close to the Earth, the average field strength varies from 5 nT at solar minimum to

20 nT at solar maximum. Again, however, this is only a rough mean, and the field strength is very variable. On a time scale of minutes, the direction is almost random. Longer averages, however, show the field lines to be following the expected spiral direction out from the Sun. Either polarity may be observed; i.e. the field lines may be directed inwards or outwards along the spiral. The magnetic field is in the form of sectors of opposing polarities which are dragged around by the solar rotation. Typically two such pairs of sectors would be found in the solar equatorial plane, with an individual field pattern stable over several solar rotations. As a sector boundary sweeps past the Earth, the wind speed may rise, reaching a maximum about a third of the way through a sector, and then falling back until the next sector boundary approaches.

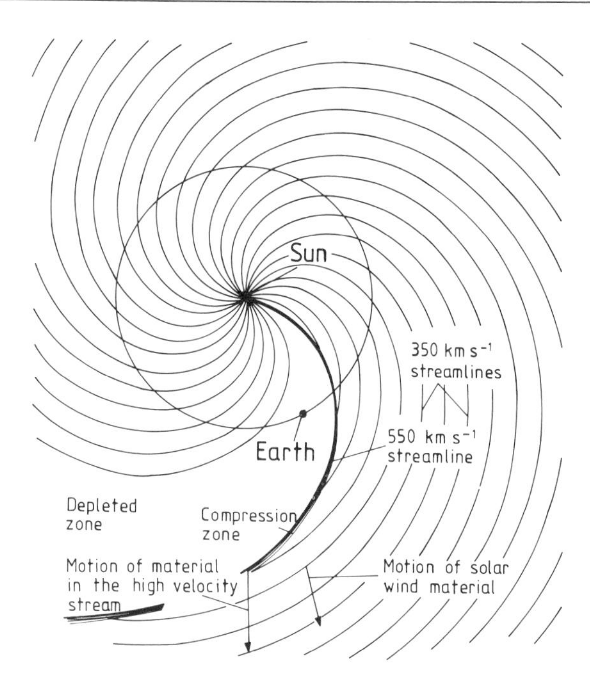

Figure 5A.1 Formation of a 'gust' in the solar wind following emission of a high-velocity stream.

The cause of the solar wind is the very high coronal temperature (§4A.5). Essentially this leads to a permanent pressure imbalance with the interstellar medium. Since the corona is thus not in hydrostatic equilibrium, it is continually expanding out into the interstellar medium, and this expansion we observe as the solar wind. The precise nature of this accelerating mechanism is considered, along with other processes which lead to stellar winds, in §5A.2.

5A.2 STELLAR WINDS

Observations

The Sun is losing mass via the solar wind at a rate of about $10^{-14} \mathcal{M}_{\odot}$ per year. Such a low loss rate would not generally be detectable in any other star. The observed stellar winds are thus many orders of magnitude stronger than the solar wind. The stars observed to have winds fall into three broad categories: the very hot stars, the very large stars, and the very young stars (*2).

Newly formed stars such as the T Tauri stars and the Herbig Ae and Be stars (§3A.5) show P Cygni line profiles in their spectra. The expansion velocities obtained from these lie in the region of 100 to 200 $\mathrm{km\,s^{-1}}$, and represent mass loss rates of 10^{-9} to $10^{-7} \mathcal{M}_{\odot}$ per year or more. The wind is very variable in its strength, both between stars of similar types, and for a single star with time. Occasionally it can even reverse so that material is accreting on to the star. It is not yet clear whether the observed outflows lead to a true mass loss from the star, or not; perhaps mass loss is occurring as jets over some parts of the star, while infall occurs elsewhere. The low velocity (50 to 200 $\mathrm{km\,s^{-1}}$), often bi-polar, outflows detected in many of the molecular clouds associated with these young stars by observation of the carbon monoxide microwave emissions, with mass loss rates of up to $10^{-2} \mathcal{M}_{\odot}$ per year, tend to suggest, however, that the observed stellar winds do indeed lead to net mass loss from the stars (Chapter 7A, §8A.1).

Relatively low velocity winds are observed for many cool supergiants and long-period variables. Although the velocities are only 10 to 100 $\mathrm{km\,s^{-1}}$, the material is escaping completely from the stars. This is established when the star is part of a wide binary system; absorption lines arising in the stellar wind may be found in the spectrum of the companion, and the expansion velocity at that distance from the star is then found to exceed the escape velocity. Some of the supergiant spectrum lines may have P Cygni line profiles, but the stellar wind lines often blend with those from the star, making their interpretation difficult. Nonetheless, mass loss rates of 10^{-9} to $10^{-5} \mathcal{M}_{\odot}$ per year for these stars are variously estimated.

Stellar winds reach their greatest developments in the very hot stars. 'Normal' O-type stars on the main sequence have mass loss rates around $10^{-7} \mathcal{M}_{\odot}$ per year. Thus, over their main sequence lifetimes, such stars may lose up to one solar mass via their stellar wind. Spectral class W stars, otherwise known as Wolf–Rayet stars (§§3A.2, 3A.6), have similar temperatures to the O-type stars, but very much more intense stellar winds. Expansion velocities of up to 3000 $km\,s^{-1}$ may occur, and mass loss rates are typically $5 \times 10^{-5} \mathcal{M}_\odot$ per year, with maximum rates four or five times larger still. Since the masses of these stars are variously estimated to be around $10\,\mathcal{M}_\odot$, such stellar winds must obviously cause radical changes in the stars in well under 100 000 years. The Of stars, which appear as intermediates in terms of their spectra between the O and W stars, appropriately have intermediate stellar winds, their mass losses typically being found to lie between 10^{-6} and $10^{-5} \mathcal{M}_\odot$ per year.

Planetary nebulae (§6A.5) have very hot central stars. The nebula itself is evidence of mass loss at some time from the star. The nebula masses lie in the region of 0.001 to $0.5\,\mathcal{M}_\odot$, and many still show expansion velocities of a few tens of kilometres per second. Mass loss rates would therefore appear to be within an order of magnitude of $10^{-5} \mathcal{M}_\odot$ per year during the formation of the nebula. Some of the central stars have spectra similar to those of the Wolf–Rayet stars, even though the stars must be of much smaller masses than is normal for that class. They thus, like normal W-type stars, have stellar winds with terminal velocities of thousands of kilometres per second. Such high velocity winds probably develop after nebula formation, and are a separate phenomenon entirely.

The star P Cygni, which gave its name to the line profile, is the prototype of a rather diverse group of stars. Its spectrum is that of a B1.5 supergiant with additional emission lines, and most of the line profiles are of P Cygni form. The line profiles suggest expansion velocities ranging from 30 to 500 $km\,s^{-1}$, increasing with distance from the star. The mass loss rate is rather uncertain and estimates range from 10^{-6} to $4 \times 10^{-4} \mathcal{M}_\odot$ per year. Similar results are found for the few other stars in the group, although the peculiarities are generally not quite so strongly developed as in P Cygni, and they tend to be somewhat cooler objects. Other possibly related systems include η Car, slow novae, and symbiotic stars.

Hot supergiants in general have stellar winds. The optical emission at Hβ suggests a velocity of only 100 $km\,s^{-1}$ or so, but ultraviolet observations reveal expansion velocities reaching thousands of kilometres per second. The mass loss rates are typically $10^{-6} \mathcal{M}_\odot$ per year, but may be higher by up to an order of magnitude. In at least one case (ξ^1 Sco), the emission is not even approximately steady, but occurs in a series of bursts at widely varying velocities.

Binary stars may lose mass, but this is usually in a rather different manner from the winds just discussed. Mass loss occurs when evolution leads one of the stars to expand and fill its Roche lobe (§8A.2). Material then spills over from one star into the sphere of influence of the other, and probably eventually accretes on to its surface. Very high rates of mass loss, or when both stars are filling their Roche lobes, may lead to mass loss from the system as a whole as a low velocity wind.

One individual star is worth a mention in this connection, and that is the star known as SS433. This has a relatively normal stellar wind, with a velocity of some 2000 km s^{-1}, and a mass loss rate around $3 \times 10^{-6} \mathcal{M}_{\odot}$ per year. Additionally, however, material is being expelled in two opposing jets at velocities of 75 000 km s^{-1} (25% of the speed of light)! One interpretation of this fascinating system is that it is a black hole/white dwarf binary system, the white dwarf being disrupted by the black hole, and a small proportion of the material falling towards the black hole being expelled at very high velocities along the black hole's rotational axis.

Theory

There are several processes which can lead to the expulsion of material from a star. As we have just seen, corotation in a binary system is one process, solar flares (§4A.3) also expel material, sometimes at relativistic velocities, while in protostars (§8A.1), angular momentum transfer from the central object to the surrounding nebula may lead to the latter's being driven outwards and lost to the system. In none of these cases, however, does the sort of large-scale, high velocity wind described in the previous section develop. The two forces which do seem to cause the majority of stellar winds are thermal pressure and radiation pressure. The first leads to coronal-type winds, such as the solar wind, the latter to the intense winds of hot stars, and rather less certainly to the winds of cool supergiants.

The thermally driven stellar winds are typified by the solar wind, and are often called coronal winds for that reason. The kinetic temperature at the base of the solar corona is about 1.5×10^{6} K, and decreases with height according to

$$T(r) = T(0)(r/R_{\odot})^{-2/7} \tag{5A.2.1}$$

where r is the distance from the centre of the Sun, $R_{\odot}$ is the solar radius and $T(0)$ is the base temperature of the corona.

Since the RMS velocity of a particle is given by

$$\bar{V}_{\mathrm{p}} = (3kT/m)^{1/2} \tag{5A.2.2}$$

and the escape velocity by

$$V_{\mathrm{e}} = (2G\mathcal{M}_{\odot}/r)^{1/2} \tag{5A.2.3}$$

where m is the particle's mass, we have for protons:

$$\bar{V}_{\mathrm{p}}/V_{\mathrm{e}} \simeq 2.2 \times 10^{-4} r^{0.36}. \tag{5A.2.4}$$

This ratio reaches a value of unity for a distance of about 20 solar radii (0.1 AU) from the Sun, and so beyond that point the bulk of the coronal particles are travelling faster than the local escape velocity from the Sun. The pressure due to the coronal material by the perfect gas law (equation

2P.1.15) is

$$P(r) \simeq 2N(r)kT(r) \tag{5A.2.5}$$

where $N(r)$ is the proton or electron number density. Thus the equation of hydrostatic equilibrium (equation 2P.5.12) with equations (5A.2.1) and (5A.2.5) becomes

$$\frac{\mathrm{d}}{\mathrm{d}r}\left[2N(r)kT(0)\left(\frac{r}{R_\odot}\right)^{-2/7}\right] = \frac{-G\mathcal{M}_\odot m_\mathrm{H} N(r)}{r^2} \tag{5A.2.6}$$

or

$$\frac{\mathrm{d}}{\mathrm{d}r}\,[N(r)r^{-2/7}] = \frac{-G\mathcal{M}_\odot m_\mathrm{H}}{2kT(0)R_\odot^{2/7}}\,\frac{N(r)}{r^2}. \tag{5A.2.7}$$

Expanding the left-hand side and rearranging gives

$$\frac{\mathrm{d}N(r)}{\mathrm{d}r} = \left(\frac{2}{7}r^{-1} - \frac{G\mathcal{M}_\odot m_\mathrm{H}}{2kT(0)R_\odot^{2/7}}\,r^{-12/7}\right)N(r). \tag{5A.2.8}$$

Integrating from the solar surface out to distance r then gives

$$\log_\mathrm{e}\left(\frac{N(r)}{N(0)}\right) = \log_\mathrm{e}\left(\frac{r}{R_\odot}\right)^{2/7} + \frac{7G\mathcal{M}_\odot m_\mathrm{H}}{10kT(0)R_\odot}\left[\left(\frac{r}{R_\odot}\right)^{-5/7} - 1\right] \tag{5A.2.9}$$

where $N(0)$ is the proton or electron number density at the base of the corona. Hence

$$N(r) = N(0)\left(\frac{r}{R_\odot}\right)^{2/7} \exp\left\{\frac{7G\mathcal{M}_\odot m_\mathrm{H}}{10kT(0)R_\odot}\left[\left(\frac{r}{R_\odot}\right)^{-5/7} - 1\right]\right\}. \tag{5A.2.10}$$

Now, from equations (5A.2.1) and (5A.2.5),

$$N(r) = \frac{P(r)}{2kT(0)(r/R_\odot)^{-2/7}}. \tag{5A.2.11}$$

Combining the last two equations we have

$$P(r) = P(0)\exp\left\{\frac{7G\mathcal{M}_\odot m_\mathrm{H}}{10kT(0)R_\odot}\left[\left(\frac{r}{R_\odot}\right)^{-5/7} - 1\right]\right\}. \tag{5A.2.12}$$

Thus at large distances from the Sun ($r \to \infty$), we find

$$P(\infty) = P(0)\exp\left(\frac{-7G\mathcal{M}_\odot m_\mathrm{H}}{10kT(0)R_\odot}\right) \tag{5A.2.13}$$

or, substituting for the quantities in the exponent,

$$P(\infty) = 2.1 \times 10^{-5}P(0). \tag{5A.2.14}$$

Now the pressure at the base of the corona (§4A.5) is about 0.04 N m^{-2}, so that the limiting pressure of the corona is about 10^{-6} N m^{-2}. The pressure of the interstellar medium averages some $10^{-12}\ \mathrm{N\,m^{-2}}$. Thus not only is the coronal material uncontained by the solar gravitational field, but it will also never reach pressure equilibrium with the interstellar medium. It therefore expands continuously away from the Sun and becomes the solar wind.

Similar winds will arise in any star with a hot corona. The cause of the high temperature of the solar corona is thought to be the deposition into it of energy from the convection region or from magnetic fields (§4A.5). Thus we may expect coronal-type winds to occur in some or all solar- and similar type stars; probably most main sequence stars (by number) have coronal-type winds, therefore.

The speed of the solar wind is highly supersonic, but this need not always be the case. Subsonic winds (stellar breezes!) may also occur. The velocity equation is

$$v(r)\left(1-\frac{k}{\mu m_{\mathrm{H}}}\frac{T(r)}{v(r)^2}\right)\frac{\mathrm{d}v(r)}{\mathrm{d}r}=-G\mathcal{M}_{\odot}\left(1-\frac{2k}{\mu m_{\mathrm{H}}G\mathcal{M}_{\odot}}rT(r)\right)\frac{1}{r^2}-\frac{k}{\mu m_{\mathrm{H}}}\frac{\mathrm{d}T(r)}{\mathrm{d}r}. \tag{5A.2.15}$$

For the essentially isothermal conditions in the lower corona, this reduces to

$$v(r)\left(1-\frac{2kT(0)}{m_{\mathrm{H}}v(r)^2}\right)\frac{\mathrm{d}v(r)}{\mathrm{d}r}=-G\mathcal{M}_{\odot}\left(1-\frac{4kT(0)}{Gm_{\mathrm{H}}\mathcal{M}_{\odot}}r\right)\frac{1}{r^2} \tag{5A.2.16}$$

where $v(r)$ is the wind velocity at radius r. Now it can easily be seen that the right-hand side of equation (5A.2.16) becomes zero when

$$r=G\mathcal{M}_{\odot}m_{\mathrm{H}}/4kT(0) \tag{5A.2.17}$$

i.e. at

$$r\simeq 2.7\times 10^{9}\ \mathrm{m} \tag{5A.2.18}$$

(about four solar radii). Thus at this point within the corona, called the critical radius, at least one of the following terms from the left-hand side of equation (5A.2.16) must also be zero for the equation to be valid:

$$v(r),\qquad \left(1-\frac{2kT(0)}{m_{\mathrm{H}}v(r)^2}\right),\qquad \frac{\mathrm{d}v(r)}{\mathrm{d}r}.$$

The complete solution of equation (5A.2.16) shows that if $v(r)$ is zero at the critical radius, then it is zero throughout the corona; i.e. there is no solar wind. Since we know this is not the case, then this possibility may be rejected. For $\mathrm{d}v(r)/\mathrm{d}r$ to be zero at the critical radius, the velocity must be at an extremum. Again, more general solutions show that for physically real situations, the required extremum must be a maximum, and the velocity less than the speed of sound. Since the solar wind is strongly supersonic near the Earth (§5A.1), this is also not acceptable for the solar case. Thus we are left with the term $\{1 - [2kT(0)]/[m_{\mathrm{H}}v(r)^2]\}$ being zero at the critical radius for the solar wind. Thus, at about four solar radii, we have

$$v = (2kT(0)/m_{\mathrm{H}})^{1/2} \simeq 160 \text{ km s}^{-1} \tag{5A.2.19}$$

for the wind speed. Now the isothermal speed of sound is given by

$$v_{\mathrm{s}} = (kT/\mu m_{\mathrm{H}})^{1/2} \tag{5A.2.20}$$

and the mean particulate mass of the coronal material is close to 0.5, since its hydrogen is almost completely ionised. Thus equation (5A.2.19) is also the expression for the speed of sound in the corona. The wind is thus subsonic for radii less than the critical radius, and supersonic further out, and it passes through the local speed of sound at the critical radius. In the more general stellar case, the critical radius is given by

$$r_{\mathrm{c}} = \frac{G\mu m_{\mathrm{H}}}{2k} \frac{\mathcal{M}_*}{T(0)_*} \tag{5A.2.21}$$

where $T(0)_*$ is the temperature at the base of the star's corona.

The complete velocity structure of a coronal wind is shown in figure 5A.2. The single transsonic solution shown there is the one which models the solar wind, and which is the most generally useful solution for coronal stellar winds. There are an infinite number of subsonic solutions, of which two only are shown on figure 5A.2. Other possible solutions are not shown as they are not physically realisable, but they include two-valued solutions, solutions supersonic everywhere, and a transsonic solution decreasing from supersonic to subsonic velocities at the critical radius.

The very high mass loss rates of $10^{-5}\mathcal{M}_\odot$ per year observed in very hot stars would require coronal temperatures of over 10^8 K if they were therm-

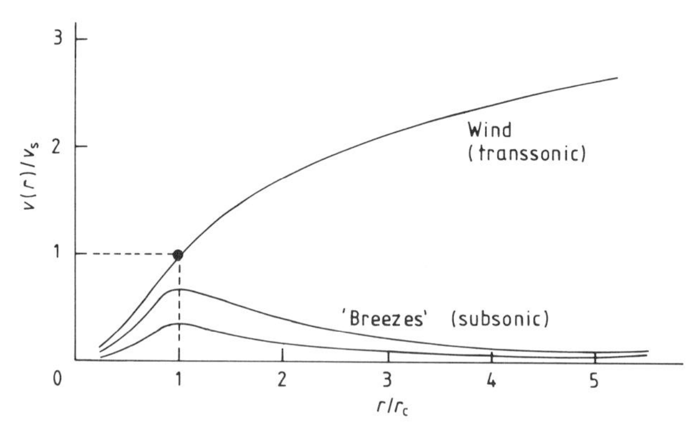

Figure 5A.2 Physically real solutions to the stellar wind velocity equation.

ally driven. Such a temperature is ruled out by other observations, such as the low levels of ionisation, and the lack of or low level of the x-ray emission. For such winds, a different source of propulsion is therefore needed. The answer appears to lie in radiation pressure.

Radiation exerts a force upon any object with which it interacts, through the change in the momentum of the photon (§2P.3). In a black body radiation field, the pressure amounts to

$$P_R = \tfrac{1}{3} aT^4 . \qquad (5A.2.22)$$

The momentum of an individual photon is

$$p = h\nu/c \qquad (5A.2.23)$$

so that if the radiation field is not isotropic, there will be a net force acting on any particle within its influence. More specifically, at distances of more than a few stellar radii from a star, the radiation field becomes essentially a radially directed outward flow. Any particle near a star will thus absorb photons from only a very small solid angle directed towards the star. In radiative equilibrium, however, it will then re-emit the energy isotropically with no net force on the particle. Hence, recognising that $h\nu$ in equation (5A.2.23) is simply the photon's energy, we have the net force acting upon a perfectly absorbing particle of radius r in a radially directed radiation field:

$$F_R = (\mathscr{F}/c)\pi r^2 \qquad (5A.2.24)$$

where $\mathscr{F}$ is the radiation flux (net energy flow within the radiation field). In the solar system, the effect of this is to sweep small dust particles away from the Sun. The radiation force on such a particle is

$$F_R = \frac{L_\odot}{4\pi D^2 c}\pi r^2 \qquad (5A.2.25)$$

where D is the particle's distance from the Sun. The gravitational force on the same particle is

$$F_G = G\mathscr{M}_\odot \tfrac{4}{3}\pi r^3 \rho / D^2 \qquad (5A.2.26)$$

where ρ is the particle's density. Thus the ratio of the radiative and gravitational forces on solar system objects is

$$\frac{F_R}{F_G} = \frac{3}{16\pi G c}\frac{L_\odot}{\mathscr{M}_\odot}(r\rho)^{-1} \qquad (5A.2.27)$$

and this is independent of the distance from the Sun. Hence for

$$F_R/F_G = 1 \qquad (5A.2.28)$$

there will be no net force on the particle, and (ignoring minor perturbations and corrections) its path will be a straight line through the solar system. Typical small meteoroids or zodiacal light particles are expected to have densities around 2500 kg m^{-3} so that the particle radius, r' (assuming complete absorption), for equality of the radiation and gravitational forces acting on it, from equation (5A.2.27), is

$$r' \simeq 2 \times 10^{-7}\ \mathrm{m}. \tag{5A.2.29}$$

Particles smaller than this will experience a net outward force and be driven very rapidly out of the solar system. Particles of this size will be lost on a time scale of a few years. Particles larger than r' will experience a reduced net gravitational force, and so a reduced escape velocity:

$$V_e' = (2G\mathcal{M}_*'/D)^{1/2} \tag{5A.2.30}$$

where $\mathcal{M}_*'$ is the effective stellar mass, and is given by

$$\mathcal{M}_*' = \mathcal{M}_* - \frac{3L_*}{16\pi Gcr\rho}. \tag{5A.2.31}$$

The net force will still follow an inverse square law, however, so that particles with velocities below the escape velocity will follow the Keplerian orbits appropriate for the effective stellar mass. The effect of this may be observed on comets' tails (figure 5A.3) where striations occur within the dust tails due to sorting according to size, as the particles follow different orbits (*3).

Particles near any star will experience a similar outward force due to radiation pressure. If enough particles exist, then their collisions with atoms in the surrounding gas, and the viscosity of that material will result in the outward force being communicated to the whole of the outer layers of the star, and hence possibly to a net flow of gas away from the star. Near hot stars, dust particles would be evaporated too quickly for such a process to lead to winds. The temperatures in the outer parts of red giants, however, may be low enough for particles to exist and for these to cause the observed stellar winds. Carbon and silicate particles are known to exist near some of these stars from their infrared emission, but it is by no means clear at the moment whether or not sufficient numbers of particles can occur close enough to the star to produce the observed mass loss rates. The reality of stellar winds produced in this way therefore remains to be demonstrated.

Radiation pressure can act directly upon the atoms or ions, rather than using dust particles as an intermediary, and this is thought to be the basis of the stellar winds in the hot stars. The process is essentially the same as that for the dust particles, except that the absorption occurs only for those wavelengths within the atom's or ion's spectrum lines. Most estimates of the number density of dust particles near stars are many orders of magnitude lower than the known number densities of the commoner atoms and ions. So the total absorption, and hence the total force acting on the material, is far larger than could be the case with dust-particle-driven winds, even though line rather than continuum absorption is involved. Thus, for

Figure 5A.3 Comet West, showing striations in the tail. (Copyright © 1977, Royal Observatory, Edinburgh.)

example, Lucy and Solomon found the radiative acceleration on triply ionised carbon atoms near an O-type supergiant from the single resonance absorption at 154.8 nm to be nearly 3 km s^{-2}. This is some 300 times larger than the gravitational acceleration towards the star at the same point!

The calculation of the total outward radiation force on material near a star is a complex problem. It obviously depends upon the strength and wavelength of individual atomic and ionic transitions, upon the relative proportions and abundances of those atoms and ions, as well as on the nature of the star's photospheric radiation field. The forces experienced by the material will also change as it starts to move away from the star. A strong absorption line in the photospheric spectrum which coincides with one in the moving material will initially cause the radiation pressure to be reduced. As the material's velocity increases, the effect of the Doppler shift will be to cause the absorbing particles to experience radiation away from the photospheric absorption line centre. Eventually they will 'see' the full photospheric continuum intensity and experience the full outward radiative force. In a similar way, an emission line from the star, or inner parts of the envelope, would lead to a high initial acceleration, which would diminish as the material's velocity increased.

The full velocity structure, in the isothermal case, of a radiatively-driven stellar wind is given by

$$v(r)\left(1-\frac{2kT(0)}{m_{\mathrm{H}}}\,v(r)^{-2}\right)\frac{\mathrm{d}v(r)}{\mathrm{d}r}=-G\mathcal{M}_*\left(1-\Gamma(r)-\frac{4kT(0)}{m_{\mathrm{H}}G\mathcal{M}_*}\,r\right)r^{-2} \tag{5A.2.32}$$

where $\Gamma(r)$ is the ratio of the radiative and gravitational accelerations at a distance r from the centre of the star. Apart from the additional term, $\Gamma(r)$, on the right-hand side, this equation is identical to equation (5A.2.16) which gave the velocity structure for an isothermal coronal wind. In a similar way, therefore, a transsonic solution for a radiatively driven wind requires the term on the right-hand side inside the brackets to fall to zero at some point:

$$1-\Gamma(r)-\frac{4kT(0)}{m_{\mathrm{H}}g\mathcal{M}_*}\,r=0\,. \tag{5A.2.33}$$

Unlike the coronal case, however, where some value of r had to exist (equation 5A.2.17) to reduce the right-hand side of equation (5A.2.16) to zero, and so allow a transsonic solution, the right-hand side of equation (5A.2.32) can never be zero when

$$\Gamma(r)>1 \tag{5A.2.34}$$

and, in this circumstance, no transsonic solution is possible. Even for lower values of $\Gamma(r)$, if the wind speed is subsonic at some radius r' and

$$\Gamma(r')>1-\frac{4kT(0)}{m_{\mathrm{H}}G\mathcal{M}_*}\,r' \tag{5A.2.35}$$

then it cannot subsequently become supersonic unless Γ is reduced in value. Since Γ is the ratio of the radiative and gravitational forces, equations (5A.2.34) and (5A.2.35) imply that *large* initial radiative forces *do not* lead to the very strong supersonic stellar winds which are observed. If a wind is already supersonic, then large values of Γ do lead to further acceleration, but if it is subsonic, then too high an applied force causes compression and turbulence, and does not lead to acceleration through the speed of sound (*4). If such a wind is to be transsonic, then the rate of acceleration must remain low until after the speed of sound has been achieved.

Two scenarios have been proposed whereby the observed supersonic winds might originate. The first is that the initial rate of radiative acceleration is low, due to the coincidence of the main absorbing lines in the stellar wind material with strong photospheric absorption lines (see previous discussion). Supersonic velocities are thereby reached before the full radiative forces start to act. The second possibility is that the initial acceleration to supersonic velocities occurs in a very thin coronal wind at the base of the observed radiatively driven wind. The x-ray luminosities of hot stars are consistent with such a proposal, but do not provide proof of its validity.

In either case the mass-loss rate for a radiatively driven wind may be found by equating the radiative momentum, L/c, to the material momentum, $\dot{\mathscr{M}}_* v_\infty$:

$$\dot{\mathscr{M}}_* \leqslant L_*/v_\infty c \qquad (5A.2.36)$$

where v_∞ is the asymptotic wind velocity, or

$$\dot{\mathscr{M}}_* \leqslant 2 \times 10^{-5} \mathscr{L}_*/v_\infty \quad (\mathscr{M}_\odot \, \mathrm{yr}^{-1}) \qquad (5A.2.37)$$

where $\mathscr{L}_*$ is the stellar luminosity in units of the luminosity of the Sun. Equations (5A.2.36) and (5A.2.37) assume only single scatterings of the photons. The Wolf–Rayet stars, with values of L_* of $10^6 \, L_\odot$ and v_∞ of $2 \times 10^3 \, \mathrm{km\,s^{-1}}$, thus have predicted mass-loss rates of $10^{-5} \, \mathscr{M}_\odot$ per year for radiatively driven winds. This is rather less than the observed rates for some of the stars. Multiple scatterings of the photons may lead to higher mass losses, but this is not yet certain. Magnetohydrodynamic (Alfvén) waves and/or transfer of angular momentum from the star to the wind by magnetic fields (cf the loss of angular momentum by protostars, §8A.1) may provide alternative or additional driving forces for the winds of such stars. But a slight question still remains at the time of writing over the Wolf–Rayet winds. The winds of other hot stars are satisfactorily explained by the single-scattering radiative winds just discussed.

PROBLEMS

5A.1 Suppose that the material in a stellar wind corotates with the star out to N stellar radii, beyond which it then escapes freely from the star. Determine the expression for the time required for a spherical, uniformly rotating star whose moment of inertia is

$$\psi \mathscr{M}_* R_*$$

to have its rotation halted, for a mass loss rate of $\dot{\mathscr{M}}$.

How long will this take for the present Sun?

Moment of inertia of a thin spherical shell $= \frac{2}{3} \mathscr{M} R^2$
$\psi = 0.04$
$\dot{\mathscr{M}} = 5 \times 10^9 \, \mathrm{kg\,s^{-1}}$
$\mathscr{M} = 2 \times 10^{30} \, \mathrm{kg}$
$N = 10$.

6P Physical Background

SUMMARY

Masers, origins of synchrotron radiation and the resulting properties of the spectrum.

Provides background material for Chapter 6A in particular.

See also transition probabilities (§3P.1), magnetic fields (§3A.8), sunspots and flares (§4A.3), H II regions (§§6A.2, 6A.3), supernova remnants (§6A.6), molecular clouds (§7A.3), diffuse galactic radio emission (§7A.4), and the formation of stars (§8A.1).

6P.1 MASERS

A maser is the microwave or radio equivalent of the more familiar laser. Technologically, though, it predates the laser, being used within low-noise radio receivers from an early stage for example. The principle of either device, or of the naturally occurring examples in the interstellar medium (§6A.2) is the same, however. It relies upon the existence of the two modes of downward transition from an excited level in an atom, ion or molecule: spontaneous and stimulated emission (§3P.1). Most such excited levels have transitions permitted under electric dipole selection rules to lower levels. In such a case, the lifetime of an electron in the excited level, due to spontaneous downward transitions, will be only about 10^{-8} seconds. More rarely, there may be no such permitted transitions. Then the only escape from the level for the electron is either via much lower probability forbidden transitions (§3P.1), or through re-excitation to an even higher level by radiation or collision. A level with only forbidden downward transitions is often called a metastable level, since in low density gases electrons may remain in it for long periods (lifetimes under spontaneous intercombination and forbidden transitions range from 10^{-3} to 10^{9} seconds).

In a low gas- and radiation-density environment, providing that there is an appropriate excitation mechanism, large fractions of an atomic, ionic, or molecular species may thus become excited to the metastable level. The exciting mechanism is often called the pump, and possible energy sources for it include absorption of radiation, collisions, and chemical reactions. It is this process which leads to the spectra of hot gaseous nebulae being dominated by forbidden lines from the ions of carbon, oxygen, nitrogen, neon, etc (Chapter 7A). In a maser we have additionally to take account of the effects of stimulated emission (hence the name: *m*icrowave *a*mplification by *s*timulated *e*mission of *r*adiation). Stimulated emission (§3P.1) is the inverse of absorption; radiation of the same frequency as that emitted by a transition induces the transition to occur in any appropriately excited atoms, ions or molecules that the photon may encounter. This second photon moves off in phase with, and in the same direction as, the incident photon. Should these two photons encounter appropriately excited particles, then two more photon emissions will be stimulated, and four in-phase, parallel photons will then move off through the medium. When there are many suitably excited particles, the process will become a chain reaction leading to the emission of a very intense, coherent, almost parallel beam of radiation. This is the familiar property of lasers, much used for lighting effects in show business etc. Spontaneous and stimulated transition probabilities vary as ν^2 and ν^{-1} respectively. Thus (equation 3P.1.16) their ratio varies as

$$\frac{\text{Spontaneous transition probability}}{\text{Stimulated transition probability}} \propto \nu^3 . \qquad (6P.1.1)$$

The importance of stimulated emission mechanisms thus decreases sharply as the frequency rises. Outside the artificial environment provided by man-made equipment, therefore, they are of significance only in the low-frequency radio and microwave regions of the spectrum.

Transitions in atoms and ions are between different energy levels for their constituent electrons. Molecules, in addition to transitions of this type, may also change their rotational and vibrational energies. These energies are quantised just like those of the electrons, and so they result in emission and absorption lines in a similar way. Thus the 1665 and 1667 MHz OH masers discussed in §6A.2 are due to rotational transitions. Vibrational transitions generally lead to much lower frequency radiation. In other respects, the processes are identical, whether they involve changes in the bound electron energies, the rotational energies, or the vibrational energies. The spontaneous transition probabilities for the transitions leading to astrophysical masers are very low indeed; the lifetime of the OH level which gives rise to the 1665 and 1667 MHz masers under spontaneous emission, for example, is over 400 years! Very high levels of population of the metastable level thus become possible even if the pump mechanism is not very efficient.

6P.2 SYNCHROTRON RADIATION

This is a special case of free–free radiation (§3P.1), usually but not necessarily from electrons. It was first observed coming from electrons in the atomic physicists' early particle accelerators (synchrotrons), hence its name. It may also be called magnetic bremsstrahlung, cyclotron, and gyro-synchrotron radiation.

In a 'normal' free–free transition (§3P.1), a free electron moving through the electric field of a nearby atom or ion reduces its relative velocity and radiates the released kinetic energy as a photon. In energy-level terms, the electron undergoes a downward transition from one level above the ionisation limit of the particle to a lower level, but which is still above the ionisation limit. The crucial factor in this process is that we have a charged particle (the electron) being accelerated; the *direction* of its velocity is altering as the electron is influenced by the particle's electric field. Any other process accelerating a charged particle can similarly lead to the emission of radiation. In particular, a magnetic field will cause charged particles moving across its direction to spiral around that direction. The particles will then radiate energy in a similar way as when they were accelerated by an atom's or ion's electric field.

The radiation from an accelerated, non-relativistic charged particle has a dipole pattern (figure 6P.1) around its acceleration vector. The emitted intensity at an angle θ to the acceleration is given by

$$I(\theta) = \frac{q^2}{16\pi^2 c^3 \varepsilon_0} a^2 \sin^2 \theta \qquad (6P.2.1)$$

where a is the acceleration and q is the charge. Consider an electron (since in practice these are the main particles producing synchrotron radiation in astrophysical situations) moving perpendicularly to a magnetic field. The moving electron constitutes an electric current in the magnetic field, and so by the Lorentz equation it experiences a force at right angles to both the field and the current (the left-hand rule of school physics):

$$\boldsymbol{F} = q\boldsymbol{V} \wedge \boldsymbol{B} \qquad (6P.2.2)$$

where $\boldsymbol{F}$ is the force, $\boldsymbol{V}$ is the electron's velocity and $\boldsymbol{B}$ is the magnetic field. The electron therefore moves in a circular path, whose plane is perpen-

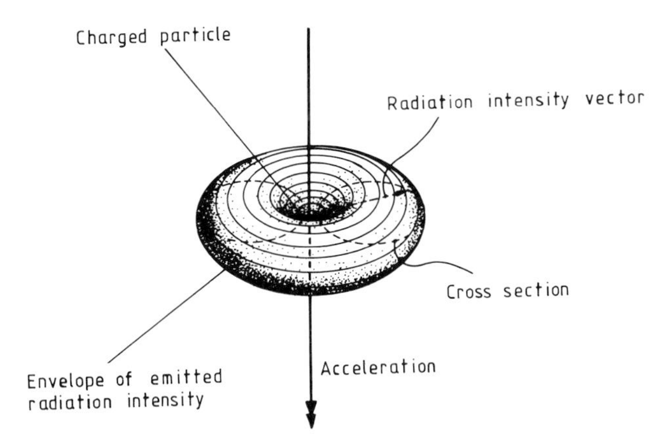

Figure 6P.1 Dipole radiation by an accelerated charged particle.

dicular to the field direction, and whose radius, the gyro radius, is given by

$$R_G = \frac{m_e v}{qB(1 - v^2/c^2)^{1/2}} \tag{6P.2.3}$$

where m_e is the electron's mass and v is the electron's velocity (or component of the velocity across the magnetic field). The acceleration is thus at right angles to the direction of motion of the electron. Now to an external observer the dipole radiation (figure 6P.1) will be affected by the aberration due to the electron's motion. Radiation emitted by the electron at an angle α' to its direction of motion will appear therefore to the external observer to be at an angle α to the direction of motion. Using the relativistic aberration formula, since only high velocities will be of significance for astrophysical situations, α is then given by

$$\alpha = \sin^{-1}\left(\frac{(1 - v^2/c^2)^{1/2}\sin\alpha'}{1 + (v/c)\cos\alpha'}\right). \tag{6P.2.4}$$

Thus the originally symmetrical emission around the acceleration vector will be concentrated into the forward direction. At 99.9% of the speed of light, for example, radiation originally emitted at 45° to the direction of the velocity will be observed only 1° away from it. Similarly the nulls, at 90° in the electron's frame of reference, will appear at 2.5° in the external observer's frame of reference. Only radiation originally emitted between 177.4° and 180° to the direction of motion by such an electron will actually appear to be emitted in a rearward direction. The observed radiation from a charged relativistic particle being accelerated by a magnetic field is thus very tightly beamed around the particle's instantaneous forward direction (figure 6P.2). The original half-power points occur at an angle of 45° to the acceleration (equation 6P.2.1). In the plane of the velocity and acceleration vectors, this corresponds to $\alpha' = 45°$, and so we find

$$\alpha_{\alpha'=45°} = \sin^{-1}\left(\frac{(1 - v^2/c^2)^{1/2}}{\sqrt{2} + (v/c)}\right) \tag{6P.2.5}$$

or, since $v/c \simeq 1$ and α is small,

$$\alpha_{\alpha'=45°} \simeq 0.4(1 - v^2/c^2)^{1/2} \text{ (rad)} \tag{6P.2.6}$$

The nulls similarly appear at

$$\alpha_{\alpha'=90°} \simeq (1 - v^2/c^2)^{1/2} \text{ (rad)}. \tag{6P.2.7}$$

Thus, to a good degree of approximation, the emitted radiation appears in a beam whose half width is $[1 - (v/c)^2]^{1/2}$ radians to an external observer. The loss of energy causes the electron to spiral inwards to tighter and tighter orbits. The velocity and acceleration vectors are thus not truly at right angles to each other. The deviation from 90° is usually very small, however, and does not significantly affect the above analysis.

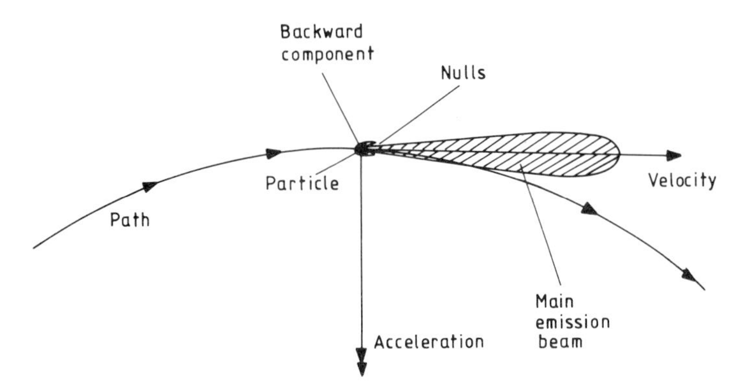

Figure 6P.2 Cross section through the synchrotron emission pattern of a charged particle accelerated in a magnetic field.

The radiation in the rest frame of the electron has an emission spectrum given by

$$I(\nu) \propto \nu^2 a(\nu)^2 \tag{6P.2.8}$$

where $a(\nu)$ is the component of the acceleration of frequency ν. The observed spectrum of the radiation from a charged relativistic particle in a magnetic field, however, is very different. An observer in the plane of an electron's motion will only be able to see its radiation while the main emission lobe is along his line of sight. This will occur for about $2(1 - v^2/c^2)^{1/2}/2\pi$ of the orbital period. The orbital period is given by the gyro frequency:

$$\nu_G = \frac{qB(1 - v^2/c^2)^{1/2}}{2\pi m_e} \tag{6P.2.9}$$

so that radiation may potentially be seen for a time interval, Δt,

$$\Delta t = \frac{2m_e}{qB} = \frac{2}{{}_c\omega_G} \tag{6P.2.10}$$

where ${}_c\omega_G$ is the non-relativistic circular gyro frequency. However, relativistic particles will be moving only a little slower than their emitted radiation. The electron will thus be close behind the first visible photons to be emitted as it emits the last visible ones. The observed pulse duration will thus be highly compressed in comparison with the prediction of equation (6P.2.10) (*1). A photon emitted at the leading edge of the main emission beam will have travelled a distance $c\Delta t$ in the time that the electron travels $v\Delta t$. Photons from the trailing edge of the main emission beam are thus emitted from a point $v\Delta t$ closer to the observer than those at the leading edge, and are a distance $(c - v)\Delta t$ behind them. The photons are hence received by the external observer only over a time interval

$$\Delta t_o = (c - v)\Delta t/c \tag{6P.2.11}$$

$$= (1 - v/c)\Delta t \tag{6P.2.12}$$

$$\simeq \tfrac{1}{2}(1 - v^2/c^2)\,\Delta t \tag{6P.2.13}$$

(using the approximation $1 + v/c \simeq 2$). Thus we have

$$\Delta t_o = \frac{m_e}{qB}\left(1 - \frac{v^2}{c^2}\right) = \frac{1 - v^2/c^2}{{}_c\omega_G}. \tag{6P.2.14}$$

The reciprocal of this interval gives the maximum Fourier component of the received radiation:

$$\nu_{max} = \frac{qB}{m_e(1 - v^2/c^2)} = \frac{{}_c\omega_G}{(1 - v^2/c^2)} = \frac{2\pi\ {}_c\nu_G}{(1 - v^2/c^2)}. \tag{6P.2.15}$$

Thus the observed synchrotron spectrum consists of high harmonics of the electron's non-relativistic gyro frequency, up to the limit ν_{max}. Overlap of the emission lines results in an apparent continuum spectrum, and a higher-precision analysis gives for the envelope of this continuum

$$I(\nu) \simeq 5 \times 10^{-25}\left(\frac{\nu}{\nu_c}\right)^{1/3} B_p\ (\mathrm{W\,Hz^{-1}}) \tag{6P.2.16}$$

for $\nu \ll \nu_c$, and

$$I(\nu) \simeq 3 \times 10^{-25}\left(\frac{\nu}{\nu_c}\right)^{1/2} \exp(-\nu/\nu_c) B_p\ (\mathrm{W\,Hz^{-1}}) \tag{6P.2.17}$$

for $\nu \gg \nu_c$ where B_p is the component of the magnetic field perpendicular to the electron's velocity vector. ν_c is a critical frequency which marks the approximate limit of significant emission, and

$$\nu_c = \frac{3}{2(1 - v^2/c^2)} {}_c\nu_G \,. \tag{6P.2.18}$$

The maximum emission occurs at about 0.29 ν_c, and the half width of the emission is about 1.5 ν_c (figure 6P.3).

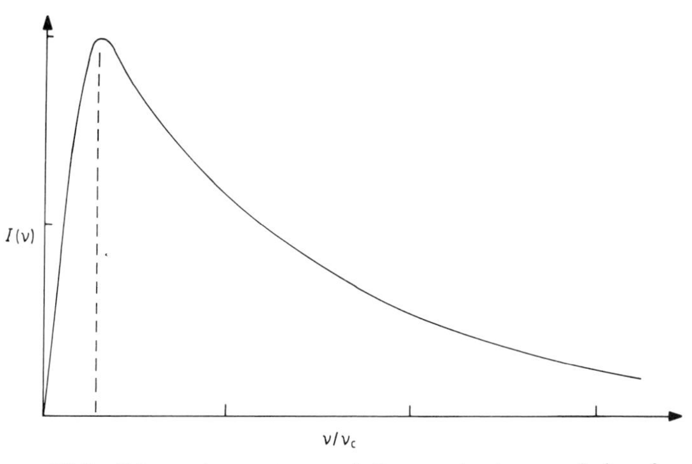

Figure 6P.3 Schematic spectrum of the synchrotron emission from a single electron.

Of course, in practice synchrotron radiation comes from electrons with varying pitch angles to the magnetic field, and with a range of velocities, and quite probably the magnetic field will vary in its strength and direction over the emission region. If we take an exponential distribution for the electron energies (a commonly occurring situation in astrophysics):

$$N_e(v) \propto E^{-\gamma} \tag{6P.2.19}$$

where E is the electron's kinetic energy, then this is much less sharply peaked than the emission from an individual electron (figure 6P.3), and we may usefully approximate an individual electron's emission as though its total energy were emitted over an interval $\Delta\nu_c$ centred on ν_c. The total emission averaged over all pitch angles for a single electron is proportional to $[1 - (v/c)^2]^{-1}$, and for relativistic particles we have

$$E = \frac{\frac{1}{2} m_0 c^2}{(1 - v^2/c^2)^{1/2}} \,. \tag{6P.2.20}$$

Thus the flux from an electron of energy $E, \mathscr{F}(E)$, is given by

$$\mathscr{F}(E) \propto E^2 \tag{6P.2.21}$$

spread over the interval $\Delta\nu_c$. From equation (6P.2.18) we have

$$\nu_c \propto (1 - v^2/c^2)^{-1} \tag{6P.2.22}$$

$$\propto E^2 \tag{6P.2.23}$$

(from equation 6P.2.20), and so

$$\Delta\nu_c \propto E\Delta E. \tag{6P.2.24}$$

The total flux at a given ν_c is thus

$$\mathscr{F}(\nu_c) = \frac{\mathscr{F}(E)}{\Delta\nu_c} N_e(v)\,\Delta E \tag{6P.2.25}$$

$$\propto \frac{E^2}{E\Delta E} E^{-\gamma}\,\Delta E \tag{6P.2.26}$$

$$\propto E^{1-\gamma} \tag{6P.2.27}$$

or, from equation (6P.2.23),

$$\mathscr{F}(\nu) \propto \nu^{(1-\gamma)/2}. \tag{6P.2.28}$$

Thus the spectrum of a real synchrotron radiation source may be expected to follow a power law. The index in the form

$$\mathscr{F}(\nu) \propto \nu^{-\alpha} \tag{6P.2.29}$$

is known as the spectral index, and we may easily see that it is related to the index of the electron energy distribution by

$$\gamma = 1 + 2\alpha. \tag{6P.2.30}$$

The value of the spectral index for astrophysical sources ranges between about 0.2 and 2.0.

The other major characteristic of synchrotron radiation which distinguishes it from thermal emission is its polarisation. This is generally elliptical polarisation, becoming closer to linear polarisation as the electron's energy increases.

Considering a single electron, its radiation is polarised with the electric vector parallel to the acceleration vector. Thus for electrons moving perpendicular to a magnetic field, the observed projection of the acceleration vector will be perpendicular to the magnetic field, and the observed radiation will be linearly polarised (figure 6P.4). When the electron has a component of its velocity along the magnetic field, the emitted beam of radiation sweeps around a cone centred on the magnetic field direction (figure 6P.5). As can be seen from the diagram, the plane of the linear polarisation rotates

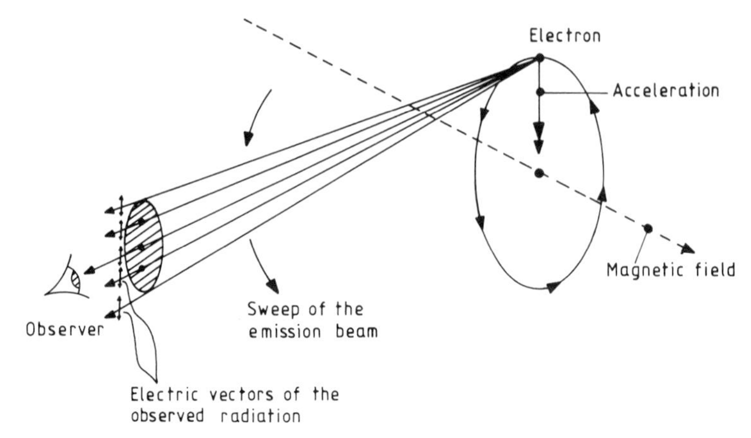

Figure 6P.4 Polarisation of synchrotron radiation perpendicular to the magnetic field.

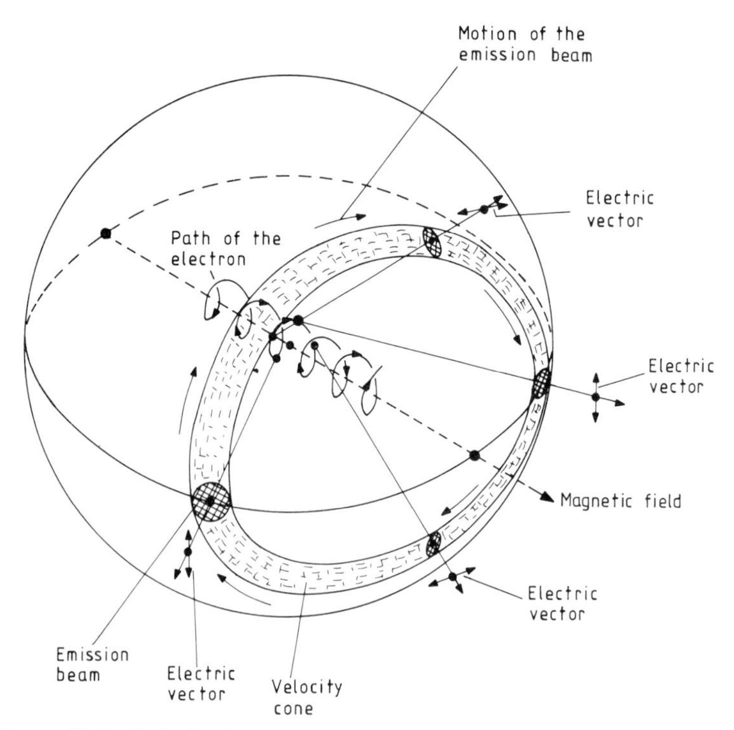

Figure 6P.5 Polarisation of synchrotron radiation from an electron with a non-zero velocity component along the magnetic field.

in space as the emission beam sweeps around this velocity cone. A single observer sees the radiation only while his line of sight intersects the emission beam just as before, but now during that interval the plane of the polarisation rotates through a small angle. In the example drawn in figure 6P.5, the rotation for the observer is clockwise. Thus there is a small degree of circular polarisation to be added to the linear polarisation; i.e. the total radiation is elliptically polarised. However, the observer will also be able to see radiation from electrons whose velocity vectors do not pass directly through the line of sight, providing that they are within $[1-(v/c)^2]^{1/2}$ radians of the line of sight. Across the emission beam, the polarisation varies as shown in figure 6P.6, according as the observed direction of the projected acceleration of the electron changes. The direction of rotation of the plane of polarisation will thus be modified by its variation across the emission beam. At high velocities, however, the effect of beams to one side of the line of sight will be cancelled by the beams on the other side, leaving just the basic effect. At low velocities such cancellation may not occur completely, leaving significant degrees of elliptical polarisation in the observed radiation.

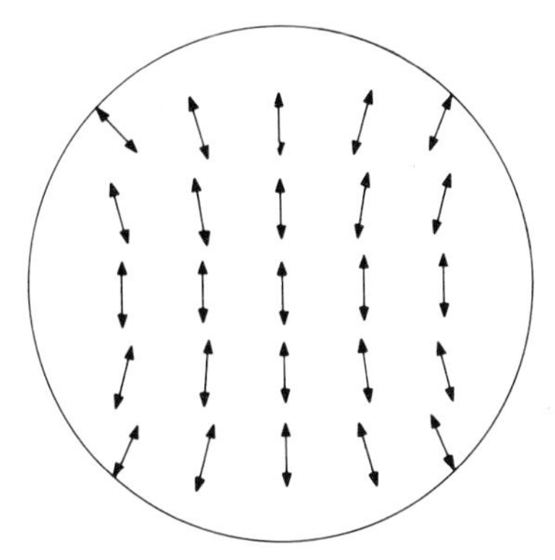

Figure 6P.6 Schematic variation in the linear polarisation across the synchrotron radiation emission beam.

For relativistic electrons, the emitted power falls rapidly as the angle to the perpendicular to the field direction increases. The observed radiation therefore originates from angles to the magnetic field near to 90°. The rotation of the plane of polarisation by the sweeping of the emission beam around the velocity cone is then very small, and so the radiation is largely linearly polarised. For electrons with the velocity distribution given by equation (6P.2.19), and in a uniform magnetic field, the observed degree of linear polarisation, π_L, is given by

$$\pi_L = \frac{\gamma + 1}{\gamma + \frac{7}{3}} \tag{6P.2.31}$$

where π_L is the usually observed parameter in astronomical polarimetry (*2). For the observed range in the values of γ from about 1.5 to 5.0 for astrophysical synchrotron radiation sources, the observed degrees of linear polarisation may thus be expected to range from 65% to 80%.

6A H II Regions and Related Objects

SUMMARY

H II regions, formation of an ionisation cavity, limitations on size, Strömgren sphere, properties of H II regions, interstellar masers, formation and evolution of H II regions, use of H II regions for determining distances, planetary nebulae and their central stars, supernova remnants.

See also stellar compositions (§§1A.2, 3A.6), stellar winds (§5A.2), masers (§6P.1), synchrotron radiation (§6P.2), interstellar medium (§7A.1), formation of interstellar dust grains (§7A.2), energy balance in the interstellar gas (§7A.3), origin of cosmic rays (§7A.4), star formation (§8A.2), and the end points of stellar evolution (§8A.2).

INTRODUCTION

In one convention, an ionised hydrogen atom is given the symbol H II (neutral hydrogen being H I; see §3P.1). A region of the interstellar material within which the hydrogen is predominantly ionised is therefore known as an H II region. Other names for the objects include Strömgren spheres, H^+ regions, and ionisation cavities.

6A.1 H II REGIONS: THEORY

A hydrogen atom with its electron in the ground state requires 2.18×10^{-18} J (13.6 eV) of energy to be added to it in order to cause it to lose its electron. The absorption of radiation with a wavelength of 91.2 nm or less will provide the required energy. H II regions are thus primarily associated with very hot stars which have significant ultraviolet fluxes.

Ionisation can occur through other processes. Thus, if the electron is in an excited level, then longer wave radiation than 91.2 nm will be able to ionise it. Collisions with other particles can also excite or ionise the electron. In practice, such effects are quite negligible for actual H II regions.

Within an H II region, therefore, hydrogen is ionised by the stellar ultraviolet photons. The protons and electrons left after this has occurred will recombine, emitting Balmer, Paschen, Pfund, etc lines and continua as well as Lyman lines, as they return to the ground state (Lyman continuum photons emitted by direct recombination to the ground state have wavelengths shorter than 91.2 nm, and so do not represent an effective absorption of the stellar photons). These emissions then render the gas visible to optical astronomers. Close to a hot star the ultraviolet flux is so large that any protons and electrons recombining will be almost instantaneously reionised. The central parts of the region are therefore almost totally ionised. The nebula may be radiation- or matter-bounded. In the former case, the gas cloud containing the H II region is sufficiently dense and extensive that the number of recombinations eventually balances the number of ionisations. An alternative description is to say that the gas cloud is optically thick to Lyman continuum radiation. There is a comparatively sharp boundary known as the transition zone or ionisation front, and over this zone the hydrogen ionisation decreases from near 100% to near zero. The H II region is then embedded in a much larger neutral hydrogen cloud (H I region). In a matter-bounded H II region, the whole gas cloud is ionised, and its outer limits are simply those of the gas cloud itself. In this situation the cloud is optically thin to the Lyman continuum photons, and some proportion of them escape entirely from the region without being absorbed.

The visible shape of a matter-bounded H II region is just that of the gas cloud. It can therefore be very complex and may be quite asymmetrical with respect to the exciting star. A radiation-bounded H II region in a uniform gas cloud of pure hydrogen would be a sphere centred on the star (hence the alternative name of Strömgren sphere). Most of the time, the shapes of even these H II regions are much more complex, due to density variations in the gas cloud and the presence of dusty obscuring matter which shadows parts of the cloud (figure 6A.1). Some of the most beautiful astronomical

---

Figure 6A.1 H II regions. Top: the Rosette nebula in Monoceros (NGC 2237–46). Note the appearance of small, dark Bok globules outlined against the bright emission regions (see §8A.1). Right: the Lagoon nebula in Sagittarius (NGC 6523, M8). (Both photographs from the Hale Observatories, USA.)

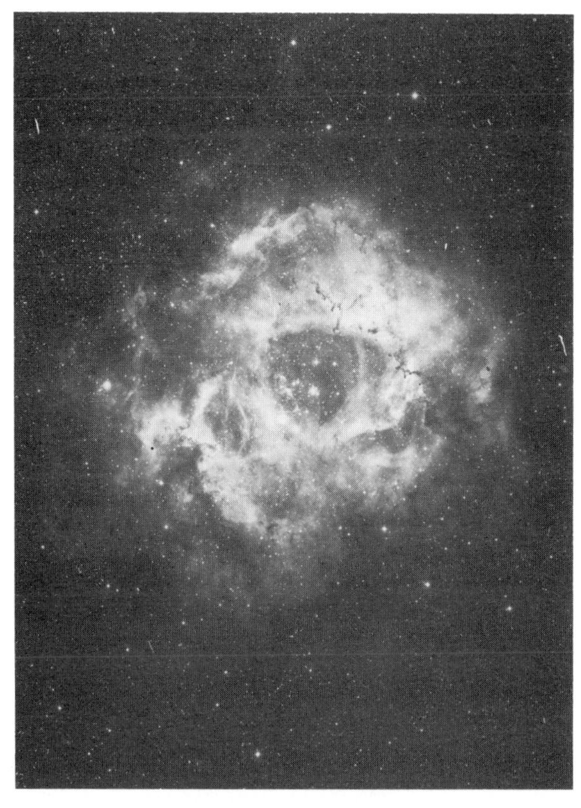

objects of all result from this interplay of matter and radiation, as may be seen from figure 6A.1.

The sizes of radiation-bounded H II regions depend upon the hydrogen density of the gas cloud, and on the number of ionising photons from the star. Its volume is just that at which the total number of recombinations to levels other than the ground state is equal to the number of ionising photons from the star. The recombination rate, Δ_n, to any level n for a hydrogen atom may be written as

$$\Delta_n = \alpha_n N_e N_p \tag{6A.1.1}$$

where α_n is the recombination rate coefficient for level n, and N_e and N_p are the electron and proton number densities. Assuming a uniform, pure hydrogen nebula, the H II region will be spherical, of radius R_S (generally called the Strömgren radius). The total number of recombinations to levels above the ground state will thus be

$$\tfrac{4}{3}\pi R_S^3 N_e N_p \sum_{n=2}^{\infty} \alpha_n \tag{6A.1.2}$$

where we assume N_e and N_p to be constant over the whole H II region, an assumption that will be justified later. This must be equal to the total stellar photon flux ν_L of photons below the Lyman limit (*1). The Strömgren radius is thus

$$R_S = \left(\frac{3\nu_L}{4\pi N_e N_p \sum_{n=2}^{\infty} \alpha_n}\right)^{1/3} \tag{6A.1.3}$$

Now Kaplan and Pikelner (*The Interstellar Medium*, Harvard University Press, 1970) have shown that

$$\sum_{n=2}^{\infty} \alpha_n \simeq 6.5 \times 10^{-15} T_e^{-0.85} \ (\mathrm{m^3\,s^{-1}}) \tag{6A.1.4}$$

where T_e is the electron temperature, while the stellar photon flux is given by

$$\nu_L = 8\pi^2 c R_*^2 \int_0^{91.2\times 10^{-9}} \frac{d\lambda}{\lambda^4 [\exp(hc/\lambda k T_*) - 1]} \tag{6A.1.5}$$

where R_* is the stellar radius and T_* is the stellar effective temperature. This has the approximate form:

$$\nu_L \simeq 5 \times 10^{42} \exp(3.2 \times 10^{-4} T_*) \ (\text{photons/s}). \tag{6A.1.6}$$

Finally, for pure hydrogen

$$N_e = N_p \tag{6A.1.7}$$

or, writing x for the fractional degree of ionisation (which will have values close to 1 for H II regions and to 0 for H I regions),

$$N_e = xN_H \tag{6A.1.8}$$

where N_H is the total number of density of neutral and ionised hydrogen, so that we obtain

$$R_S \simeq 1.2 \times 10^{19} x^{-2/3} N_H^{-2/3} T_e^{0.3} \exp(1.07 \times 10^{-4} T_*) \text{ (m)}. \tag{6A.1.9}$$

Now, as discussed later, [O III] emission limits the electron temperature of the H II region to about 10^4 K, and with $x = 1$, we have

$$R_S \simeq 6200\, N_H^{-2/3} \exp(1.07 \times 10^{-4} T_*) \text{ (pc)}. \tag{6A.1.10}$$

The hydrogen number densities of actual H II regions vary from 10^7 to $3 \times 10^9\ \text{m}^{-3}$, so that an O5 star whose effective temperature is about 50 000 K would have H II regions with radii between 28 and 0.6 pc. For a B0 star, with an effective temperature of 32 000 K, the range would be from 4 to 0.09 pc.

The transition zone between the H II and H I regions is thin in comparison with the size of the H II region. The photoionisation cross section for neutral ground-state hydrogen is about $6 \times 10^{-22}\ \text{m}^2$ at the Lyman limit, falling to $1.2 \times 10^{-22}\ \text{m}^2$ at 50 nm, and to $7 \times 10^{-25}\ \text{m}^2$ at 10 nm. For a density of 10^7 atoms per cubic metre, unit optical depths are reached over linear distances of 0.005, 0.03 and 5 pc. For densities of $3 \times 10^9\ \text{m}^{-3}$, the distances are a factor of 300 smaller still. If the material within the H II region were totally ionised, then these distances would be a good guide to the thickness of the transition zone. In fact, the HII region contains small proportions of neutral hydrogen, and these atoms provide the main source of opacity to the Lyman continuum photons. For equilibrium conditions, the number of ionisations must balance the number of recombinations throughout the whole of the region, and so

$$\sum_{n=2}^{\infty} \alpha_n N_H^2 x(r)^2 = \frac{\nu_L}{4\pi r^2} \exp[\overline{\tau(r)}]\, \bar{\sigma}\, [1 - x(r)]\, N_H \tag{6A.1.11}$$

where $x(r)$ is the fractional ionisation at distance r from the centre of the star, $\bar{\sigma}$ is the average ground-state photoionisation cross section for hydrogen, weighted by the number of available photons at each wavelength. Reasonably representative values are $2 \times 10^{-22}\ \text{m}^2$ for O5 stars, and $10^{-22}\ \text{m}^2$ for B0 stars. $\overline{\tau(r)}$ is the average optical depth for ionising radiation from the star to radius r. Since the star is generally much smaller than the H II region, to a good degree of approximation we may write

$$\mathrm{d}\bar{\tau} = (1 - x) N_H \bar{\sigma}\, \mathrm{d}r \tag{6A.1.12}$$

and

$$\overline{\tau(r)} \simeq \int_0^r (1 - x) N_H \bar{\sigma}\, \mathrm{d}r. \tag{6A.1.13}$$

Thus we find, combining equations (6A.1.11) and (6A.1.12),

$$r^2\,\mathrm{d}r = \frac{\nu_\mathrm{L}}{4\pi x(r)^2 N_\mathrm{H}^2 \sum_{n=2}^{\infty} \alpha_n} \exp[-\overline{\tau(r)}]\,\mathrm{d}\bar{\tau} \qquad (6\mathrm{A}.1.14)$$

or, from equations (6A.1.3), (6A.1.7) and (6A.1.8), and assuming $x(r)$ and N_H to be constant,

$$\int_{R_*}^{r} r^2\mathrm{d}r = \int_0^{\tau} \tfrac{1}{3} R_\mathrm{S}^3 \exp[-\overline{\tau(r)}]\,\mathrm{d}\bar{\tau}. \qquad (6\mathrm{A}.1.15)$$

For r large compared with the stellar radius, we then find

$$r = \{1 - \exp[-\overline{\tau(r)}]\}^{1/3} R_\mathrm{S} \qquad (6\mathrm{A}.1.16)$$

or

$$\overline{\tau(r)} = -\log_e[1 - (r/R_\mathrm{S})^3] \qquad (6\mathrm{A}.1.17)$$

and this varies as shown in figure 6A.2. As would be expected, the optical depth approaches infinity as the Strömgren radius is neared.

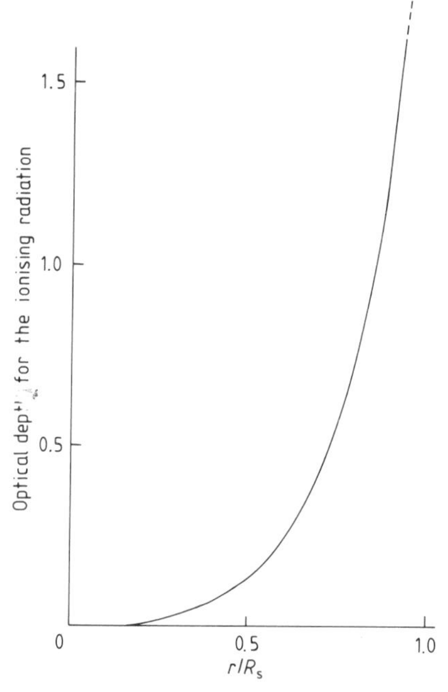

Figure 6A.2 Optical depth for the ionising radiation in a pure hydrogen H II region.

Equations (6A.1.11) and (6A.1.17) may be combined to give

$$1 - x(r) = \frac{4\pi r^2 \sum_{n=2}^{\infty} \alpha_n x(r)^2 N_{\mathrm{H}}^2}{\nu_{\mathrm{L}} \bar{\sigma} N_{\mathrm{H}} [1 - (r/R_{\mathrm{S}})^3]}. \tag{6A.1.18}$$

Using equation (6A.1.3), we then find

$$x(r) = 1 - \frac{3r^2}{R_{\mathrm{S}}^3 \bar{\sigma} N_{\mathrm{H}} [1 - (r/R_{\mathrm{S}})^3]} \tag{6A.1.19}$$

which is valid for values of $x(r)$ close to unity. A more convenient form of the relation is

$$x(r) = 1 - \frac{3}{R_{\mathrm{S}} \bar{\sigma} N_{\mathrm{H}}} \frac{R_{\mathrm{S}}/r}{[(R_{\mathrm{S}}/r)^3 - 1]}. \tag{6A.1.20}$$

Thus the percentage of ionised hydrogen atoms near an O5 star varies as

$$x(r) = 100 - 0.18 \frac{R_{\mathrm{S}}/r}{[(R_{\mathrm{S}}/r)^3 - 1]} \ \% \tag{6A.1.21}$$

for a total hydrogen number density of $10^7\ \mathrm{m}^{-3}$, and as

$$x(r) = 100 - 0.028 \frac{R_{\mathrm{S}}/r}{[(R_{\mathrm{S}}/r)^3 - 1]} \ \% \tag{6A.1.22}$$

for a total hydrogen number density of $3 \times 10^9\ \mathrm{m}^{-3}$. For a B0 star, the corresponding formulae are

$$x(r) = 100 - 2.5 \ \frac{R_{\mathrm{S}}/r}{[(R_{\mathrm{S}}/r)^3 - 1]} \ \% \tag{6A.1.23}$$

and

$$x(r) = 100 - 0.37 \frac{R_{\mathrm{S}}/r}{[(R_{\mathrm{S}}/r)^3 - 1]} \ \% \tag{6A.1.24}$$

and these are shown in figures 6A.3 and 6A.4. It can now be seen that the assumption of almost complete ionisation of hydrogen throughout the H II region that has been used several times is well justified. The value of $x(r)$ is thus nearly constant at near 1.0. Taking the transition zone to lie between values of $x(r)$ of 0.95 and 0.05 (95% and 5% levels of hydrogen ionisation), we find the edge of an H II region around an O5 star to have a thickness of 0.1% and 1% of the Strömgren radius for particle densities of $3 \times 10^9\ \mathrm{m}^{-3}$ and $10^7\ \mathrm{m}^{-3}$ respectively. The corresponding linear thicknesses are 6×10^{-4} pc (2×10^{13} m, 100 AU), and 0.28 pc (8×10^{15} m). For a B0 star, the corresponding figures are 2% and 14%, or 0.002 pc and 0.6 pc respectively. These are comparable with the unit optical depth distances in neutral hydrogen calculated above, but are probably overestimates of the actual thicknesses of the transition zones. This arises through the use of

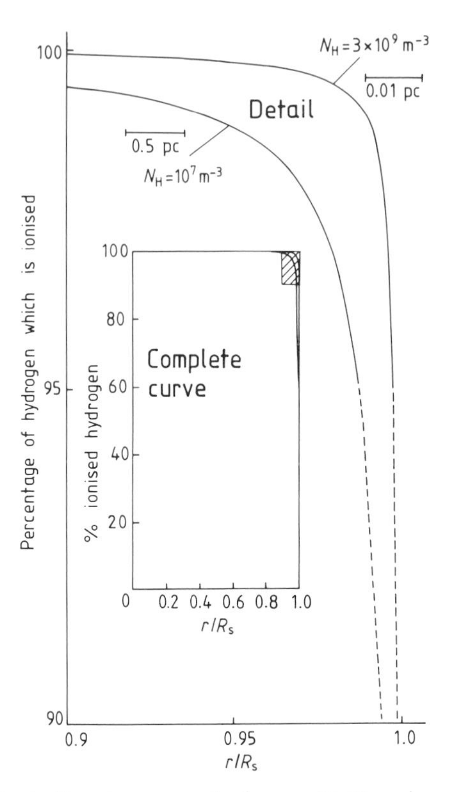

Figure 6A.3 Ionisation in a pure hydrogen H II region around an O5 star.

equations (6A.1.21) to (6A.1.24) which become inaccurate in the transition zone through being based upon the assumption of a value of 1.0 for $x(r)$ (the curves are shown dotted on figures 6A.3 and 6A.4 for that reason). The breakdown of the assumption, however, arises through increasing number of neutral atoms. The opacity will thus increase more rapidly than predicted across the transition zone, and it will be a much sharper region than the above figures suggest.

This analysis for a pure hydrogen nebula is complicated in reality by the presence of other atoms, and in one respect simplified by their presence. The latter effect has already been applied. It is that the kinetic temperature of the atoms, ions and electrons within the H II region is almost constant at about 10 000 K. The spectrum of an H II region (figure 6A.5) is composed largely of emission lines. Some of these are due to hydrogen, and arise from the recombination of protons and electrons. These are not the most prominent lines, however. Most other lines in the spectrum and all the strong

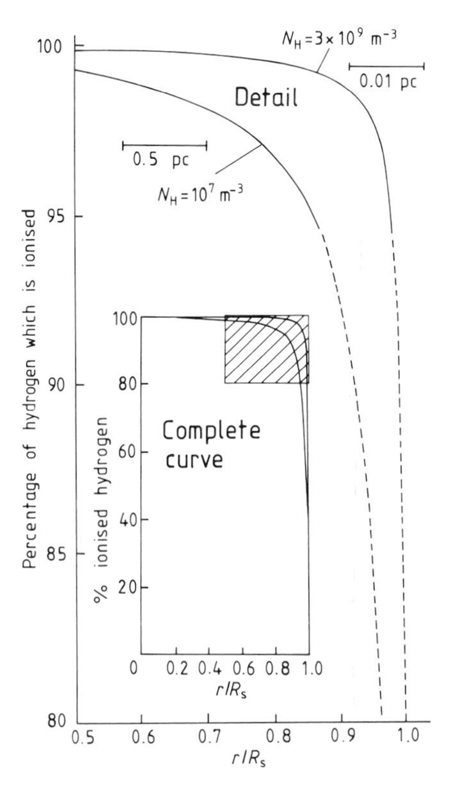

Figure 6A.4 Ionisation in a pure hydrogen H II region around a B0 star.

lines are due to forbidden transitions (§3P.1), mainly from ionised oxygen, neon, sulphur and iron. The collision of these ions with the high-energy electrons in the H II region excites their electrons. These then tend to accumulate in metastable levels, which can thereby become very highly populated. The only transitions down from such metastable levels are forbidden under electric dipole selection rules, but permitted under other lower probability interactions. The large populations of the metastable states lead to intense emission in these normally very faint lines. In particular, the [O III] lines at 495.9 and 500.7 nm, and the [O II] lines at 372.6 and 372.9 nm (figure 6A.5) can become sufficiently strong to act as a thermostat for the material. The electrons' kinetic energy, and through collisions that of the atoms and ions as well, is converted into photons which can be lost from the nebula, thereby limiting the temperature to a maximum of about 10 000 K and a mean value of about 6700 K. This constancy of the electrons' kinetic temperature was one of the assumptions underlying the calculation of the Strömgren radius, and is thus of considerable importance in enabling H II regions to be understood.

The other effect of the presence of additional species of atoms is that each can form its own version of the Strömgren sphere. For a gas composed purely of a single element, the analysis is identical to that for hydrogen except for the different ionisation limit, and perhaps the need to take accound of more than one level of ionisation. In a mixed situation, however, the nebula may be optically thick for some atoms or ions, and optically thin for others, there is competition for the ionising photons, and the photons emitted by the recombination of some ions and electrons may still be sufficiently energetic to ionise other species of atom or ion. Only helium, however, is sufficiently abundant for its ionisation to be able to affect the electron density of the H II region. It is present in the gas cloud to the extent of about 10% by number, and its ionisation potentials are 24.6 and 54.4 eV. These energies require photons of wavelengths shorter than 50.4 and 22.8 nm respectively in order to ionise the particles. Even O5 stars emit only about 0.02% of their radiation shorter than the second limit (10^{46} photons per second) and so the contribution from doubly-ionised helium is unimportant. The neutral helium, even with an ionisation potential twice that of hydrogen, can produce significant effects. For temperatures of the exciting star above 40 000 K there are sufficient photons between 91.2 and 50.4 nm to ionise the hydrogen completely. The bulk of the photons below 50.4 nm are therefore available to ionise the helium. The H II and He II regions will thus be approximately coincident. For temperatures of the central star below 40 000 K, the hydrogen and helium will be competing for the shortwave photons. These are therefore likely to be absorbed before all the photons between 91.2 and 50.4 nm are absorbed by the hydrogen acting alone. The He II region in this circumstance will be smaller than the H II region, and the ionisation cavity will be stratified with somewhat greater electron number densities in its central regions compared with the outer portions.

Atoms other than hydrogen and helium are too low in abundance to affect the electron number density by their ionisation to any significant extent. However, as we have seen, much of the optical radiation may come from these heavier atoms, and so their ionisation structures can be observed. The ionisation potentials of neutral oxygen and nitrogen from their ground states are 13.6 and 11.3 eV respectively. These are close to the 13.6 eV for hydrogen and so the O II and N II emission occurs out to the limits of the H II region. Their second ionisation potentials at 35.1 and 29.6 eV, and the first for neon at 21.6 eV, are closer to helium's first ionisation stage. These ions therefore tend to occur only in the inner parts of the H II region, unless the star is very hot, when they may extend throughout its volume. Some observational evidence for this stratification comes from taking photographs in the light of different ions, when different shapes and

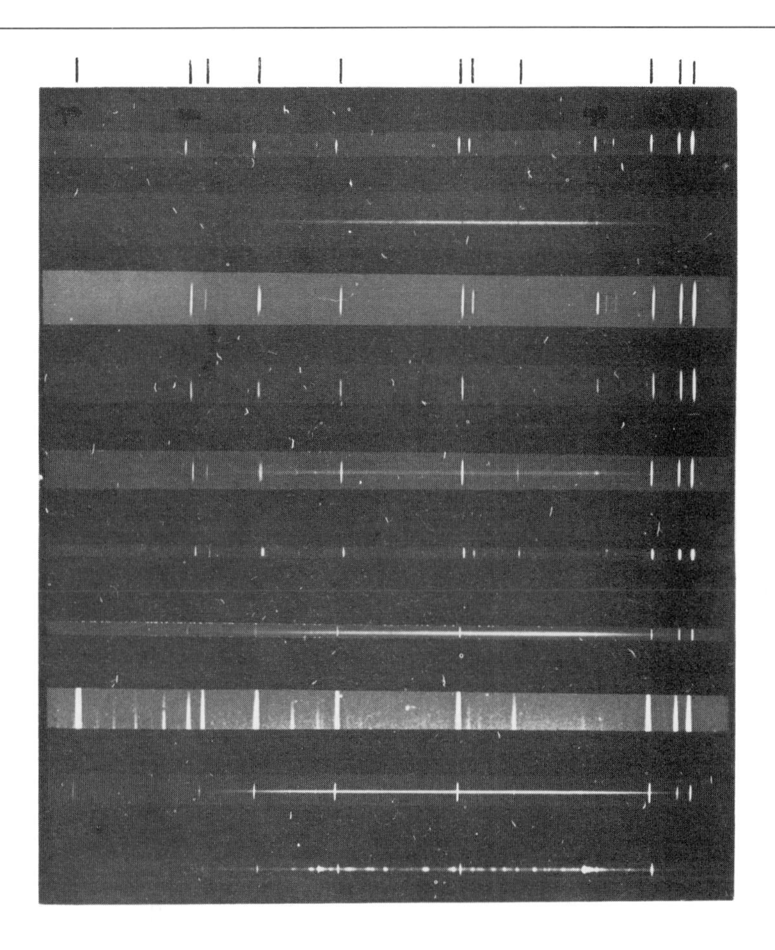

Figure 6A.5 Spectra of planetary nebulae (P) and H II regions (H). From top to bottom: NGC 7027 (P); NGC 2392 (P); NGC 7662 (P); NGC 7009 (Saturn nebula) (P); NGC 6543 (P); NGC 6572 (P); NGC 6826 (P); M8, NGC 1976–7 (H) (Orion nebula); IC 418 (P); BD + 30° 3639 (P). (Reproduced by permission of the Royal Astronomical Society.)

sizes are found for the nebula which are roughly in line with the above expectations.

H II regions for O5 and B0 stars have been used as examples in the above discussion. These are just about the observed limits for the occurrence of H II regions. The hottest known stars at about 52 000 K are of class O4, but these are very rare indeed. Stars cooler than B0 hardly possess a directly observable H II region at all. A B1 star, for example, emits only 0.5% of the ionising photons of a B0 star. In a gas with a hydrogen number density of 10^7 m^{-3}, its H II region would have a maximum radius of only 0.6 pc; more typically the radius would be about 0.1 pc. Given the typical distance of H II regions of 5000 pc, such a star would be surrounded by an H II region 4″ across, scarcely distinguishable from the star's seeing disc on a photograph.

6A.2 H II REGIONS: PHYSICAL PROPERTIES

Some of the physical properties of H II regions will have become apparent already. In particular, their physical sizes have been discussed at some length. The theory is in good agreement with the observations, to the point where the angular size of the H II region and the spectral type of the exciting star can be used to determine their distance (§6A.4, Appendix I). The shapes of H II regions are sometimes approximately spherical, as predicted. Often, however, they differ markedly from such a shape and are shot through with light and dark lanes and nodules to produce some of the most beautiful sights in astronomy (figure 6A.1, for example). These shapes simply mirror the contortions and density variations of the original gas and dust cloud, together with the distribution of the exciting stars within the cloud when there are more than one of these. The reddish hue of some of the nebulae arises from the Balmer Hα emission following recombination of the protons and electrons to the $n = 3$ level. The whitish regions are areas where the forbidden oxygen emission is strong, while the bluish regions sometimes observed are due to light scattered from the central star.

The temperature of H II regions has also been mentioned in §6A.1, and the reasons for its limitation to less than 10 000 K discussed. The actual temperature may be found in several ways. Forbidden-line intensity ratios are sensitive to temperature when the originating levels for the lines have markedly different energies. The intensities depend upon the level populations, which in turn depend upon the temperature (in TE and LTE, Boltzmann's equation gives the relationship; the situation here is far from either situation but there will still be some dependence of level population on excitation energy and temperature). Amongst the prominent lines in H II region spectra, those from O III and N II are most suitable for this purpose.

Thus, for example, transitions from the 2.7 and 5.4 eV levels in O III produce lines at 436.3, 495.9 and 500.7 nm, the first originating from the higher level and the other two from the lower level. We then find

$$\frac{I(495.9) + I(500.7)}{I(436.3)} = 8.32 \exp(3.92 \times 10^4 T^{-1}) \left(1 + 4.5 \times 10^{-10} \frac{N_e}{T^{1/2}}\right)^{-1} \tag{6A.2.1}$$

where $I(\lambda)$ is the intensity of the line of wavelength λ. For the conditions in H II regions, the bracketed factor on the right-hand side of this equation differs from unity by less than 1%, so that we find

$$T \simeq \frac{3.92 \times 10^4}{\log_e\{[I(495.9) + I(500.7)]/I(436.3)\} - 2.12} \text{ (K)}. \tag{6A.2.2}$$

Similarly, for the [N II] lines at 575.5, 654.8, and 658.3 nm we have

$$T \simeq \frac{2.5 \times 10^4}{\log_e\{[I(654.8) + I(658.3)]/I(575.5)\} - 2.0} \text{ (K)} \tag{6A.2.3}$$

although in this latter case there may be some inaccuracy at the highest levels of electron density.

The ratio of line intensity to that of the nearby continuum is also temperature-dependent. In the optical region the deduced temperatures are of low accuracy due to the uncertainty in the relative contributions to the continuum of free–free, free–bound, and two-photon emissions, scattering, and of weak lines. The radio region provides more reliable results. The radio lines arise from transitions between highly excited levels in hydrogen and helium following their recombinations with electrons. The radio continuum is due to free–free emission (§3P.1). The resulting estimates lie close to 9000 K for the electron temperatures. The radio line intensities give generally similar values. The line widths are due to Doppler broadening, and so can lead to estimates of the kinetic temperature of the emitting gas. This tends to be rather more variable, ranging from 7000 to 13 000 K.

The radio continuum as a whole depends upon both the electron temperature and number density, and there are two different regimes: the optically thick and optically thin wavelength regions. The radio emission varies according to the Rayleigh–Jeans approximation to the Planck equation (equation 2P.3.5):

$$B_\nu(T) \simeq 2kc^{-2}\mu^2 T\nu^2 \tag{6A.2.4}$$

or, since the refractive index is close to unity,

$$B_\nu(T) = (2kT/c^2)\nu^2. \tag{6A.2.5}$$

The radio emission is largely due to free–free radiation as the electrons interact with the ionised hydrogen and helium nuclei (§3P.1). Its absorption coefficient, $\varkappa_{\mathrm{ff}}$, is then

$$\varkappa_{\mathrm{ff}} = 10^{-14} \frac{N_e N_i}{T_e^{1.5} \nu^2} \log_e(5 \times 10^7 T_e^{1.5}/\nu) \qquad (6A.2.6)$$

where N_i is the ion number density, and the equation is in SI units. For the temperatures of H II regions and for wavelengths from a few millimetres to a few metres, this simplifies to

$$\varkappa_{\mathrm{ff}} \simeq 2 \times 10^{-13} \frac{N_e N_i}{T_e^{1.35}} \nu^{-2.1}. \qquad (6A.2.7)$$

Thus the radio absorption coefficient of the gas is approximately dependent upon the inverse square of the frequency. When the frequency is high enough for the nebula to be optically thin, the total emission at any frequency varies according to the product of the absorption coefficient and the emission. The two frequency dependences cancel out in this case, and the radio brightness is left nearly independent of the frequency of observation. At lower frequencies the nebula is optically thick, and its emission simply follows the Rayleigh–Jeans equation, varying as the square of the frequency. Thus scanning the radio spectrum of an H II region from low to high frequencies should reveal an initial increase in brightness, which fairly abruptly levels out. The frequency at which this 'knee' occurs is where the nebula becomes optically thin. The surface brightness in the flat part of the spectrum is given by

$$B_\nu(T) \propto E T^{-1/2} \qquad (6A.2.8)$$

where E is a quantity called the emission measure, and is given by

$$E = \int N_e^2 \, \mathrm{d}s \qquad (6A.2.9)$$

integrated across the nebula. In the optically thick region, as we have seen, the brightness is

$$B_\nu(T) \propto T\nu^2. \qquad (6A.2.10)$$

Thus, potentially, the radio continuum observations can give both the temperature and electron density of the nebula.

The results of all these, and other, methods of determining the temperatures within H II regions are in moderately good agreement with each other and with the theoretical upper limit due to the thermal balance. The average temperature is about 8000 K, and some specific examples are given below.

H II region	T_e (K)
NGC 1976 (Orion)	7000–8900
M17 (Omega)	6000–7000
M20 (Trifid)	5000–10 000
M8 (Lagoon)	6500–8000

Some evidence of a reduction from 10 000 K at the centre of the region to perhaps 5000 K near the transition zone has been found. Much smaller scale variations are found by the measurements of radio interferometers. 'Clumps' a few hundredths of a parsec across are found in NGC 1976, for example. The temperature variations, however, are small; the major effect in such cases is the variation in density of the gas (see below).

The density of H II regions can be found from the electron density since, as we have seen, the hydrogen is almost completely ionised. The electron density is determinable from the turnover point of the radio continuum emission, once the temperature is known (see above). The electron density may also be found from the forbidden line intensity ratios. Unlike their use to determine temperature, however, the levels must be selected to be close together in energy terms. At low densities, the populations of the levels, and hence the resulting line intensities, will be proportional simply to the statistical weights of the levels, while, as the density increases, an additional dependence upon the transition probabilities from the levels will become important as collisional de-excitation becomes more significant. [O II], [O III] and [S II] provide the lines mainly used to determine densities, with the [O II 372.9]/[O III 372.6] and [S II 671.6]/[S II 673.1] ratios varying from 1.5 at low densities to 0.35 at high densities. The value of the ratio depends in a complex manner upon both electron density and temperature, so that no simple formulae can be given on the lines of equations (6A.2.2) and (6A.2.3) for the densities.

The electron densities have a much wider range than the temperatures, both between nebulae and within individual nebulae. Examples for individual nebulae are:

H II region	*Electron density* (electrons/m^3)
NGC 1976 (Orion)	1.7×10^9
M17 (Omega)	5.7×10^8
M20 (Trifid)	7.2×10^7
M8 (Lagoon)	2.1×10^8

with the upper and lower limits being around 3×10^9 and 10^7. The corresponding total densities, assuming complete ionisation of the hydrogen and singly ionised helium, are thus

H II region	*Total density* (kg m^{-3})
NGC 1976 (Orion)	3.5×10^{-18}
M17 (Omega)	1.2×10^{-18}
M20 (Trifid)	1.5×10^{-19}
M8 (Lagoon)	4.3×10^{-19}

and the range is from 2×10^{-20} (10^7 atoms/m^3) to 6×10^{-18} (3×10^9 atoms/m^3).

Knowing the size and density we may easily find the mass of the H II region. It generally lies between about 1 and 10^4 solar masses, with specific examples being

H II region	*Mass* ($\mathscr{M}_{\odot}$)
NGC 1976 (Orion)	10
M17 (Omega)	300
M20 (Trifid)	150
M8 (Lagoon)	20

At large distances the H II regions may be obscured to optical observations by the interstellar extinction (§7A.2). They may still be found, however, from their radio continuum emission. The optically thin radio emission is almost, but not quite, constant with frequency. Equations (6A.2.4) and (6A.2.7) do not cancel out the frequency dependence exactly. There remains a variation of the form

$$B_{\nu} \propto \nu^{-0.1} . \qquad (6A.2.11)$$

This enables H II regions to be distinguished from other non-thermal radio sources, even when not optically detectable. Objects such as quasars, radio galaxies, and supernova remnants have a radio continuum spectrum (§6P.2) which varies approximately as

$$B_{\nu} \propto \nu^{-0.7} . \qquad (6A.2.12)$$

Thus, for example, the optically invisible nebula Sgr B2 may be distinguished from the non-thermal source at the galactic centre, Sgr A, by the differences in their radio continua.

The distances to H II regions may be estimated from their radial velocities, the radio lines being used for the optically invisible nebulae. The Doppler shifts, combined with a model for the rotation of the Galaxy, then give the distance. In this way they are found to be closely associated with the spiral arms of the Galaxy.

The luminosity of an H II region is governed by the energy in the ionising photons from the central star (or stars). From equation (6A.1.6), and taking the effective wavelength of the ionising radiation to be 66 nm for an O5 star, and 75 nm for a B0 star, we find total luminosities of 1.4×10^{32} W (350 000 $L_{\odot}$) and 4×10^{29} W (1000 $L_{\odot}$) for optically thick H II regions around such stars. In many cases dust in the surrounding H I cloud absorbs the remaining radiation from the star and re-emits it in the infrared. The combined luminosity of the H I and H II regions may then reach 10^7 $L_{\odot}$ for the hottest exciting stars.

The composition of H II regions is rather uncertain from direct observations because of the comparatively few elements producing detectable spectrum lines. Those elements which are observed, hydrogen, helium, nitrogen, oxygen, neon and sulphur, however, all have abundances close to their cosmic values (§1A.2). Cool interstellar gas clouds (H I regions) are observed to be depleted in the high-melting-point elements, from the strengths of the lines which their elements and molecules produce in the spectra of the more distant stars (Chapter 7A). Since infrared emission reveals the presence of dust in such regions, it is likely that the missing elements have condensed to form the dust. Overall the element abundances in H I regions are thus probably close to the cosmic distribution. Since most H II regions develop from H I regions as their exciting stars form (§3A.3), it is likely therefore that the composition of H II regions is also close to the cosmic values. Possible exceptions to this, such as the lithium abundance, have been mentioned in §3A.6.

A relatively recently discovered and striking phenomenon associated with some H II regions is the occurrence of maser emission at radio wavelengths (§6P.1). About one in four H II regions have such emission. Mostly it is due to the OH molecule, occasionally to H_2O, and more rarely to SiO and other more complex molecules. The main OH emission frequencies are at 1665 and 1667 MHz (wavelengths of 180.18 and 179.96 mm). They consist of many components with widths of a fraction of a kilometre per second spread over a few tens of kilometres per second. The strengths of the components vary markedly on time scales of a year or so. The emission regions are very small: several with sizes of 10^{-5} pc (2 AU) tending to occur over a region perhaps 10^{-3} pc across. The luminosity can be comparable with that of the Sun, but since it is emitted over such a small spectral region, the brightness temperature can reach 10^{15} K! The maser emission is usually strongly elliptically polarised, some of the components reaching 100% circular polarisation. The H_2O masers are generally much smaller and shorter lived than the OH masers, and when both occur near the same H II region, there is usually little physical association between them. All the masers, however, are almost invariably associated with small bright knots within a larger H II region. Such objects are often called compact H II regions. Their sizes are typically 0.01 to 0.1 pc, and their densities up to a thousand times those normally found in H II regions. Very often there is an association with strong compact infrared sources as well.

A possible scenario to account for the masers may be as follows. The maser occurs in the H I region just outside the ionisation cavity. Dust in the H I region shields the maser molecules from the remaining stellar radiation which would otherwise dissociate them. The nature of the energy input to the maser (the 'pump') is unclear. Most probably it is infrared radiation; as mentioned above there is often a close relationship between masers and infrared sources. Collisions, ultraviolet radiation, and chemical reactions are other possibilities. Compact H II regions are expanding (§6A.3), since they are not in hydrostatic equilibrium. The expanding ionisation front is likely to destroy a typical maser source in about 10 000 years. H II regions larger than about 0.1 pc thus do not have associated masers.

In this connection, it is perhaps worth mentioning that some masers are associated with cool stars, mostly the Mira variables, long-period variables, and short-period semi-regular variables. The sources appear comparatively large (up to 3×10^{-4} pc), and the emission is rarely strongly polarised. The maser intensity variations often correlate well with those of the stars, suggesting that a radiative pumping mechanism is involved. A possible model has the maser action occurring in a large (0.01 pc) expanding envelope around the star. The observed emission comes only from the nearest and furthest portions of this shell, since in other regions velocity gradients will preclude maser emission along our line of sight. The frequently observed splitting of the emission into two components separated by 10 to 50 $\mathrm{km\,s^{-1}}$, and the phase delay of 20 to 30 days between variations of the blue- and red-shifted components (due to the flight time across the shell) provide strong support for such a model.

Somewhat similar phenomena to H II regions may result from stellar winds (Chapter 5A). These can inflate hot bubbles inside gas clouds with a superficial resemblance to H II regions. Channelling by denser portions of the nebula, by an equatorially concentrated envelope around the star, or by the particle emission occurring in jets can lead to bi-polar or more complex shapes for such regions. An expanding bubble with a shock transition to the interstellar medium may also develop as a result of the postulated magnetically driven outflows from protostars (§8A.1).

6A.3 H II REGIONS: FORMATION AND EVOLUTION

H II regions do not come into existence fully formed, but develop with time around newly formed hot stars. Dense H I regions are sites of star formation. The final stage, when the star develops significant levels of ionising radiation, occurs relatively rapidly ($< 10^5$ years). During the time that the star's temperature is increasing, and its ionising radiation is growing, it will already have achieved an energy output close to its final luminosity. If the H I region contains significant quantities of dust, then this will be driven outwards by the radiation pressure of the longwave radiation. Velocities of 10 $\mathrm{km\,s^{-1}}$ or more may be reached so long as the dust density is too low for significant coupling to the gas of the H I region. Eventually the dust accumulating in this expanding wave front may become dense enough to drag the gas with it. The gas and dust together then expand at about 1 $\mathrm{km\,s^{-1}}$. When the star achieves its final temperature, therefore, it is surrounded by a dust-free region a few tenths of a parsec across with the normal H I region gas density, then by a region of low-density gas, and a high-density gas and dust front which is still slowly expanding, before the unaltered H I region is reached. Without the dust, the H I region effectively extends up to the star's surface until the onset of the ionising component of the star's radiation.

The dust shell would be optically thick in the visible part of the spectrum, and so the star and developing H II region would only be apparent as a strong infrared source embedded in an H I region. The frequent association of strong infrared sources with actual H I regions, and the infrared emission from dust around developed H II regions argue strongly that most H II regions do form inside dusty H I regions. In either case, however, the effect is that a cloud of cool neutral gas suddenly has an intense source of radiation switched on inside it. A small H II region develops around the star, probably in material rather denser than average for the cloud, due to the local concentration which produced the star initially. The radius of this region will not initially be the appropriate Strömgren radius, because the available photons have to ionise the neutral gas, as well as balancing the recombinations. The ionisation front expands into the H I region at many tens or hundreds of times the speed of sound, until it nears the Strömgren radius. Typically this may take from 100 to 10 000 years in clouds of high and low densities respectively. The temperature within the developing H II region will be 100 to 1000 times that of the surrounding H I region. Furthermore, both ions and electrons will be available to provide the gas pressure. Thus the gas pressure inside the H II region is several hundred times that of the H I region around it. The material in the H II region is therefore not in hydrostatic equilibrium, and it expands outwards, pushing the material in the H I region ahead of it. The velocity of expansion is about the speed of sound in the hot gas, some 10 km s^{-1}. Since the ionisation front moves outwards much more rapidly than this until it approaches the Strömgren radius, the H II region effectively develops in static gas. When the front approaches the size of the Strömgren sphere, however, its expansion velocity slows to less than the speed of sound. The physical expansion of the material then becomes significant. It leads to a reduction in the density of the H II region, and, therefore, from equation (6A.1.3), to an increase in the Strömgren radius. Assuming a simple expansion of the region by a factor x, the particle number density would decrease by x^3, and the Strömgren radius increase by a factor x^2. The ionisation front therefore follows closely on the heels of the expanding edge of the hot bubble, at roughly the speed of sound.

The rate of expansion of the H II region into the H I region is thus at about the speed of sound for the material in the H II region. Since the temperature of the H I region is about 1% that of the H II region, the speed of sound there is about 10% of that inside the H II region. The H II region is therefore expanding highly supersonically with respect to the surrounding material. The interaction region therefore develops into a strong shock front, and the material ahead of the shock front will remain at rest until the front reaches it.

Eventually, the expansion should reduce the density in the H II region to the point at which there is pressure equilibrium with the H I region, and the expansion should come to a halt. This is likely to require some 10^7 to 10^8 years, however, and the main sequence lifetime of an early O-type star is only about 10^6 years. A stable state for an H II region is therefore never reached, and all observed H II regions are still expanding. The lifetime of an H II region is thus that of the main sequence stage of its exciting star: 10^6 to 10^7 years. This is short compared with the time scale of 10^9 years or so for changes in the Galaxy. H II regions in our own and other galaxies thus delineate the regions of star formation. Since they can be detected in the radio region even when invisible optically, they have also provided one of the methods used to map the structure of the Milky Way Galaxy.

6A.4 H II REGIONS AS DISTANCE YARDSTICKS

H II regions may be used to determine distances in two different ways. As mentioned in the previous section, they provide a means of mapping out our own Galaxy, when the Doppler shifts of their radio lines are combined with a model for the rotation of the Galaxy.

The second use is to determine the distances of other galaxies. They are bright enough and large enough to be resolved on photographs of nearby galaxies (figure 6A.6). Now the Strömgren radius for the hotter stars in low-density gas is about 30 pc (§6A.1). The largest H II region for a single star is thus likely to be 100 pc or less in diameter. If several exciting stars are involved, as is often the case, then the upper limit might rise to 200 pc. Taking this as the size of the largest H II region to be found in a galaxy, then it becomes a simple matter to determine the distance of the H II region, and hence of the galaxy containing it, from the observed angular diameter. For example, if the largest H II regions in a galaxy are observed to be 10″ across, then assuming a true size of 200 pc gives the distance to the galaxy as 4 Mpc. Other sizes of H II regions can be used if the spectral type of their central star can be determined. The appropriate physical radius is then given by equation (6A.1.10).

6A.5 PLANETARY NEBULAE

Many of the physical processes associated with H II regions also take place in planetary nebulae. But their structures, and, more particularly, their origins, are quite different from those of the H II regions.

The name of these objects arises from the fancied similarity of appearance of the nebula to a planet, when viewed through a small telescope. They are generally much smaller and fainter than H II regions, so that in at least

Figure 6A.6 A complex H II region in the Large Magellanic Cloud, the Tarantula nebula (30 Doradus complex). (Copyright © 1979, Royal Observatory, Edinburgh.)

half the known examples, they are unresolved on a direct image, and appear star-like rather than planet-like. Their true nature in such cases is obtained from their spectra. Those nebulae which are angularly resolved often show quite complex shapes, though generally with more spherical symmetry than H II regions (figure 6A.7).

The material forming the planetary nebula originates from the central star, possibly through a dust-driven radiative stellar wind (§5A.2), possibly

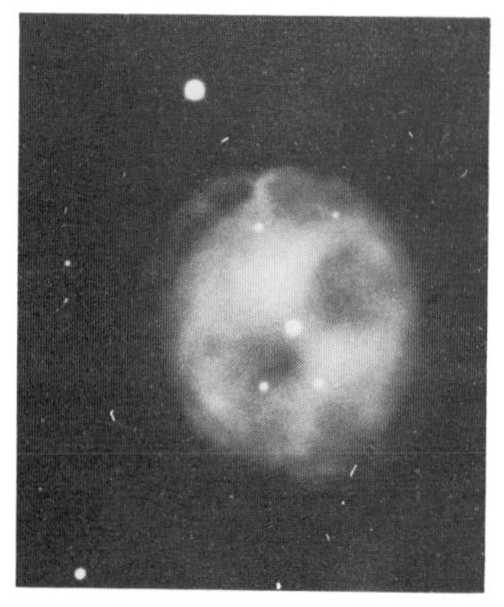

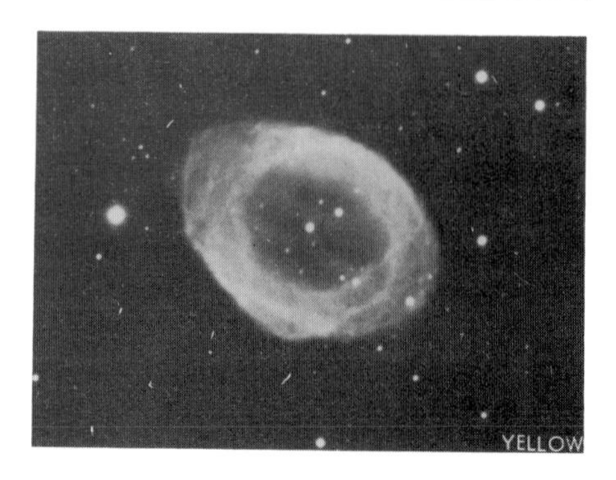

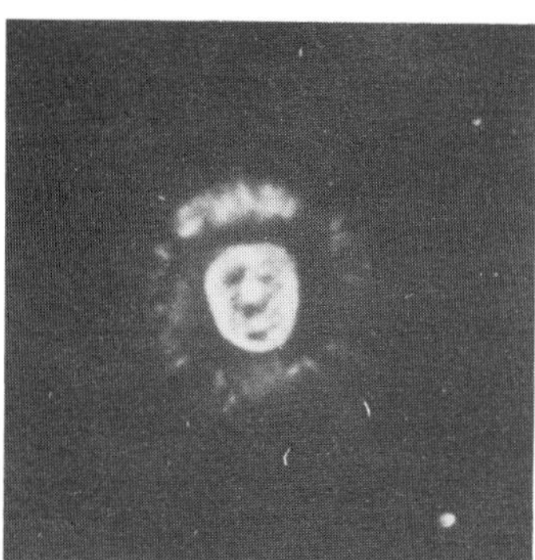

Figure 6A.7 Planetary nebulae. Upper left: the 'Owl' nebula in Ursa Major (M97, NGC 3587). (Photograph from the Mount Wilson and Palomar Observatories.) Lower left: the 'Eskimo' nebula in Gemini (NGC 2932). (Photograph from the Hale Observatories.) Right: the 'Ring' nebula in Lyra (M57, NGC 6720). (Photograph from the Hale Observatories.)

from an evolving star in a binary system overflowing its Roche lobe (§8A.2). The central star is old, evolved and very hot. Thus the gas forming the planetary nebula will generally be of low mass and be expanding unrestrictedly into the interstellar medium. The expansion velocity may be up to 60 $\mathrm{km\,s^{-1}}$. There is no surrounding neutral hydrogen cloud of any significance, though there may be a remnant of a slow (10 $\mathrm{km\,s^{-1}}$) stellar wind from the possible red giant precursor of the system. The material will normally be ionised throughout, and the visible extent of the nebula limited by the greatest distance reached by the expanding gas, rather than by the Strömgren radius. Many planetary nebulae have the appearance of fairly sharply bounded rings (actually spheres). This may arise from the compression of the material between the red giant wind, and a more recently developed high-velocity (1000 $\mathrm{km\,s^{-1}}$) Wolf–Rayet-type wind (§§3A.6, 5A.2).

The central stars are very hot: temperatures of 10^5 to 2×10^5 K are found in some cases. Their spectra are those of early O-type stars, Of stars,

Wolf–Rayet stars (§§3A.2, 3A.5), or in some cases are featureless with no discernible spectrum lines, just a pure continuum. Their immediate precursors are thought to be Mira-type variable stars. The masses of the central stars are very uncertain, but seem likely to be rather less than that of the Sun, on the average. Their luminosities, compared with those of the stars producing H II regions, are low: 0.1 to 100 times the solar luminosity. Their radii are estimated to range from one per cent to twice the solar radius, with the average about half the present size of the Sun. It is generally accepted that these central stars are old, just at the end of their nuclear burning phases, and about to collapse down to white dwarfs.

The spectra of planetary nebulae are similar to those of developed H II regions. There is a radio continuum which increases in intensity with frequency initially, but then flattens out as the nebula becomes optically thin, with superimposed radio recombination lines. There is a large infrared emission from dust particles, and an optical and ultraviolet spectrum composed mostly of emission lines (both forbidden and allowed) superimposed upon a weak continuum due to free–bound transitions.

The temperatures and densities of the nebulae may be found in similar manner to those of H II regions: from forbidden-line intensity ratios, radio continuum turn-offs, etc. The results give temperatures between 6000 and 20 000 K, a wider range than found in H II regions. Electron densities a few times 10^9 m^{-3} are usually found: towards the top end of the range for H II regions, and with a much lower dispersion. In one or two cases electron densities up to 10^{11} m^{-3} may occur. The luminosity of the nebula is low, mirroring the low luminosity of the central star. Their integrated absolute magnitudes range from $+2$ to -2 (10 to 400 solar luminosities). Their physical size is usually a few hundredths of a parsec, but can reach half a parsec occasionally. Their surface luminosity is low compared with that of H II regions, and tends to decrease with increasing physical size. The mass of ionised gas is variously estimated at 0.001 to 1 solar mass, with the total mass being perhaps a few times larger. Condensations and stratification with the nebula (§6A.1) can lead to unusual shapes and internal structure. But generally the gas appears spherical and centred on the star.

Since the central star is an evolved object, the material of the nebula may be expected to differ in composition from the cosmic average, due to processing in the nuclear reactions inside the star. The Wolf–Rayet-type central stars in particular are low in hydrogen, and overabundant in helium, carbon, nitrogen, and oxygen (§3A.6). However, the few directly determined abundances for the nebulae do not differ much from solar values. Thus it may be that the nebula originates only from material previously forming the outer layers of the central star, and that it is the removal of this material which reveals the processed core. Nonetheless, some change in the composition from the cosmic values probably does occur, and when this mixes with the interstellar medium (at a rate of $5\mathscr{M}_{\odot}$ per year for all planetary nebulae in the Galaxy) it is probably the main cause of the increase in the abundance of the medium-mass elements with time (§1A.2).

6A.6 SUPERNOVA REMNANTS

The final stage in the lives of some massive stars, the tremendous disrupting explosion known as a supernova, was described in §3A.5. After the explosion has occurred, we are left with the supernova remnant, or SNR for short. Strictly, for at least some supernovae, there are two components to the remnant: the expanding cloud of gas, and the central neutron star or black hole. Only the former is normally included in the usual meaning of the term, SNR, and that is what is now to be considered. Neutron stars (pulsars) and black holes have been discussed in §§2P.1, 2A.1, 3A.5, and 3A.9, and are not associated with all of the gaseous SNRs.

A SNR may be expected to be detectable for a period of some two or three hundred thousand years after the explosion. Given the estimated rate of supernovae in our Galaxy of two or three per century (§3A.5), there should be several thousand SNRs in our Galaxy. That only about a hundred are actually known is probably due to their low surface brightnesses, and to the obscuration of the more distant examples. The majority of those which are known are identified from their radio emissions. The appearances of SNRs may resemble those of H II regions, although more often they have a filamentary, spherical shell-like appearance (figure 6A.8). The two types of object can be distinguished unequivocally, however, from the nature of their radio emissions. Recombination lines appear in the radio spectra of H II regions, but not in those from SNRs. Furthermore, the emission from a SNR is often strongly polarised, while that from H II regions is not, and has a spectral index (§6P.2) of 0.5 ± 0.2, while the spectral index of H II region emission is about -2 at low frequencies, altering to $+0.1$ at higher frequencies as the gas becomes optically thin. The difference arises from the mechanisms producing the radiation: synchrotron emission from SNRs (§6P.2), and thermal emission from H II regions (§6A.1).

The forms of SNRs are largely governed by three factors: whether or not it possesses a central pulsar to provide a continuing energy source, its age, and whether the originating supernova was of type I or II (§3A.5). The most famous SNR of all is the Crab nebula (M1) in Taurus. It is visible on clear moonless winter nights in the northern hemisphere to anyone with a 0.2 m or larger telescope. It results from a supernova which was seen in 1054 AD, and contains the pulsar NP 0531. It differs from most other SNRs, however, in appearing to be a filled sphere rather than a thin shell (figure 6A.9). The Crab nebula and a few similar SNRs are sometimes called *plerions* to highlight this difference. It may be that the possession of the pulsar is the cause of this disparity since it can act as a continuing source of electrons to power the synchrotron emission. Cas A (figure 6A.8) is more typical of the appearance of a SNR. This was the result of a supernova some three centuries ago which was not observed at the time. It has the appearance on its radio 'photograph' of a thin spherical shell, with much fine structure.

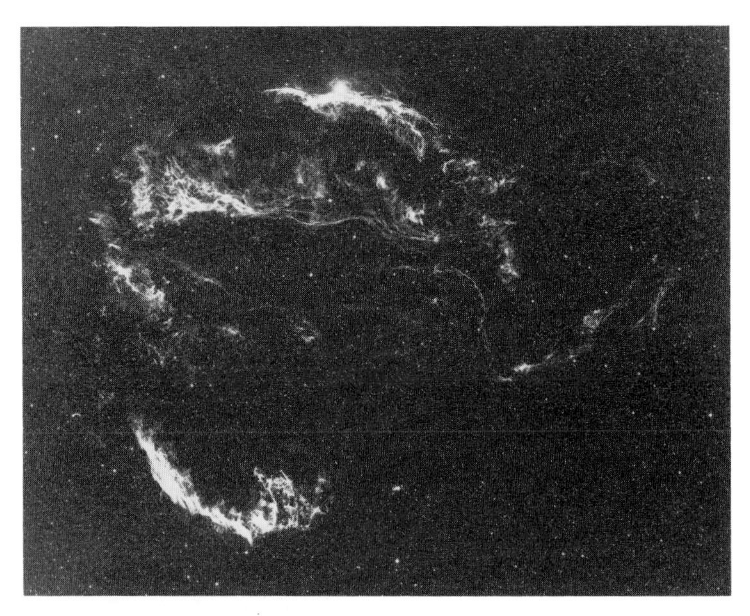

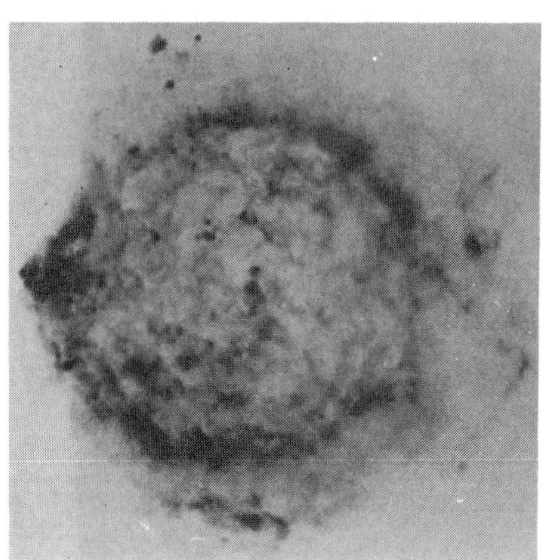

Figure 6A.8 Supernova remnants. Top: the Cygnus loop (NGC 6960–92). (Photograph from the Hale Observatories.) Bottom: a negative radio 'photograph' of Cas A. (Reproduced by permission of the Mullard Radio Astronomy Observatory, Cambridge, UK.)

The radiation from SNRs originates in synchrotron emission. The source of the relativistic electrons, however, seems to change as the remnant ages. Young SNRs are very intense radio sources; the apparent brightness of

Cas A at 100 MHz is comparable with the quiet Sun, though it is six hundred million times further away. In this first phase, the relativistic particles, and probably the magnetic fields as well, arise directly from the supernova explosion. Very high energies for charged particles can be achieved in a chaotic situation by multiple reflections from moving magnetic fields (a process often called the Fermi mechanism, §7A.4). The fine structure may result from local regions of high compression. Later, except possibly for the plerions where the source of particles may be a neutron star, the radio emission seems likely to arise from the acceleration of electrons passing through the supersonic shock front as the expanding SNR meets the interstellar medium, allied to the concomitant compression of the magnetic field.

Figure 6A.9 The Crab nebula (M1, NGC 1952). (Reproduced by permission of the Royal Astronomical Society.)

A few SNRs are observable outside the radio region. In particular, the Crab nebula (figure 6A.9) emits detectable radiation from the infrared to the gamma ray region. Most of this appears to be by synchrotron radiation, as shown for example by the high degree of optical polarisation of the nebula (figure 6A.10). Some of the visible light, however, is due to recombination following ionisation of the gas by the ultraviolet synchrotron emission. The SNR of Tycho's supernova of 1592 shows an almost pure Balmer line emission spectrum in the optical region. Since the nearby interstellar hydrogen would have been ionised if the supernova had had a hot star as its precursor, this remnant must have come from a type I supernova. In older SNRs, such as the Cygnus loop (figure 6A.8), the kinetic energy of the electrons falls to 100 eV or so. Ions of carbon, nitrogen, and oxygen can then exist, and their ultraviolet and optical line emissions dominate the SNR's shorter-wave emissions.

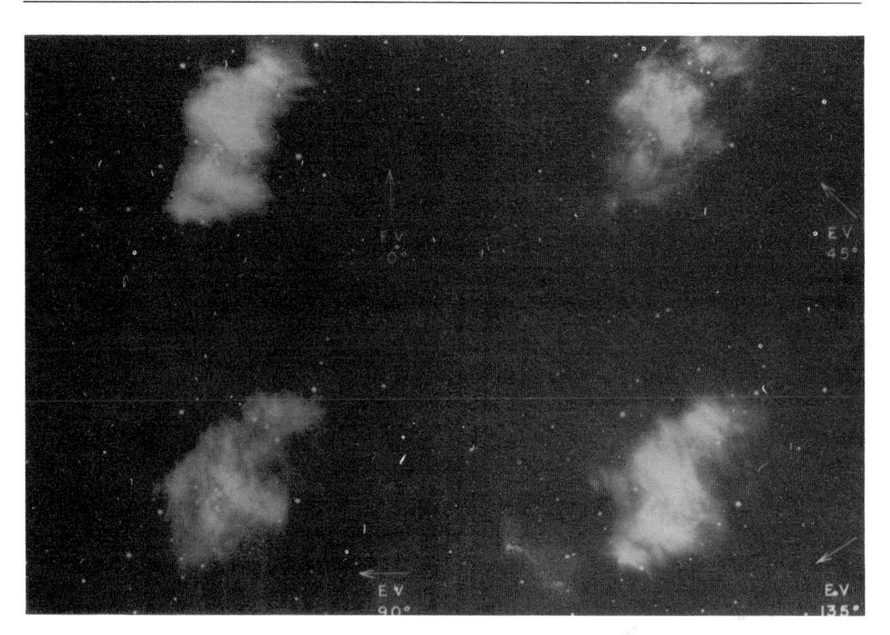

Figure 6A.10 The Crab nebula (NGC 1952) in polarised light. The arrow gives the direction of polarisation for each photograph. (Mount Wilson and Palomar Observatories.)

The physical properties of SNRs obviously vary considerably, both within individual nebulae, and from one nebula to another. The expansion velocity is typically in the region of 10 000 km s^{-1}. Individual particle velocities can be much higher, though: within a fraction of one per cent of the speed of light. The kinetic temperature is correspondingly high: well over 10^6 K until cooling by line emissions from the heavier atoms becomes effective. The electron densities of the filaments and denser portions are probably within an order of magnitude of 10^9 m^{-3}. Their physical extent obviously starts at a value comparable with the size of a star, and then increases with time. The Crab nebula is now about 4 pc across, as is Cas A. The remnants of Tycho's (1572) and Kepler's (1604) supernovae are now about 9 and 5 pc respectively. The Cygnus loop, a much older remnant, is over 40 pc across, and several others over 30 pc are known. Some hot bubbles in the interstellar medium may be even older remnants and can be hundreds of parsecs across. The total mass of a SNR is probably less than 4 or 5$\mathcal{M}_{\odot}$ on theoretical grounds (§3A.5). Observationally, the mass of the Crab nebula has been estimated to be about one solar mass, giving partial confirmation of this suggestion.

The evolution of a SNR may be divided into several phases. During the first of these, the density of the hot gases exploding outwards at around 10 000 km s^{-1} is so much higher than that of the surrounding interstellar medium that the expansion is essentially into a vacuum. At about one parsec in size, perhaps a century after the occurrence of the supernova, the interstellar material swept up by the expanding remnant reaches densities at which it starts to impede the expansion. This leads to the development of a strong and turbulent shock zone, with synchrotron emission from the accelerated electrons as already described. The expansion velocity will slowly decrease until at about 100 km s^{-1} the third phase commences with the increasing optical line emission from the heavier elements. Finally, after some few hundred thousand years, the remnant will merge with the interstellar medium.

When a SNR merges with the interstellar medium, it changes it in two important ways. First the material of the SNR is highly processed and rich in heavy elements. Although the total rate of addition of SNR material to the interstellar medium is only 1% of that from planetary nebulae (§6A.5), about 0.1$\mathcal{M}_{\odot}$ per year, they make the major contribution to the elements heavier than iron. Secondly, the energy dissipated from the SNR probably provides an appreciable fraction of the kinetic energy and turbulence of the interstellar gas.

SNRs are also of wider significance in two further respects which are discussed in detail later. Thus they may be sources of cosmic rays, particularly of the electron component, through the interaction of the charged particles with the turbulent magnetic fields in the initial stages of the expansion and/or in the chaos behind the shock fronts at a later stage (§7A.4). Also, the ram pressure of the expanding SNR on the surrounding interstellar medium may precipitate star formation (§8A.1), as indeed may the rather slower expansions of H II regions.

PROBLEMS

6A.1 How many B0 stars would be required to produce a Strömgren sphere of the same size as produced by a single O5 star?

6A.2 A single O5 star and a single B0 star are observed to have developed

Strömgren spheres of the same size. Explain how this can occur in the light of your answer to question 6A.1.

6A.3 Show that the time for a sphere of uniform density (mass $\mathcal{M}$ and radius R) to collapse under gravity, assuming all other forces to be negligible (the free-fall time), is

$$t_{\mathrm{ff}} = \left(\frac{\pi^2 R^3}{8G\mathcal{M}}\right)^{1/2}.$$

(*Hint*: remember $(\mathrm{d}/\mathrm{d}x)(\mathrm{d}y/\mathrm{d}x)^2 = 2(\mathrm{d}y/\mathrm{d}x)(\mathrm{d}^2y/\mathrm{d}x^2)$.)

Hence estimate the time for the collapse of the relativistically electron-degenerate core of a supernova precursor to a baryon-degenerate configuration, if its mass is one solar mass.

$\mathcal{M}_\odot = 2 \times 10^{30}$ kg
$G = 6.67 \times 10^{-11}$ J m^2 kg^{-2}.
See also figure 2A.11.

7A The Interstellar Medium

SUMMARY

Thermal instability of the interstellar medium, interstellar dust particles; their effects on observations, their possible natures and modes of origin, interstellar absorption and emission lines, molecules, giant molecular clouds, masers, bi-polar outflows, protostars, primary cosmic rays, solar cosmic rays, secondary cosmic rays.

See also spectrum line formation (§3P.1), forbidden lines (§3P.1), atomic and ionic populations (§3P.4), interstellar absorption (§3A.1), supernovae (§3A.5), magnetic fields (§3A.8), stellar winds (§5A.2), masers (§6P.1), H II regions (§§6A.1, 6A.2, 6A.3), planetary nebulae (§6A.5), supernova remnants (§6A.6) and star formation (§8A.1).

INTRODUCTION

If we define the outer edge of a star as its photosphere, then everything beyond that point must be the interstellar medium. We have therefore, on this definition, already encountered several aspects of it: stellar winds, coronae, planetary nebulae and H II regions. Most astronomers would probably consider the first three of these were more validly taken to be a part of the star rather than the interstellar medium, however. The difficulty of defining the interstellar medium precisely simply reflects the fact that there is not truly a rigid division between it and a star's outer layers. Nonetheless, well away from stars (without trying to define the meaning of this too precisely), we clearly reach a region which is quite separate from the stars. The major influences on this very low density material are then very highly diluted stellar radiation, the microwave background radiation, cosmic rays, etc, and it is these which are the concern of this chapter.

7A.1 BASIC PROPERTIES

The interstellar medium is extremely variable in its properties. Average values are thus only of limited use, except as very rough guides. They are just briefly summarised here to enable the later, more detailed discussions to be placed in context.

The interstellar medium is composed of two components: gas and dust particles. The gas is largely hydrogen and has a mean density of about 10^6 atoms/m^3 (one atom per cubic centimetre). The kinetic temperature of the gas has three regimes: around 50 K in the cold molecular hydrogen clouds, over 10 000 K in the intercloud medium, and an additional component of atoms, ions and particularly electrons at 10^6 K in some regions. Dust particles with a size around 100 nm occur within the gas at a rate of about 1 in 10^5 m^3 (10 000 per cubic kilometre). Their compositions (§7A.2) are thought to include graphite, silicates and ices, so that their masses average

Figure 7A.1 The region of the Horsehead nebula in Orion. Fewer stars are seen in the lower half of the photograph because of absorption by intervening interstellar dust. (Copyright © 1983, Royal Observatory, Edinburgh.)

a little over 10^{-18} kg. The dust particles hence form rather less than 1% of the interstellar medium by mass. The effect of these few particles upon astronomical observations is profound, however, at least in the optical region. Their absorption and scattering limit observations within the galactic plane to only a few thousand parsecs. The interstellar material is concentrated into the disc of the Galaxy in a layer less than 200 pc thick. Its total mass is about 10^{40} kg ($10^{10}\mathcal{M}_\odot$ or about 10% of the total mass of the Milky Way Galaxy).

Figure 7A.2 The Pleiades star cluster and associated nebulosity in Taurus. (Photograph from the Hale Observatories, USA.)

Thermal instability divides the interstellar medium into cool dense regions and hot rarefied regions (§7A.3), known as clouds or nebulae and the intercloud medium respectively. The nebulae in turn may be grouped into absorption, emission, and reflection nebulae according to their optical behaviour. The absorption nebulae are up to a million times the average density of the interstellar medium, and they contain abundant quantities of dust. In the optical region, their presence is detected by their obscuration of more distant stars (figure 7A.1). Extinctions of 100 magnitudes per parsec may be reached. Some absorption clouds, often rather hotter, larger, and denser than 'normal' and with radio emission from molecules are probably sites of star formation (§§7A.3, 8A.1). The reflection nebulae occur when a dusty cloud is illuminated by a nearby star. The starlight is in fact scattered rather than reflected by the dust. The nebula is often bluish in colour, therefore, partly due to the greater efficiency with which shorter-wave radiation is scattered, and partly to the illuminating star often being very hot (figure 7A.2). B-type stars are probably the commonest illuminating sources, since they are bright enough to provide sufficient photons, but without evaporating the dust grains, as might a hotter star. The spectrum of a reflection nebula is just that of the illuminating star, but with the continuum biased towards the shorter wavelengths. Emission nebulae have already been discussed (Chapter 6A), and comprise such objects as H II regions, planetary nebulae and SNRs etc, where the material is sufficiently energetic to radiate in the optical of its own accord.

7A.2 DUST GRAINS

The presence of dust particles in the interstellar medium is revealed by many types of observation. One of the most important is the reduction in the apparent brightness of stars which is known as interstellar extinction. This arises from scattering and absorption of the starlight by the grains to roughly equal degrees. The scattered light reappears as a general diffuse glow of the interstellar medium, while the absorbed light is generally re-radiated at infrared and microwave wavelengths. The degree of extinction, A, is measured by the number of stellar magnitudes (§3A.1) by which the star has been dimmed for each parsec along the line of sight. Its value is strongly wavelength-dependent. Since the particle size (100 nm) is only just smaller than the wavelength of optical and near ultraviolet radiation, the dependence of the scattering cross section upon wavelength is roughly inverse:

$$E_\lambda \propto \lambda^{-1} \tag{7A.1.1}$$

where E_λ is the fractional proportion of energy lost from the starlight at wavelength, λ:

$$A_\lambda \simeq 0.0008\ \lambda^{-4/3}\ (\mathrm{mag\,pc^{-1}}) \tag{7A.1.2}$$

for λ in microns. There are significant departures from this at wavelengths of 0.2, 3, 10, and 30 microns (figure 7A.3). These are attributed, not without some disagreement, to the composition of the grains: graphite (0.2 μm feature), ices (3 μm feature), silicates (10 μm feature) and MgS (30 μm feature).

The extinction is revealed in two main ways: via differential spatial effects, and by differential spectral effects. Any given interstellar dust cloud is obviously limited in its physical extent. If it is relatively uniform, then the stars behind it will be dimmed by a constant factor. Plots of the number of stars per unit area of a given apparent magnitude (*1) against that magnitude for two parts of the sky, one within the cloud's confines, and one separated from it, will coincide for the stars nearer than the cloud, and be separated by the absorptivity of the cloud for those which are more distant (figure 7A.4). Such a plot is often known as a Wolf diagram. It may be used to delineate the extent of the cloud when this is not obvious (as is usually the case, except for clouds silhouetted against bright nebulae or the densest parts of the Milky Way). It can also be used to estimate the thickness and distance of the cloud (Appendix I).

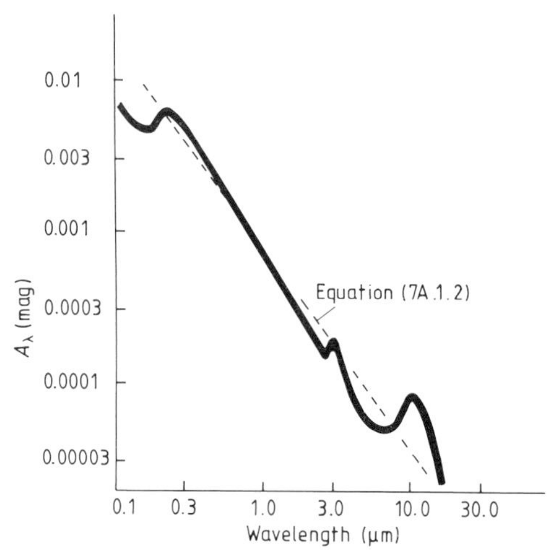

Figure 7A.3 Interstellar extinction as a function of wavelength (mean values per parsec).

The scattering of the light from stars behind the nebula is wavelength-dependent (equation 7A.1.2), with the shorter wavelengths being scattered more than the longer ones. As a result, a distant star appears redder than it should. The spectrum line strengths, however, relative to the *nearby* continuum, are unaffected by the scattering, and so the star's spectral type may still be determined. Comparison with the continuum of an unreddened star of the same spectral type will then reveal the extent of the reddening. Should the average interstellar extinction be known, the star's distance may then be estimated.

Photometry (§3A.1) is more seriously affected by the interstellar extinction, as would be expected. We may usefully define the colour excess, E, as the actual colour index minus the unreddened index. Thus in the *UBV*

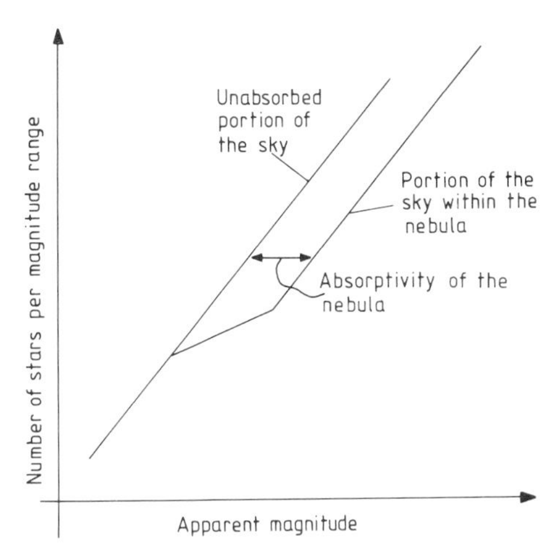

Figure 7A.4 Schematic Wolf diagram of an interstellar absorption nebula.

photometric system we have

$$E_{U-B} = (U - B) - (U - B)_0 \tag{7A.1.3}$$

and

$$E_{B-V} = (B - V) - (B - V)_0 \tag{7A.1.4}$$

where the subscript 0 denotes the unreddened values. A consequence of the form of the spectral dependence of the extinction is that these colour excesses vary in similar ways, and their ratio is therefore almost constant. The reddening ratio

$$E_{U-B}/E_{B-V} \tag{7A.1.5}$$

has a value of 0.72 and is almost independent of the star's spectral type and distance. For the B-type stars, there is an empirical relationship between the intrinsic $(B - V)$ colour index and the colour factor, Q:

$$(B - V)_0 \simeq \tfrac{1}{3} Q \tag{7A.1.6}$$

where

$$Q = (U - B) - (E_{U-B}/E_{B-V})(B - V) \,. \tag{7A.1.7}$$

Thus, again, the star's distance may be found when the average absorption is known.

A third effect of the dust particles is to introduce linear polarisation into the starlight. This reaches a maximum in the centre of the optical region

(550 nm), and amounts to about 1% linear polarisation per magnitude of extinction.

Now polarisation may easily be introduced into reflected radiation, but almost the only process which could lead to polarisation of transmitted radiation under the circumstances applying to the interstellar medium would require the dust particles to be non-uniform and aligned. The mechanism causing the polarisation would then be a differential scattering efficiency for different directions of the electric vector of the radiation, arising from the non-spherical particle shapes. Needle shapes are usually postulated, with an aspect ratio of about three; photons with their electric vectors parallel to the long axis are then preferentially scattered. Such a particle will always produce a polarising effect by itself, but in a random assemblage, the effect of one particle will be cancelled out by another, leaving no net polarisation. Only for a non-random distribution can there be any net polarisation from a large number of particles. The generally accepted hypothesis to induce such a state postulates the existence of an interstellar magnetic field with a strength of around 10^{-10} to 10^{-9} T. The needle-shaped particles, almost certainly rotating rapidly around their minor axis, align themselves with their rotation axes parallel to the magnetic field.

There are several lines of evidence which tend to confirm such a model. First the direction of polarisation of the radiation from individual stars shows a Galaxy-wide pattern consistent with the existence of a Galactic magnetic field. Secondly radio astronomy can detect the magnetic field more directly in two different ways. Polarised radio waves from synchrotron radiation sources have the plane of their polarisation rotated by the Faraday effect. This occurs when a plasma (such as the interstellar medium) is permeated by a magnetic field, the degree of rotation being given by

$$\omega = 8.1 \times 10^3 \lambda^2 \int_0^D N_e B \, dx \quad (7A.1.7)$$

where ω is the rotation of the plane of polarisation (in radians), D is the star's distance (pc), N_e is the plasma electron density (m^{-3}) and B is the strength of the component of the magnetic field along the line of sight (T). The wavelength dependence of the effect enables observations at different wavelengths to lead to an estimate of the field strength. The quantity

$$8.1 \times 10^3 \int_0^D N_e B \, dx \quad (7A.1.8)$$

is known as the rotation measure and typically has values in the region of 0 to 100 rad m^{-2} for lines of sight close to the galactic plane. Combined with the dispersion measure (§7A.3):

$$\int_0^D N_e \, dx \tag{7A.1.9}$$

it gives the magnetic field strength along the line of sight weighted by the plasma density. The second line of evidence from radio observations comes from the direct effect of the Galactic magnetic field upon the emission of the radiation. The Zeeman effect (§3P.1) causes a splitting of a line into components separated by

$$\Delta\nu = 1.40 \times 10^{10} B \text{ (Hz)}. \tag{7A.1.10}$$

The neutral hydrogen line at 1420 MHz (§7A.3) will thus be split into components separated by about 14 Hz by a magnetic field of 10^{-9} T. As with the detection of stellar magnetic fields (§3A.8), the components will not be separately visible, but the imbalance of the polarisation in the wings of the blend will still allow the magnetic field strength to be estimated.

The overall results of these measurements and theories suggest an interstellar magnetic field parallel to the plane of the Galaxy and running around it with a strength of about 2×10^{-10} T, for the region within a few kiloparsecs of the Sun. This would be just about adequate to align the interstellar dust particles in the required manner.

The dust particles also reveal their presence by their own direct emissions. Most H II regions and many planetary nebulae (Chapter 6A) have infrared excesses and/or are associated with strong nearby infrared sources. Almost certainly the emission in such cases comes from dust particles. The particle density can be very high, completely obscuring the central star on occasion. The dust grains are heated to several hundreds or even a thousand degrees, re-emitting the radiation primarily between 1 and 30 microns. A similar situation occurs near some cool supergiants; their infrared excesses may amount to 10 or 20% of their total energy. In this case, however, the particles are thought to have originated in the outer layers of the star and may therefore differ to some extent from 'standard' interstellar dust grains. Away from stars, the dust is still heated by the radiation that it absorbs, but this is so weak that temperatures of only a few tens of degrees Kelvin are reached. The re-emission then occurs at wavelengths around 0.1 mm, and is very faint and difficult to observe.

The composition of the grains is unknown. There are several hints about their nature, however. The extinction curve (figure 7A.3) has several features which, as already mentioned, are attributed to the presence of graphite, ices, silicates and MgS in the dust grains. A consequence of the scattering component of interstellar extinction is the diffuse light from the Galaxy, since this is just starlight scattered by the interstellar grains. It is very feeble, and the effects of many faint stars pollute the observations. The indication, however, is that the grains have an albedo of 60 to 70% in the optical and ultraviolet, dropping to about 35% around 220 nm, perhaps due to graphite. There is also a feature near 220 nm in the emission from nebulae around some Wolf–Rayet stars, and it has again been suggested that this is due to graphite. Graphite, perhaps with a contribution from iron-based particles, can explain the longwave infrared emission from planetary nebulae. Silicates have been suggested as possibilities to explain features near 10 microns in T Tauri and Herbig Ae and Be stars. The former also have features at 3.1 and 20 microns, perhaps attributable to ices (frozen gases such as water, methane, ammonia etc).

A rare class of cool supergiant variables, the R Corona Borealis stars, behave rather as 'inverse novae'. Their optical brightnesses suddenly decrease by up to 10 or 12 magnitudes and then more slowly return to normal. They are rich in helium and carbon, and depleted in hydrogen. The explanation for their abrupt variations is thought to lie in the formation of graphite particles in the outer, cooler parts of the atmosphere. These then partially obscure the star until radiation pressure has blown them well out into the interstellar medium. Thus again graphite appears as a possible composition for the interstellar dust. Silicate particles might originate in a similar manner from an oxygen-rich star.

Somewhat less directly, an indication of the potential grain composition can be obtained from the abundance of elements in the interstellar medium (§7A.3). It is found to be depleted in almost all the heavier elements, relative to hydrogen. Carbon, for example, is less than 20% as common in the interstellar medium as in stars generally, while oxygen, silicon and iron have 25%, 2% and 1% of their stellar abundances respectively. This depletion of the interstellar gas is generally attributed to the partial condensation of such elements into the dust grains. Thus the 'missing' elements from the interstellar gas should give a clue to the probable composition of the dust grains. On this hypothesis, the major grain forming elements would be as given in Table 7A.1. Additionally, various hydrogen compounds such as H_2O and NH_3 are to known to occur in the interstellar gas (§7A.3), and these are likely to be present in the grains in frozen form as ices.

Combining these various lines of evidence enables a model for the interstellar grains to be developed. Probably there are at least three major types of composition: graphite, silicon carbide, and silicates (particularly aluminium and magnesium silicate). These provide a good fit to the ultraviolet and optical data. The sizes required for such grains are around 30, 80, and 50 nm respectively. These particles then form the cores for the condensation of the volatiles, particularly water, ammonia and methane ices. The total size of the composite grains reaches perhaps 200 to 300 nm along their longer axes. Close to stars, the ice coating evaporates leaving just the core. Variations in the relative proportions of the types of core can explain the variations to be found in the extinction curves for different stars. Although this model fits the results quite well, and many astronomers accept it as correct, it cannot yet be regarded as definitive; the observational constraints are insufficient to eliminate many other possibilities.

The manner of formation of the grains is unclear, but again plausible guesses may be made. First we strongly suspect that particles are condensing in the outer regions of stars such as cool supergiants, R CrB stars etc. The rate of production by such means appears, however, to be too low to explain the observed particle density, perhaps by as much as a factor of ten. An additional site might then be the dense molecular clouds (§7A.3).

Table 7A.1 Major components of interstellar dust grains

Element	Number of atoms in the interstellar dust grains (relative to 100 oxygen atoms)
Oxygen	100
Carbon	54
Nitrogen	15
Iron	7.8
Silicon	6.5
Magnesium	5.2
Sulphur	1.6
Aluminium	0.5
Calcium	0.4
Nickel	0.4
Sodium	0.3
Chromium	0.1
Manganese	0.1
Phosphorus	0.1
Others	<0.1

Once a core has formed and been moved by radiation pressure to a cool environment, the ice coating may be built up by accretion of individual atoms and molecules. In dense, cool clouds ($>10^9$ atoms/m^3, <100 K), a size of 100 to 200 nm might be achieved in 10^9 years. Thus most grains must be expected to have acquired a significant ice mantle. A heavy atom in such a cloud would have a mean lifetime before accretion of a few times 10^8 years. In dense molecular clouds, where the density might be 10^{10} atoms/m^3, the lifetime might be as low as 10^6 years. Since the heavy elements are depleted, but not totally consumed as ought to be the case with such short time scales, some mechanism to destroy the grains must also be occurring, in addition to the processes leading to their formation.

There are several possible destruction mechanisms. First, there is direct evaporation due to the internal temperature of the grain. This can probably occur near H II regions, and becomes significant at grain temperatures in excess of 40 or 50 K. Secondly, there is the process called *sputtering*, and this is the ejection of atoms or molecules from the grain by the impact on it of other atoms or ions moving at thermal velocities. Probably kinetic energies of over 100 eV are needed before this becomes a significant means of grain destruction. A kinetic temperature of 10^{10} K or so would thus be needed, and so sputtering is probably only significant in strong shock fronts, such as at the outer edge of a SNR. Shock waves probably also provide the energy for the third destruction process. This is the direct collision of dust particles with each other. Sufficient energy to evaporate the ices may be released at collisional velocities as low as 1 km s^{-1}, and even iron-based cores will be destroyed at velocities over 10 km s^{-1}. The passage of a single shock front through a dense dust cloud may destroy several per cent of the existing grains by this process. Additionally, glancing impacts may impart sufficient angular momentum for the rotational forces to disrupt the grain, especially the low-strength icy mantles. Two final external processes which may be important are the absorption of high-energy photons and the impact of cosmic rays. Ultraviolet photons may reduce the lifetime of molecules adsorbed on to the surfaces of the grains to less than a century, even in the depths of interstellar space. Near to stars, of course, the whole of the icy mantle will be evaporated very rapidly. Cosmic rays may destroy grains completely; a 10^8 eV cosmic ray particle carries sufficient energy to evaporate even the most refractory particle. Normally, however, the cosmic ray would pass through the grain, losing only a small fraction of its energy, and leading to the loss of a few atoms or molecules in a similar manner to sputtering. Some or all of these processes may be aided by chemical reactions amongst the constituents of the grain. Radicals accumulating in the icy mantle could react explosively following a small increase in temperature due to one or other of the above processes, and cause complete loss of the mantle.

7A.3 INTERSTELLAR GAS

The bulk (99%) of the interstellar material is in the form of gas. As we have already seen (§7A.2), this is largely hydrogen and helium with the heavier elements depleted with respect to their solar abundances. Its presence, however, is much less obvious to an optical astronomer than that of the less abundant dust particles, its major effect being the introduction of a few absorption lines into the spectra of the more distant stars. In the radio region, by contrast, it has several significant effects: 21 cm lines, molecular lines, dispersion of pulsar signals, etc, which have led to much improved understanding of its distribution and properties over the last couple of decades.

We turn to the optical and ultraviolet interstellar lines first: their existence has been noted since the beginning of the century. δ Orionis is a binary system with an orbital velocity of over 100 km s^{-1}. Its spectrum lines thus change their wavelengths by 0.03% periodically over an interval of just under six days. In 1904, however, it was noticed that some of the lines in the spectrum were not following this pattern, but remained at a fixed wavelength. These arose within the gas over the 500 pc along the line of sight to the star. Many stars have since had similar lines found in their spectra. In the visible spectrum the calcium H and K lines and the sodium D lines are by far the strongest, and may sometimes be detected in the spectra of stars only 50 pc away from us. Iron and potassium lines can also be found on occasion. Most of the interstellar lines, however, occur in the ultraviolet, where over twenty elements have been detected.

Many of the atoms are found to be ionised. At first sight, this might seem surprising in the low-energy environment of interstellar space. The density of the medium is so low however, that once an atom does become ionised, its probability of meeting a free electron with which to recombine is very small indeed, and it remains ionised for a long period. Thus calcium exists in the interstellar medium primarily in the singly ionised state with perhaps 10% doubly ionised and less than 1% as neutral atoms. The low-energy environment does, however, mean that both atoms and ions are almost exclusively in their ground states. The main exception to this occurs near stars when metastable levels may become highly populated leading to forbidden-line emission from nebulae and to masers (§6P.1 Chapter 6A). Thus almost all interstellar lines are absorptions from the ground state, and the large energy gap to the first excited level which occurs for most atoms means that such lines tend to occur in the ultraviolet. The strongest line is due to hydrogen: Lyman α at 121.6 nm. Ionisation of hydrogen renders the interstellar medium almost totally opaque at wavelengths shorter than 91.2 nm (the Lyman limit), the extinction reaching 20 magnitudes per parsec or more. Thus the shortwave ultraviolet region of stellar spectra is obscured to a very large extent by the effect of this and other ionisation edges.

In addition to these relatively straightforward interstellar absorption lines, there are two other types of feature. The first is the molecular absorption lines. In the optical and ultraviolet regions these are due to H_2, OH, CO, CN, and CH. The much more extensive radio molecular lines are discussed later in this section. Again, only transitions from the lowest energy levels are observed. Rather more problematical are the diffuse absorption features. The strongest of these is at 443 nm and has a width of over 4 nm compared with the typical width of 0.01 to 0.02 nm for an atomic or ionic line. Over 40 such features are now known with the next strongest being found at 578, 618, and 628 nm. Their origin is unknown. Their strengths correlate well with the interstellar grain density, however, and so they are probably due to compounds on or in the grains.

The 'normal' or sharp interstellar lines have widths corresponding to velocities for the atoms of only a few km s^{-1}. A few of the lines (e.g. Lyman α) are saturated, so that their profiles are much broader and unrelated to the velocity structure of the gas (§3A.2). Generally the radial velocity found from the interstellar lines will be due to the rotation of the Galaxy. Sometimes, however, a cloud along the line of sight may have a significant intrinsic velocity. It will then produce its own absorption lines, shifted from those of the rest of the interstellar gas. The observed interstellar lines thus may be divided into several components; four or five components separated by up to 100 km s^{-1} are possible if several such clouds intervene.

The most powerful technique for studying the interstellar gas is based upon the radio emission from neutral hydrogen atoms. In the ground state, the electron can have its spin in the same direction as, or opposed to, the spin of the nucleus. The electron energy is higher in the former situation than in the latter by some six millionths of an electron volt. A transition between the two levels results in an emission or absorption line at 1420.405 751 79 MHz (21 cm). The lifetime of the higher level against spontaneous transitions is about 10^7 years, and roughly three-quarters of the hydrogen atoms are in that state at any given time. Most transitions (in both directions) are induced by collisions, leading to an actual lifetime for the excited state of 100 to 500 years. The density of neutral hydrogen is about 10^6 atoms/m^3 in the spiral arms of the Galaxy, and about one-third of that between the arms. At 21 cm, therefore, the spiral arms stand out almost like beacons, and can be picked up by very modest radio telescopes. When two or more spiral arms lie along the line of sight, their differing radial velocities allow the separation of their individual contributions. In this way, much of the Galaxy has been mapped out in recent years. When a strong extragalactic radio source lies behind a neutral hydrogen cloud, then 21 cm absorption lines are found and can be studied, in a similar way to the direct emission lines.

Perhaps one of the more surprising discoveries of the last two decades has been the presence of molecules in the interstellar medium. The simple bi-atomic molecules detected at optical wavelengths such as H_2, CO, and CH (see above) are perhaps not too unexpected. The OH and H_2O masers associated with some H II regions etc (§6A.2) have also already been mentioned. Radio line observations have now detected very complex organic molecules in considerable quantities in some clouds, however. The number of such molecules known at the time of writing (not including molecules which differ only by containing different isotopes of the same atom) is over sixty. The most complex of the molecules found so far comprises thirteen atoms. Since conditions in the interstellar medium differ so greatly from the terrestrial environment, many of these molecules are identified initially only from calculation, and not from experiment.

The molecules are not distributed randomly throughout the interstellar medium, but are concentrated into cool, dense regions known as molecular clouds (figure 7A.5). Most of these occur between 4 and 8 kpc from the centre of the Galaxy. Their densities range between 10^8 and 10^{11} particles/m^3, their dimensions up to a few tens of parsecs, and their total masses up to $10^6 \mathcal{M}_\odot$. They are thus the largest coherent single objects within the Galaxy. Their internal temperatures may be 100 K at the lowest densities, decreasing to 10 K as the density increases towards the upper limit. They contain large quantities of dust, and are often closely associated with H II regions. They are generally accepted as sites of current star formation.

Figure 7A.5 The Taurus molecular cloud. (Reproduced from *Photographic Atlas of Selected Regions of the Milky Way* by E E Barnard (1927), by permission of the Carnegie Institute of Washington.)

The molecules are detected via their rotational transitions, which usually occur in the 1 to 100 mm waveband. The carbon monoxide emission bands at 0.65 and 2.6 mm are particularly strong due to its high abundance (0.01% that of H_2), and these emissions form a useful tracer for the molecular clouds. Several molecules expected to occur in the clouds, such as H_2 and N_2, are symmetrical, and transitions from their ground states to their first excited levels are forbidden. They do not therefore have any significant rotational emission, since the energy density in interstellar gas clouds is rarely sufficient to excite the molecules directly to the second excited level. H_2 has been detected in the optical region (see above) and in the ultraviolet, whereby upwards of 50% of the hydrogen is found to exist in molecular form. It must generally be expected to predominate both in the clouds and throughout the interstellar medium. The other symmetrical molecules have yet to be detected. The rotational emission of the molecules is almost certainly powered by collisions; the molecule collides with a hydrogen atom or molecule to initiate its rotation, and then undergoes a spontaneous downward transition, usually in less than a few hours, and emits the observed radiation. Many of the molecules are ionised, such as CH^+, HCO^+, N_2H^+ etc, or are species difficult to produce on the Earth such as the radicals OH, CH etc, or are complex molecules such as $HC_{11}N$. They are thus of interest to chemists as well as to astronomers.

The two major problems associated with understanding and explaining the existence of interstellar molecules are how they form, and why they are not rapidly dissociated after formation. The answer to the second point at least is clear: the molecules are shielded from the stars by the dust grains (§7A.2). As we have seen, the grains, though only a small component of the material, are very efficient at absorbing shortwave radiation. Photons in the optical and ultraviolet regions are therefore almost totally converted into infrared and microwave continuum radiation, with too low an energy then to dissociate the molecules. The formation of the molecules poses more of a problem. It is possible to postulate reaction routes whereby the molecules may form in interstellar space. The low density of the gas means, however, that the reactions must be two-body; the probability of three-body collisions will be vanishingly small. A simple reaction of two atoms will not be stable, and the resulting molecule will dissociate back to the atoms again in a small fraction of a second. By involving ions, or complex molecules, stable end products may result through the dissipation of the reaction energy into routes other than dissociation. Thus one method of forming the CH molecule might be

$$\left.\begin{aligned}&C = C^+ + e^-\\&C^+ + H_2 = CH_2^+\\&CH_2^+ + e^- = CH + H.\end{aligned}\right\} \qquad (7A.3.1)$$

A possible route to the formation of water molecules is:

$$\left.\begin{aligned}&H_2 = p^+ + H + e^-\\&p^+ + O = O^+ + H\\&O^+ + H_2 = HO^+ + H\\&HO^+ + H_2 = H_2O^+ + H\\&H_2O^+ + H_2 = H_3O^+ + H\\&H_3O^+ + e^- = H_2O + H.\end{aligned}\right\} \qquad (7A.3.2)$$

The original ionisations in each case are produced by cosmic rays, ultraviolet photons etc.

Quite complex molecules can be built up in this way, and moderately good agreement with the observed molecular abundances achieved. An alternative or additional process may involve the dust grains. We have seen (§7A.2) that it is likely that material, particularly ices, will condense on the surfaces of refractory grains to form a thick mantle. Furthermore, it is likely that occasionally atoms and molecules, sometimes even the whole mantle, may be evaporated back into space by a variety of processes. The grains are at a low temperature, but their very much higher density (by a factor of some 10^{24} compared with the interstellar medium), will increase the probability of interactions and remove the restriction to two-body reactions. Additionally, some of the disruptive processes such as spallation, cosmic rays, and ultraviolet photons, are likely to leave highly reactive radicals on the surface. These may then more easily take part in chemical reactions with other nearby atoms and molecules, leading at least some of the time to the formation of the complex molecules which are observed.

High-angular-resolution observations of the infrared and microwave CO and H_2O emissions have recently resulted in the identification of an important new class of objects. These are strong outflows of material, often in the form of bi-polar jets, from proto-stellar condensations within molecular clouds. They appear to be associated with proto-stars of one or more solar masses and may themselves contain between a tenth and a hundred solar masses of material. The velocity of the material is typically 100 km s^{-1}, implying mass loss rates of up to $10^{-2}\mathcal{M}_\odot$ per year, and mechanical luminosities of up to several thousand times the solar radiative luminosity. Their evolutionary lifetimes are estimated to lie between 1000 and 100 000 years, so that their rate of formation must roughly equal that of all stars of one solar mass and above. Such outflows may therefore be associated with the formation of all the larger stars (§8A.1).

The outflows have been detected as close to their central objects as a few astronomical units, and in some cases can be found to extend to as much as a parsec outwards. They are closely linked to maser emission (§6A.2) and to Herbig–Haro objects. These latter are small, bright patches of luminosity occurring in some interstellar clouds. Their spectra contain emission lines of common elements, and are characteristic of shock-induced emission. They may originate when the supersonic outflow impinges on denser parts of the surrounding material. The H_2O masers also seem to be associated closely with the interaction of the outflows with this stationary or slowly moving material.

The origin of the outflows is still unknown. Normal radiatively driven stellar winds (§5A.2) would be too weak by a factor of 100 to 1000 to provide the observed mass loss rates. Possible alternative explanations include (*a*) a change in the stellar angular momentum with the onset of deuterium burning, (*b*) rotational instabilities, or (*c*) transfer of angular momentum by magnetic or viscous linkages from the proto-star to the nebula. The bi-polar nature of the outflow may be an intrinsic result of the accelerating mechanism. Alternatively, an outflow from a proto-star which was initially isotropic would be likely to become bi-polar as it interacted with the equatorially concentrated remnants of the collapse of the original cloud fragment. This would constrict the outflow in the plane of the equator, leaving the material to escape primarily in the directions of least resistance, around the rotational poles of the proto-star.

If, as seems likely, many proto-stars develop this type of behaviour, then the resulting input of turbulent energy to the remaining portions of the gas cloud may be sufficient to delay their gravitational collapse. The observed lifetimes of the clouds are uncomfortably long without such a process occurring, so that there is some support for the probable universality of bipolar outflows associated with star formation. The jets may also influence other nearby proto-stars in the cloud, possibly disrupting some of them. They may also act to compress other parts of the cloud, so initiating star formation elsewhere (§8A.1). The importance, or even reality, of these processes remains to be demonstrated, however, at the time of writing.

The temperature of the interstellar gas ranges from 10 to 10^6 K, and this very wide variation may be largely understood in terms of the heat balance of the medium. The heat sources are primarily shortwave stellar radiation and cosmic rays. The loss mechanisms are molecular and atomic line emission and continuum radiation from the dust. The molecular transitions are almost all due to rotational transitions and are driven by collisions with other atoms and molecules. The number of collisions, and hence the loss of energy via the molecules, varies as the square of the density. Thus a slight increase in the density leads to a much larger increase in the rate of loss of energy from the material. This in turn lowers the temperature and pressure, causing collapse of that region of the interstellar medium, leading to a further increase in the density and energy loss, and so on. The interstellar medium therefore tends to separate into cool dense regions, and hot rarefied regions. We may identify these as the diffuse clouds with a temperature of about 100 K, and the general intercloud medium at some 10 000 K. The very cool clouds arise from the shielding effect of the dust. This reduces the effect of the heat sources and further enhances the efficiency of the energy sinks. The very hot (10^6 K) regions are thought to arise as the last phase of a supernova remnant (§6A.6) before it merges completely with the interstellar medium.

In addition to its intrinsic significance and interest, the interstellar gas is important for its effect on the observation of other astronomical objects. Extinction has already been discussed in some detail (§7A.2), and Faraday rotation of the plane of polarisation mentioned in connection with the galactic magnetic field. The other two main effects are the dispersion of pulsar signals and scintillation. In an ionised gas (plasma), the velocity of electromagnetic radiation of frequency ν is

$$v(\nu) = [1 - (\nu_p/\nu)^2]^{1/2} c \qquad (7A.3.3)$$

for $\nu \gg \nu_G$, where ν_G is the gyro frequency (equation 6P.2.9) and ν_p is the plasma frequency, the frequency of oscillation of the gas following a disturbance in its electrical neutrality, given by

$$\nu_p = (e^2 N_e / 4\pi^2 \varepsilon_0 m_e)^{1/2} . \qquad (7A.3.4)$$

For typical interstellar conditions, we have $\nu_G \sim 1$ Hz, $\nu_p \sim 1$ MHz, and so we find at the normally observed frequencies, which lie in the region of tens of megahertz to gigahertz:

$$v(\nu) \simeq c / [1 + (e^2 / 8\pi^2 \varepsilon_0 m_e)(N_e/\nu^2)] . \qquad (7A.3.5)$$

If a signal travels a distance D through the interstellar medium, then the travel time for radiation of frequency ν will be

$$t(\nu) = \frac{D}{c}\left(1 + \frac{e^2}{8\pi^2 \varepsilon_0 m_e} \frac{N_e}{\nu^2}\right) \qquad (7A.3.6)$$

for constant density, or

$$t(\nu) = \frac{D}{c} + \frac{e^2}{8\pi^2 \varepsilon_0 m_e c \nu^2} \int_0^D N_e \, dx \qquad (7A.3.7)$$

if the density varies along the path. If we have a sharp pulse, such as the output from a pulsar, then the frequency dependence in equations (7A.3.6) and (7A.3.7) will cause it to be observed arriving at the Earth at different times as the frequency of observation changes. The relative delay of one signal with respect to the other is given by

$$\Delta t = \frac{e^2}{8\pi^2 \varepsilon_0 m_e c} (\nu_1^{-2} - \nu_2^{-2}) \int_0^D N_e \, dx \qquad (7A.3.8)$$

where ν_1 and ν_2 are the two observed frequencies. Thus when a pulsar (§3A.5) is observed, a particular pulse (which may generally be identified by its detailed profile) timed at two or more frequencies will provide a measure of this delay and allow the determination of $\int_0^D N_e \, dx$ along the line of sight to that pulsar. This quantity is known as the dispersion measure, and in units of pc m^{-3} its value ranges in practice from a few times 10^6 to 4×10^8. If the distance to the pulsar is known, then it may be used to estimate a value for N_e. Alternatively a value for N_e may be assumed (typically about 5×10^4 m^{-3}) for the intercloud medium, and the pulsar's distance estimated.

The fluctuations with time in the electron density of the interstellar medium along the line of sight to a distant radio source cause scintillation of the sources in a manner analogous to the 'twinkling' of stars in the visible. The search for this effect led to the discovery of the first pulsar (§3A.5). Scintillation due to the interstellar medium is noticeable only on sources of very small angular diameters, and amounts to some 10^{-3} arcseconds (cf 1″

for the optical scintillation due to the Earth's atmosphere). The time scale of the variations is around 10^3 seconds, and the size of the cells within the interstellar medium about 10^9 m (10^{-7} pc).

7A.4 COSMIC RAYS

Primary cosmic rays

Almost all astronomical knowledge has been gained from the information carried by electromagnetic radiation. Only within the solar system has direct sampling been possible. The only direct source of information on the Universe outside the solar system (since neutrinos and gravity waves remain to be found with certainty) lies in the cosmic rays. These are not electromagnetic radiation in the normal sense, but very highly-charged energetic particles. They permeate most of the Galaxy, and possibly the intergalactic space as well. The Earth is continuously being bombarded by them. Thus in addition to their having a certain intrinsic interest, as our only unequivocal samples of the Universe outside the solar system, they have an importance far beyond that.

These particles are called the primary cosmic rays. There are also solar cosmic rays and secondary cosmic rays. Both are discussed later in this section. Briefly, the solar cosmic rays are high-energy particles from the Sun, usually but not always charged, and with energies towards the lower end of those of the primary cosmic rays. The secondary cosmic rays are subatomic particles and photons produced in vast numbers by the interaction of the higher energy primary and solar cosmic rays with atoms in our own atmosphere.

The composition of the primary cosmic rays is mostly protons (84%) and helium nuclei (14%). About half the remaining particles are electrons, and the remainder (1%) the heavier nuclei (figure 7A.6). Nuclei heavier than iron are found, but only very rarely; they are less than 10^{-5} times as frequent as the iron-group nuclei. A few positrons are to be found, at about 1% of the number density of the electrons. Antiprotons also occur at a rate of one to about two thousand protons. Such numbers, however, are consistent with their formation by the collision of other cosmic ray particles with the interstellar gas nuclei, and thus they do not necessarily imply the existence of 'anti-stars' etc. At very high energies, gamma rays may form a significant component of the primary cosmic rays.

Two features of the abundance curve are particularly noticeable: the beryllium group (helium-3, lithium, beryllium, boron) are about two and a half million times more abundant in the cosmic rays than in the cosmos generally (figure 1A.11), and the iron-group nuclei are overabundant by

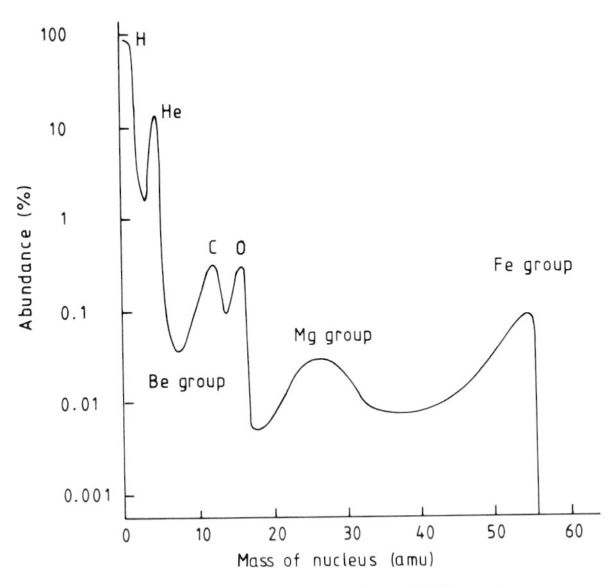

Figure 7A.6 Abundance by number of nuclei in primary cosmic rays.

about a factor of one hundred. The first of these anomalies has two contributing mechanisms. The more significant is probably spallation. This is rather different from the spallation reactions involved in the destruction of interstellar dust grains (§7A.2). Here it is the collision of the heavier cosmic ray nuclei (particularly carbon, oxygen etc) with interstellar hydrogen. The heavy nuclei fragment in such interactions into the lighter beryllium-group nuclei. The observed ${}^3_2\mathrm{He}/{}^4_2\mathrm{He}$ ratio of about 0.1 suggests that about 3% of the heavier elements have suffered such collisions. Now the mean interaction cross section of carbon, nitrogen, oxygen, and neon with protons is about $1.2 \times 10^{-30} \mathrm{m}^2$, so that a nucleus must pass through a column density of about 1.4×10^3 kg m^{-2} of hydrogen on average before collision. These nuclei must thus have passed through some 40 kg m^{-2} of hydrogen in order to produce the required ${}^3_2\mathrm{He}/{}^4_2\mathrm{He}$ ratio. Given the average number density of interstellar hydrogen of some 10^6 m^{-3} (§7A.3), the distance travelled by the average cosmic ray particle within the Galaxy must be some 10^6 pc, and its average flight time about 3×10^6 years. Thus we either have an estimate of the lifetime of a primary cosmic ray particle, or, more probably, an estimate of its containment time in the Galaxy. The second possible reason for the overabundance of the light nuclei in the cosmic rays is likely to be their destruction within stars and the consequent reduction of their cosmic abundances (§§1A.2, 3A.6). The factors contributing to the high abundance of the heavier elements are less clear.

Probably it arises through the origin of primary cosmic rays within objects rich in such elements, such as white dwarfs, neutron stars and supernovae.

Each component of the cosmic ray flux has its own energy spectrum, but with the exception of the electrons, they are all very similar (figure 7A.7). For energies higher than about 10^{12} eV, the data are inferred from observations of the secondary cosmic rays so that the nature of the original particle is unknown; only its energy can be determined. Since the lower energy data suggest that 84% of the particles are protons, no great loss of accuracy occurs in continuing the proton curve into the total particle curve at high energies. There are several features of the curve worth highlighting.

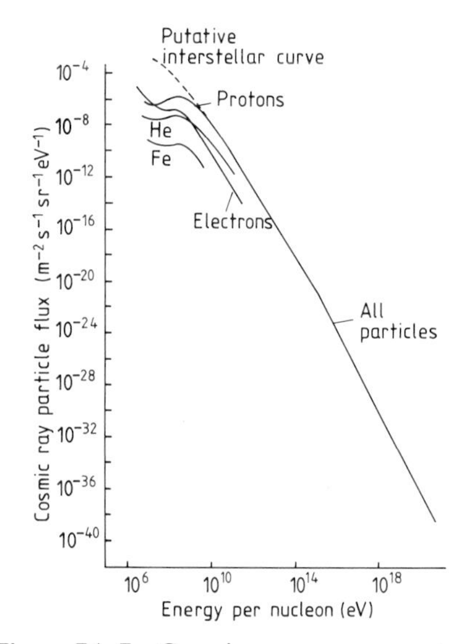

Figure 7A.7 Cosmic ray energy spectra.

There is first the enormous range in the flux, at least 33 decades. The optical equivalent would be a solar telescope which was also capable of detecting stars of magnitude + 56! Three regions may be distinguished: (*A*) up to 10^{10} eV, (*B*) 10^{10} to 10^{14} eV and (*C*) more than 10^{14} eV.

Region *A* is affected by the interplanetary magnetic field and changes rapidly with time. Diurnal and eleven-year cycles can be traced in its behaviour (see later discussion). A tentative extrapolation of the proton curve to show its form in interstellar space is given as the dotted extension on figure 7A.6.

In region *B*, the curve has a relationship

$$F \propto E^{-2.6} \tag{7A.4.1}$$

where F is the particle flux and E is the particle energy, while in region C, it is

$$F \propto E^{-3.1}. \qquad (7A.4.2)$$

An exponential law of this form is predicted by the Fermi acceleration mechanism (see later discussion). The data, however, become more uncertain as the higher energies are approached, since they are based upon relatively few particles. Thus even the difference in the exponents of equations (7A.4.1) and (7A.4.2) is not firmly established. If it does exist, then it could arise through the differing containment times of the particles within the Galaxy. Few, if any, particles with energies over 10^{20} eV have been detected. Partly this must be due to their extreme rarity even if the trend of the curve in figure 7A.6 continues unabated. However, there is reason to expect a cut-off in the energies at about 10^{21} eV. At this energy, the photons in the 2.7 K microwave background radiation are Doppler shifted into gamma rays of some 10^{8} eV or more. A cosmic ray proton of 10^{21} eV will therefore appear to be continuously bombarded by 10^{8} eV gamma rays, and they will interact to produce pions:

$$p^{+} + \gamma \rightarrow p^{+} + \pi^{0} \rightarrow p^{+} + \gamma + \gamma \qquad (7A.4.3)$$

or

$$p^{+} + \gamma \rightarrow \mathrm{n} + \pi^{+} \rightarrow \mathrm{n} + \mu^{+} + \nu_{\mathrm{e}} \rightarrow \mathrm{n} + \mathrm{e}^{+} + \nu_{\mathrm{e}} + \nu_{\mathrm{e}} + \tilde{\nu}_{\mu} \qquad (7A.4.4)$$

and so the proton will rapidly lose energy. The reality of the process is evidenced by the 10^{8} eV gamma ray diffuse galactic emission coming from the neutral pion decay.

Of the other particles, the heavier nuclei are observed only within the range where solar influences are important. The shape of the curves, and their behaviour with time, have similar characteristics and explanations to those of the protons (see later discussion). The curve for the electrons, and the few positrons, does differ noticeably from those of the other particles. Only recently have distant spacecraft observations enabled the interstellar component to be separated from terrestrial, solar and jovian contributions. The flux decreases much more rapidly with energy than that of the other particles up to the limit of observation at about 10^{13} eV:

$$F \propto E^{-3.0}. \qquad (7A.4.5)$$

The reason for this, and also for the imbalance between the proton and electron numbers, lies in the loss of energy by the electrons in the form of synchrotron radiation (§6P.2). The radiation emitted as the electrons spiral around the Galactic magnetic field appears as the diffuse Galactic radio emission between 1MHz and 1GHz. The rate of energy loss is given by

$$\frac{\mathrm{d}E}{\mathrm{d}t} = \frac{-e^{4}B^{2}}{24\pi^{2}\varepsilon_{0}^{2}m_{\mathrm{e}}^{2}c^{3}(1 - v^{2}/c^{2})} \qquad (7A.4.6)$$

where B is the magnetic field strength. For an electron with an energy of 10^{12} eV in a magnetic field of strength 10^{-9} T, we have

$$\mathrm{d}E/\mathrm{d}t = -11 \text{ eV s}^{-1} \tag{7A.4.7}$$

so that only a few thousand years suffice to slow such an electron to sub-relativistic speeds. Cosmic ray electrons will also radiate at frequencies up to the gamma rays through free–free radiation during a close passage with an interstellar nucleus. The average energy loss is given by

$$\frac{\mathrm{d}E}{\mathrm{d}t} = \frac{e^4 Z(Z+e)\log_e[191(e/Z)^{1/3}]N}{8\pi^2\varepsilon_0^3 m_e c^2 h(1 - v^2/c^2)^{1/2}} \tag{7A.4.8}$$

where Z is the nuclear charge of the interstellar atoms and N is the number density of the interstellar atoms. Thus, for interstellar conditions (§7A.3),

$$\mathrm{d}E/\mathrm{d}t \simeq -5 \times 10^{-4} \text{ eV s}^{-1}. \tag{7A.4.9}$$

Thus free–free emission is not generally significant, but it may remove a large portion of an electron's energy in particular cases, and contribute to the diffuse gamma ray emission from the Galaxy. Other energy loss processes for the cosmic ray electrons, such as the inverse Compton effect and ionisation of the interstellar gas, are much less significant than the free–free emission, and may be ignored.

The degree of isotropy of the cosmic ray flux is difficult to establish. The rotation of the Earth combined with zenithal concentration of the secondary cosmic rays (see later discussion) means that any given detector array sweeps out a band of the sky some twenty degrees wide and centred on a declination equal to the array's latitude. However, the observed direction will differ from the true direction of the cosmic ray particle due to deflection by the Earth's magnetic field. The deflection is up to 90° for a 10^{10} eV particle, and 10° for a 10^{11} eV particle. Only particles with energies of 10^{12} eV or above arrive essentially undeflected. Furthermore, the true arrival direction at the Earth will have been changed in much less predictable ways by the interplanetary and Galactic magnetic fields. Despite these and other difficulties, up to about 10^{14} eV, there appears to be a 0.05% anisotropy, perhaps due to the local flow of the interstellar medium. For energies of 10^{15} to 10^{16} eV, there appears to be a slight intensification towards the Galactic plane, attributed by some workers to the origin of cosmic rays in the sources of the gamma ray bursts. Between 10^{17} and 10^{19} eV, a deficit of several tens of per cent occurs towards the northern part of the Galaxy. This becomes a strong excess as the energy rises above 3×10^{19} eV, and this may be due to cosmic rays coming from the Virgo supercluster. Even for a true isotropic distribution, a small anisotropy will be induced into the observations by the Earth's motion through space. This will amount to about 0.1% at 10^{14} eV, and it has probably recently been detected.

The major objectives of the study of primary cosmic rays are to ascertain their source or sources, and to determine the acceleration mechanism capable of producing such enormous energies. We will see later that some of the low-energy cosmic rays originate in solar flares. The mechanism in this case might be a pinch effect as the angle between the magnetic field lines decreases (§4A.3). Presumably other solar-type stars have a similar behaviour, and certainly some cool dwarfs are known to exhibit flare activity. Perhaps more active stars might eject particles with energies one or two orders of magnitude higher than those of the Sun. 10^{11} eV or so, however, probably represents the upper limit to the possible contribution to the cosmic ray flux by relatively normal stars. A possible means of further acceleration of charged particles has been proposed by Fermi. It is based on the reflection of charged particles by magnetic fields of sufficient intensity. Imagine a charged particle and such a magnetic reflector moving towards each other at velocities v and V respectively, as seen by an external observer (figure 7A.8). Then conservation of momentum within the moving frame of reference of the magnetic reflector will lead to a gain of energy by the particle as seen by the observer. The particle's velocity after reflection, v', is given by

$$v' = \frac{v + 2V + vV/c}{1 + (2vV/c^2) + V^2/c^2} \tag{7A.4.10}$$

and the gain of energy by

$$\Delta E = E\frac{2V(v + V)}{c^2 - V^2} \tag{7A.4.11}$$

where E is the energy of the particle before reflection. If the magnetic reflector should be moving in the same direction as the particle as seen by the observer, then there will be a loss of energy in a similar manner. If a particle were between two moving reflectors approaching each other, then it would

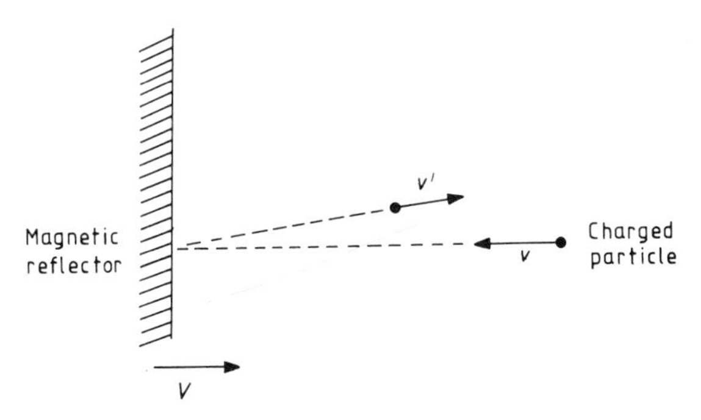

Figure 7A.8 Physical situation for the Fermi acceleration mechanism.

bounce back and forth, gaining energy each time. The potential final energy would be limited only by the number of reflections and the energy loss mechanisms, provided that the particle did not become sufficiently energetic to penetrate the reflectors. In a mixed situation where a number of reflectors may be involved, some adding energy and some removing it, the particle will still have a net gain of energy. This arises because the time intervals between reflections when a particle is gaining energy are shorter on average than those when it is losing energy (figure 7A.9), and so the rate of energy gain is higher than the rate of energy loss. The energy spectrum of a number of such particles is given by

$$F \propto E^{-1-(c^2/2\bar{n}V^2)} \tag{7A.4.12}$$

where $\bar{n}$ is the mean number of reflections per particle. This is of the same form as the observed middle- and high-energy cosmic ray spectrum (equations 7A.4.1, 7A.4.2). The Fermi mechanism was originally envisaged as operating by reflections between interstellar gas clouds. Such a process is too slow for sufficient energy gain during the three-million-year retention

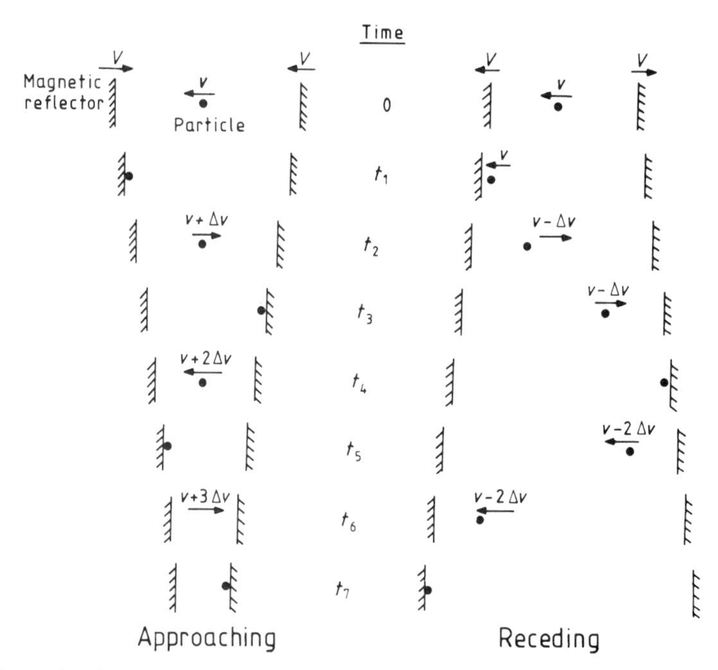

Figure 7A.9 Schematic illustration of how the rate of energy gain by the Fermi mechanism can exceed the rate of energy loss. The particle between the approaching magnetic reflectors undergoes four reflections in the time that the particle between the receding reflectors undergoes three reflections.

time of cosmic rays within the Galaxy, and is too likely to suffer significant energy losses precluding the particle from reaching the higher energies. It may more probably be found to occur in and around the shock waves, magnetic, turbulent, and other interactions to be found in supernovae, supernova remnants, pulsars, AM Her stars, dwarf novae etc. It could there lead to high energies for the particles very quickly, so that their path lengths and opportunities for losing energy are small.

Whether or not Fermi's mechanism provides the energy of the primary cosmic rays, we have still to find the type of objects from which they originate. Supernovae are currently favoured for several reasons: (*a*) They are the most violent and energetic regularly occurring local events. (*b*) The diffuse galactic gamma ray emission shows peaks which may be associated with the Crab and Vela supernova remnants, and less certainly with other SNRs. (*c*) The diffuse Galactic radio emission shows a feature, the north Galactic spur, which may be the result of a recent supernova. (*d*) The overabundance of the iron-group nuclei is to be expected from supernova material. (*e*) Sufficient energy is available to provide the observed flux. The energy density of primary cosmic rays near the Earth is about 10^6 eV m^{-3}, and the diffuse gamma ray and radio emission from the Galaxy support the view that a similar density holds throughout the disc of the Galaxy. Thus the total energy stored in cosmic rays within the Galaxy is around 10^{47} J. We have seen that the lifetime of a cosmic ray particle within the Galaxy is about three million years, so that energy must be supplied at a rate of 10^{33} W to maintain a steady cosmic ray population. This is less than one per cent of the mean energy release from supernovae.

Nevertheless, this is only circumstantial evidence, so other or additional sources may be needed. There is a strong probability that the operation of the Fermi mechanism in supernova remnants contributes to the medium-energy cosmic rays, and there is also the possibility of corotation with neutron stars accelerating particles to relativistic velocities. There may be links with the postulated sources of gamma ray bursts such as the Gemini gamma ray source, Geminga, or with Cyg X-3, and with black hole candidates such as Cyg X-1.

Most cosmic rays are trapped by the Galactic magnetic field for several million years. But if they are not destroyed in some interaction, they are likely eventually to leak out of the Galaxy. Liouville's theorem states that in the absence of collisions, the phase-space density of such particles is constant:

$$\mathrm{d}\rho/\mathrm{d}t = 0 \tag{7A.4.13}$$

where ρ is the number of particles at time t with spatial coordinates in the range $x \to x + \Delta x$, $y \to y + \Delta y$, $z \to z + \Delta z$, and momenta between $p_x \to p_x + \Delta p_x$, $p_y \to p_y + \Delta p_y$, $p_z \to p_z + \Delta p_z$, so that eventually the density of the particles in intergalactic space must approach that within the Galaxy. This does not yet seem to be the case for most particles, though the very highest energies ($>10^{19}$ eV) may be approaching the equilibrium situation. If the situation is even remotely approaching equilibrium, then the bulk of the particles will be extragalactic in origin.

Before concluding this discussion of the primary cosmic rays, it is worth noting the coincidence between a number of different energy densities in the interstellar medium, as first pointed out by Hoyle.

Effect	*Energy density* ($\times 10^{-13}$ J m^{-3})
Primary cosmic rays	1.6
Starlight	0.7
Interstellar gas turbulence	0.5
Galactic magnetic field	4
2.7 K microwave background radiation	0.4

The interactions between the cosmic rays, the interstellar gas, and the Galactic magnetic field seem likely to be sufficient for a rough equipartition of their energies. The significance, if any, of the other coincidences, however, remains to be found.

Secondary cosmic rays

We have seen that the passage of a primary particle through about 1.4×10^3 kg m^{-2} of hydrogen is sufficient on average for it to undergo a collision with a nucleus. The equivalent figure for air is about 800 kg m^{-2}. Now the mass of a vertical 1 m^2 column of the Earth's atmosphere is about 10^4 kg. So a primary cosmic ray particle will normally collide with a nucleus within the top 10% or so of the atmosphere, i.e. at a height of 30 to 60 kilometres. During such a collision, especially at the higher energies, it matters little whether the primary particle is a proton or a heavier nucleus. The nucleons in a heavier nucleus will interact essentially as individuals, rather than as members of an assemblage. We need only consider, therefore, the effects of high-energy proton impact.

Consider first a medium-energy proton of 10^{14} eV or so. Upon its first interaction with a nucleus, it will produce the whole gamut of subatomic particles. The details of the reaction are of considerable interest to the subatomic physicists, since the current energy limits of their accelerators are some 2×10^{14} eV. For our purposes, however, the significant reaction products are the nucleons and the π mesons (pions). The original particle may retain a significant fraction of its energy after the first interaction, and so go on to further collisions. Furthermore, the energies of the nucleons produced in the early collisions are likely to be sufficient for them to cause interactions of their own. Thus a burst of nucleons is rapidly produced in a narrow cone around the direction of the original particle. This is termed the nucleon cascade. Eventually the energies of the particles fall to the point at which they can no longer produce further particles during their interactions. They then lose energy via ionisation of the atmospheric gases, and the cascade is gradually absorbed. The pions meanwhile behave very differently. They also spread out in a cone around the original trajectory, but as they are unstable particles, they decay. Neutral pions have a lifetime of only 8×10^{-15} seconds, and decay according to

$$\pi^0 \rightarrow \gamma + \gamma \qquad (7A.4.14)$$

$$\pi^0 \rightarrow e^- + e^+ + \gamma \qquad (7A.4.15)$$

producing mostly high-energy gamma rays. The gamma rays may then produce positron–electron pairs, and these in turn produce further gamma rays:

$$\gamma \rightarrow e^+ + e^- \qquad (7A.4.16)$$

$$e^{\pm} \rightarrow e^{\pm} + \gamma \qquad (7A.4.17)$$

resulting in the electron–photon cascade or soft component of the secondary cosmic rays. As the energies of the particles or photons decrease, ionisation, the Compton effect, etc, again gradually lead to absorption by the atmosphere. The charged pions (π^+, π^-), have a much longer lifetime, 2.6×10^{-8} s, and quite different decay modes:

$$\pi^+ \rightarrow \mu^+ + \nu_e \qquad (7A.4.18)$$

$$\pi^- \rightarrow \mu^- + \nu_e . \qquad (7A.4.19)$$

The neutrinos are of no further concern here, but the μ mesons (muons) form the muon shower, or hard component of the secondary cosmic rays. Although the muon lifetime is only 2.2×10^{-6} s, the relativistic time dilation, given by

$$\text{Observed lifetime} = (E/mc^2) \times 2.2 \times 10^{-6} \text{ (s)} \qquad (7A.4.20)$$

where E is the muon energy and m is the muon rest mass, stretches their existence for a terrestrial observer to the point where most of them survive for the flight time of 10^{-4} s to the surface of the Earth. Some of the muons, however, do decay, to produce more electrons:

$$\mu^+ \rightarrow e^+ + \nu_e + \tilde{\nu}_\mu \qquad (7A.4.21)$$

$$\mu^- \rightarrow e^- + \nu_e + \tilde{\nu}_\mu . \qquad (7A.4.22)$$

The whole complex of particles: nucleons, photons, electrons, pions, muons, etc, is known as a cosmic ray shower or extensive air shower, and they form the secondary cosmic rays.

For energies of 10^{14} eV or more, a large fraction of the shower particles survives to be observed at ground level. As the energy of the primary particle decreases, however, fewer secondary particles are produced, and the absorption phase of the shower development starts earlier. Thus, by 10^{10} eV, few secondary particles other than neutrons survive to ground level. At very high energies a process similar to that for a 10^{14} eV particle is assumed to occur, but on an even more extensive scale. Since there are no laboratory data for such energies, however, it is possible that additional or alternative processes may be occurring.

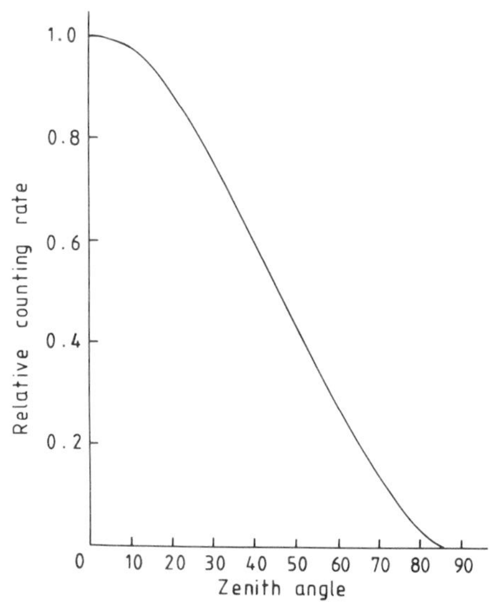

Figure 7A.10 Relative counting rate of a directional cosmic ray detector as a function of zenith angle.

There is a strong dependence of the shower strength upon the air mass along its path. The counting rate of directionally sensitive detectors therefore decreases markedly as the zenith angle of the detector increases (figure 7A.10). This is known as the zenithal concentration of the secondary cosmic rays. Other effects include the latitude effect (figure 7A.11) due to the greater ease of penetration of the low-energy primary particles near the magnetic poles, and the peaking of the shower intensity at a height of 10 to 15 km above sea level (figure 7A.11).

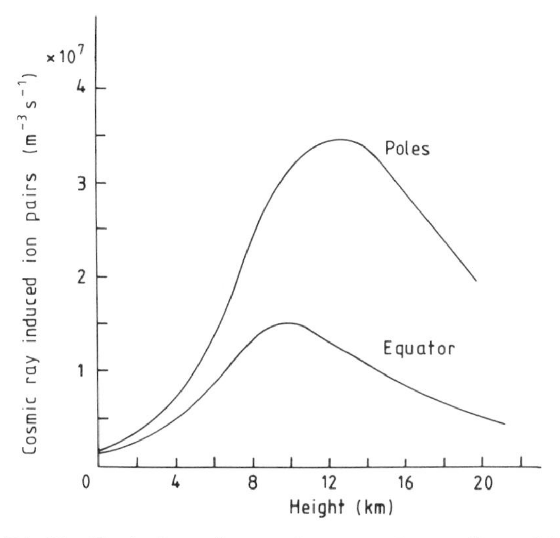

Figure 7A.11 Variation of secondary cosmic ray flux with height.

Solar cosmic rays

The Sun influences cosmic rays in two quite separate ways. First it emits low-energy cosmic rays itself, and secondly its magnetic field influences the flux of the Galactic cosmic rays.

Solar particle emission

By solar cosmic rays, we mean the particles emitted by the Sun at relativistic velocities. Thus solar wind (§5A.1) particles, etc, are not included. The majority of the solar cosmic rays are associated with flares (§4A.3); a few related events such as x-ray bursts may also contribute on a minor scale. The flux is thus very highly variable. Few of the solar cosmic rays produce effects detectable at ground level. But occasional very dramatic events can occur: in February 1956, for example, the normal cosmic ray background increased by over a factor of 50 at ground level in a few minutes, due to solar cosmic ray emission. The solar particles may more commonly signal their presence by the production of radio fade-outs (polar cap absorption events), and aurorae. Most studies of solar cosmic rays, however, have to be undertaken from spacecraft.

The composition of the solar cosmic rays mirrors that of the Sun at energies above about 10^7 eV. At lower energies, the heavier elements are enriched by up to a factor of fifty. This is probably due to the greater ease of escape of the partially ionised heavy ions from the solar magnetic fields compared with that for the completely ionised lighter nuclei. Some examples of the energy distributions of solar cosmic rays are shown in figure 7A.12. An idea of their variability may be obtained from the comparison of the average curves with those for specific major flares. Unlike the primary cosmic rays, electrons and protons are present in equal quantities, and the electron spectra are shown in figure 7A.13. Solar cosmic rays, as may be seen from the comparison of figures 7A.7, 7A.12, and 7A.13, are of low energy compared with the Galactic cosmic rays, but can reach peak fluxes a million or more times higher. Neutrons are also found with energies up to 10^8 eV coming from the Sun, but not from the Galaxy.

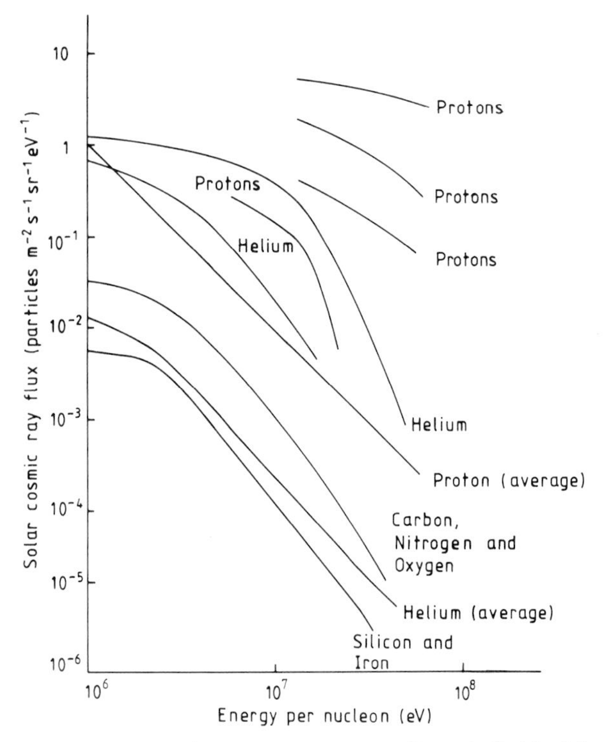

Figure 7A.12 Solar cosmic ray energy spectra (from individual flares and averages).

The burst of particles from a flare is known as a solar proton event, and such events display a reasonably consistent pattern of behaviour. About three to thirty minutes after the visible occurrence of a flare, a sharp increase in the cosmic ray flux is noticed. The delay in the particle arrival time is probably due to their long and tortuous paths around the inner portions of the interplanetary (i.e. extended solar) magnetic field. Recently, however, a neutron component also delayed to a similar extent has been observed. The neutrons cannot be directly accelerated but must be produced by collisions between the high-energy charged particles and atoms high in the solar atmosphere. Thus it may be that the delay time actually arises from the time required for the acceleration of the particles, since the neutrons will not be deflected by the magnetic fields. For about the first hour, the particles arrive preferentially from the direction of the Sun. They are guided by the magnetic field, and so do not move radially, but spiral outwards to arrive at the Earth from a direction about 10° ahead of the Sun. The particle intensity thereafter slowly decreases back to the normal background over the next twelve hours or so. Simultaneously their distribution becomes more isotropic as the particles are scattered and trapped in the inner solar system. (N.B. not all flares produce observable proton events, since the magnetic fields may guide the particles completely away from the Earth's vicinity.)

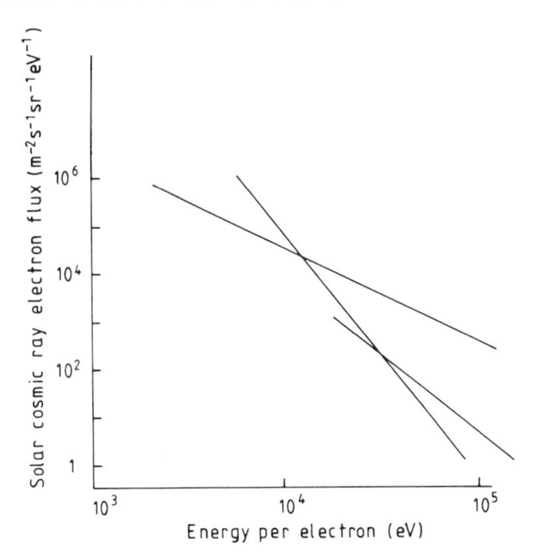

Figure 7A.13 Solar cosmic ray electron energy spectra (from individual flares).

Proton micro-events have also been recognised. Their intensity does not rise above 2×10^{-4} protons $m^{-2}\,s^{-1}\,sr^{-1}\,eV^{-1}$. These can be very common: one or more per day for major active regions on the Sun. They may be associated with x-ray bursts and other activity as well as with flares.

The origin of the solar cosmic rays is obviously closely tied to the mechanism producing solar flares. Unfortunately (§4A.3), that mechanism is not understood. Most theories of flares use the reconnection of magnetic fields near X- or Y-shaped magnetic neutral points to produce intense electric currents, which subsequently produce the observed effects of the flare. The theories mostly fall down by predicting time scales for the processes a thousand or more times longer than those observed. It is therefore difficult to make reasonable models to explain the production of the cosmic rays. One possibility, however, is that charged particles are caught between two closing magnetic fields, and squeezed out from between them at many times the closing velocity by the Lorentz forces (cf tiddlywinks, or soap squeezed between the fingers!).

The Sun is not the only source of high-energy particles in the solar system. Electrons with fluxes and energies high enough to produce observable effects near the Earth are occasionally found to come from Jupiter. Their origin seems to be in the interactions between the planet, its satellite Io, its trapped radiation belts, and its magnetosphere.

The long-term stability of the solar particle emission is difficult to establish. The intensity obviously varies with the solar cycle, and with variations in this cycle such as the Maunder minimum (§4A.3). The ancient fluxes of heavy nuclei may be estimated from the ionisation damage tracks which they produce in meteorites and lunar samples. It is very difficult to relate such data to specific times in the past, but within quite wide limits, the flux averaged over millions of years appears to be constant. Occasional very intense emission events or periods cannot be ruled out, however, and indeed are postulated as one explanation for the extinction events in the fossil record. The particle emission during such an active period, if it were to coincide with a reversal of the Earth's magnetic field, might deplete the ozone layer, allowing dangerous shortwave ultraviolet radiation to penetrate to ground level, causing mutations and extinctions.

Solar cosmic ray modulation

The intensities of the cosmic rays with energies less than about 10^{12} eV are strongly affected by the extended solar magnetic field (otherwise known as the interplanetary magnetic field). The basis of the effect is that the particles will be channelled by the magnetic field, unless the curvature of the field is stronger than the curvature of the particle's paths around the field lines. The radius of the particle's path is just the Larmor or gyro radius (equation 6P.2.3):

$$R_G = \frac{m v_p}{ZB(1 - v_p^2/c^2)^{1/2}} \tag{7A.4.23}$$

where m is the mass of the particle, Z is the charge of the particle, v_p is the component of the particle's velocity perpendicular to the magnetic field and B is the field strength. This has a value of about two million kilometres for a 10^{10} eV proton in interplanetary space. Thus if a region of the solar magnetic field has a radius of curvature less than about a million kilometres, low-energy cosmic rays will separate from its guiding effect at a point and in a direction that will depend in detail upon the properties of the field and the particle. Such a region of the interplanetary magnetic field effectively therefore acts as a scattering centre for the cosmic rays.

Now the large-scale pattern of the magnetic field is of inner and outer disordered zones, with a region from perhaps 0.1 to 2 AU which has a roughly radial structure and which corotates with the Sun. This radial zone often divides into sectors of predominantly uniform polarities (§4A.5).

These ideas provide sufficient background to yield qualitative understanding of the three main solar modulation effects: (*a*) the low-energy flux reduction and its 11-year cyclic variation; (*b*) the diurnal variation and (*c*) the Forbush decrease.

The first of these is shown in figure 7A.7 where the observed curve for the Galactic primary cosmic ray protons diverges from that expected to exist in interstellar space at energies below about 10^{10} eV. The other particles also have a similar flux reduction, and behave in a similar way. The reduction in the intensity is due to the scattering centres formed by the solar surface magnetic field irregularities as these move outwards through the solar system. Since the magnetic distortions are greater near centres of activity, there is a greater density of scattering centres, and hence a greater reduction in the low-energy flux, at solar maximum than at solar minimum. Suitable measures of the cosmic ray fluxes have only been made over about two solar cycles. The data are thus not very complete, but a reduction of about 40% in the low-energy Galactic cosmic ray flux seems to be indicated between solar minimum and maximum. Strong supporting evidence for this comes from the changes in the abundance of ${}^{14}_{6}C$ with time. This is produced by the action of the cosmic rays upon atmospheric nitrogen. Its abundance is strongly anticorrelated with sunspot number, as would be expected if the magnetic fields have been affecting the cosmic rays for at least the last three centuries (see §4A.3 for a more complete discussion of the variations).

The diurnal effect is much smaller: about 0.5%. It is due to the channelling of the low-energy cosmic rays along the local field lines. These are nearly radial to the Sun near the Earth, and the Earth's daily rotation then leads to the observed variation.

The Forbush decreases can be quite large: 10% or more. The initial decrease occurs quite rapidly, in a few tens of minutes or so, and is followed by a return to 'normal' over a few hours or days. The cause is the passage of one of the scattering centres past and around the Earth, during which time the Earth is directly shielded from the cosmic ray particles.

PROBLEMS

7A.1 If a supernova which reaches a peak absolute magnitude of -19 leads to the destruction of the interstellar dust grains by sputtering throughout about 100 $\mathcal{M}_\odot$ of the interstellar medium, estimate from the projected supernova rate (§3A.5) the rate of destruction of dust grains in the whole Galaxy. Assuming the interstellar medium to have a mean density of 10^7 hydrogen atoms per cubic metre, and to be formed into a disc approximately 20 000 pc across and 200 pc thick, estimate the lifetime of an average dust particle under this loss mechanism.

$$m_H = 1.67 \times 10^{-27} \text{ kg} .$$

8A Stellar Formation and Evolution

SUMMARY

Formation of stars from the interstellar medium, planetary systems, main sequence, post main sequence evolution for low mass stars, medium mass stars, high mass stars, and stars in close binary systems, end points of stellar evolution.

See also stellar masses (§§1A.1, 2A.1), stellar compositions (§§1A.2, 3A.6), virial theorem (§2P.5), energy sources (§2A.2), energy transfer (§2A.3), luminosity, size and density (§§2A.4, 2A.5, 3A.1, 4A.1), radiative transfer (§3P.2), variable stars (§3A.5), supernovae and remnants (§§3A.5, 6A.6), rotation (§3A.7), magnetic fields (§3A.8), black holes (§3A.9), solar neutrino problem (§4A.4), stellar winds (§5A.2), masers (§§6P.1, 6A.2), H II regions (§§6A.1, 6A.2, 6A.3), planetary nebulae (§6A.5), interstellar medium (§§7A.1, 7A.2, 7A.3), and cosmic rays (§7A.4).

INTRODUCTION

With time scales for evolutionary changes (see note 1 to the Preface) in stars generally in the range 10^5 to 10^{10} years (apart from supernovae, and one or two other exceptions), it is obviously impossible for such changes to have been observed over the three- or four-century existence of modern astronomy. Nonetheless, it is still possible to deduce from the data discussed in the foregoing chapters the probable nature of the evolutionary changes. A frequently used analogy, originally due to Herschel, consists of imagining asking a biologist to deduce the lifecycles of trees, shrubs and other plants in a primaeval forest after a fifteen-minute walk through it. No actual changes would be observed during such a walk, but upon noticing seeds, seedlings, saplings, trees, decaying fallen trees and branches, etc, a plausible scenario could be deduced to link them together. The biologist could not be certain, however, of the validity of the deduced lifecycles, no matter how much supporting circumstantial evidence might accumulate. The astronomer trying to deduce stellar lifecycles is in a rather similar position. Much of the directly observed and inferred properties of stars and the interstellar medium can be tied into a plausible set of linked phenomena, generally called stellar evolution. In many ways biological systems are more complex and less predictable than physical ones. So the astronomer can probably be more confident of his deductions than could the biologist in the analogy. It would thus be very surprising if the picture were widely wrong. Nonetheless, in the following discussions it should be remembered that this is a deduced scenario, and that at least some of the finer details must be expected to change as our understanding improves.

There are many problems of both detail and principle with most aspects of this topic. They have often been mentioned in earlier parts of the book, and where particularly important are highlighted in the subsequent discussions in this chapter. They usually underlie leading research areas and are items of greatest contemporary interest. Some of the outstanding problems at the time of writing are gathered together here before the main sections of this chapter to try and obviate any impression that may be generated later of omniscience or complacency.

First, the widely accepted evolutionary scenario which is considered in the remainder of this chapter is not without its alternatives. The most important of these envisages an explosive origin for stars. The collapse of a 10^6 solar mass gas cloud might have a core of 10^2 to 10^4 solar masses which could reach stellar densities at some stage of the collapse. Such a mass would be unstable, as we have already seen (§2A.1), and would explode. The fragments might then form the stars of a galactic cluster. Another alternative, due to Ambartzumian, suggests that stars originate from masses of material of nuclear densities which revert to normal densities, becoming visible in the process.

It is clear that many processes known to be occurring in stars, gas clouds, and the interstellar medium have yet to have their effects on stellar formation and evolution taken into account. Among the more significant we may identify the following.

(i) Mixing of products of nuclear reactions into the outer layers of the star. FG Sge, for example, with its excess barium and strontium, may be the result of such a process, and planetary nebulae nuclei may exhibit mixing into the intermediate layers which has been revealed by the loss of the surface layers of the star. There may also be non-convective mixing of various layers of the star, rotation, diffusion and magnetic fields being the main candidates for such a process. Ap stars and objects such as Przybylski's star are suspected of being examples of the operation of the last two effects. The importance of differential rotation, if any, in this context remains to be determined. Tidal effects in close binary systems could also lead to unusual mixing patterns. The effect of the complete convective mixing of the whole star in very low-mass stars has still to be elucidated in detail.

(ii) The role of interstellar dust particles in aiding the collapse of gas clouds by providing nuclei for the condensation of hydrogen molecules.

(iii) The effect of the interstellar magnetic field on cloud collapse and star formation.

(iv) The effect of rapid rotation, mass loss, etc on stellar evolution.

(v) Uncertainties in the mixing length in convective regions, especially in the envelopes of low-mass stars, leads to corresponding uncertainties in their models.

(vi) Neutrino processes, especially cooling in white dwarfs, may have much more significance than they are presently given.

(viii) Most modelling of evolutionary sequences relies on the assumption of quasi-static changes. For many occasions such as the initiation of a new nuclear burning stage, shell flashes, etc, this is far from valid.

Close binary stars can have very complex interactions (§8A.2). Apart from the qualitative outlines of the probable effects, however, definitive models of these systems remain to be developed. The origins of binary star systems, and the possibly related problem of the origin of planetary systems, still have to be determined with certainty. The answer to this problem may reveal why a majority of stars in the disc of the Galaxy belong to binary systems, but in the galactic halo the majority are isolated single stars. There is also the problem of whether stellar masses determined from binary systems (§1A.1) can be taken as representative of those of similar but single stars.

The compositions of most stars resemble that of the Sun (§1A.2); some stars, however, differ markedly from the solar pattern. Wolf–Rayet stars, of both types, seem likely to be the cores of evolved massive stars which have become visible through their outer layers being stripped off in a violent stellar wind, their compositional peculiarities then just being due to the nucleosynthetic reactions preceding their present states. In other cases, such as the R, N and S spectral types, and the overabundance of ${}^{3}_{2}He$ in some stars, the explanation for peculiarity remains unknown. White dwarfs have great uncertainties over their compositions, little observational evidence exists (§1A.2), and the theoretical models range from pure helium to pure iron spheres.

Understanding the processes underlying the origin of stars has a long way to go before it becomes satisfactory. Thus the recently discovered bi-polar outflows from many proto-stellar objects have still to be explained, and the duration of the accretion phase and whether it can coexist with outflows in the form of jets is still unknown. The rate of formation of stars may just be a statistical average, or some feedback system on the lines of

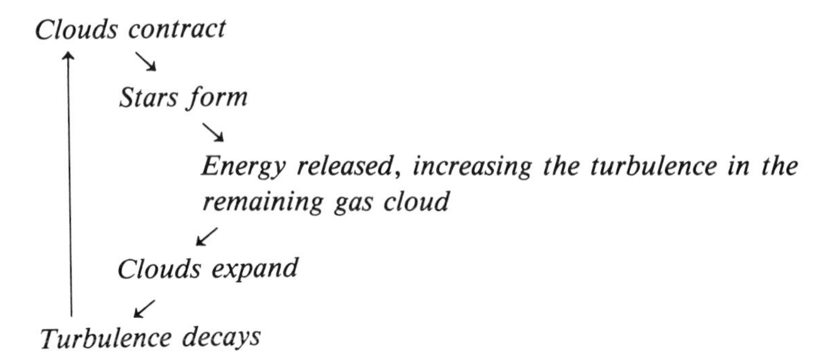

may be involved. In a wider context, star formation is presumably intimately interconnected with galaxy formation. Therefore problems such as why some clusters of galaxies (e.g. the Coma cluster) contain predominantly elliptical galaxies, whilst others have a mix of elliptical, spiral and irregular galaxies, may well have a bearing upon stellar evolutionary problems and/or receive explanations from the solutions to stellar evolutionary problems.

Problems with individual stellar types abound. Thus pulsating stars such as Cepheids have a significant discrepancy between their masses estimated from pulsation theory, and those estimated from stellar evolutionary theory. The reason for the double-beat Cepheids, and the balance between fundamental and first-harmonic oscillations in pulsating stars more generally, has yet to be found. In dwarf novae, the cause of the outbursts, and the reason for their occasional cessation in the Z Cam stars has still to be determined. The shock-bounce theory of type II supernovae leaves much to be desired in the reality of its predictions. Type I supernovae may originate in the collapse of a white dwarf in a close binary system as accretion takes it over the Chandrasekhar limit, but this has yet to be firmly established. The relationship between neutron stars, pulsars and supernova remnants also needs a satisfactory theory. For even the best known star, the Sun, we still need explanations for the solar neutrino problem, the origin of flares, and the cause of the various cycles of activity. The energy source of the corona, long thought to be acoustic energy, now appears perhaps to be magnetic in origin, thereby possibly providing an explanation for the wide x-ray variability amongst stars of otherwise similar characteristics. The origin of planetary nebulae, and the relationship, if any, of their central stars when these have Wolf–Rayet-type spectra to the more normal Wolf–Rayet stars, are both uncertain. The list could be prolonged almost indefinitely, but will be stopped here before it takes over the whole book!

In a wider context, there are intriguing questions such as whether the rough equipartition of energies of the components of the interstellar medium (§7A.3) is of significance or not, and whether accretion discs around black holes can provide the energy for some of the compact high-energy sources in Seyfert galaxies and quasars, etc. By no means finally, but the last item to be mentioned here, there is the problem of the 'missing mass' in the Universe. This latter problem arises through the apparent requirement for the stability of clusters of galaxies for them to contain up to 100 times the amount of matter that is actually visible in the form of galaxies. One possible form in which this might be found is as 'brown dwarfs', or very low-mass, low-luminosity stars. The stellar mass distribution (§1A.1) depends on $\mathcal{M}^{-2.35}$ for stars over about half a stellar mass, but this appears to fall to a dependence as $\mathcal{M}^{-1.35}$ at the low-mass end of the distribution. If the higher mass distribution were to continue to the low masses, then the vast majority of stars would be similar to the companion of VB8: only a little more massive than Jupiter, and with temperatures of a few hundred to a thousand or so degrees, and detectable (with present techniques) only out to a few parsecs.

The remainder of this chapter summarises the main ideas of star formation and evolution, and links these where possible with the data in the preceding chapters, to try to present a coherent pattern. It is not intended to be a totally thorough and complete treatment of the subject. The interested reader is referred to more specialised books and to the research journals, if such a treatment is required.

8A.1 STAR FORMATION

We may usefully consider the aging process of stars in several separate stages, the first being the formation of the star from the interstellar medium.

The main current, if not only, sites of star formation appear to be the giant molecular clouds (§7A.3). These are concentrated in the Galactic disc, particularly, but not exclusively, into the spiral arms. There are perhaps several thousand of these giant molecular clouds in the Galaxy, the Orion nebula being a small nearby example. Typical giant molecular clouds have masses in the region of 10^5 to $10^6 \mathcal{M}_\odot$, and sizes of 20 to 60 pc. The density averages some 10^9 hydrogen molecules per cubic metre. Within such a cloud, small denser regions may be expected to develop to the point at which they become self-gravitating. It may be that turbulence and random variations will suffice to cause this, but more probably an external initiating mechanism will be needed to compress the molecular cloud material. Candidates for such a trigger include shock waves and ionisation fronts from supernovae, H II regions, planetary nebulae etc, collisions between molecular clouds, and a density wave spreading out from the centre of the Galaxy following a Seyfert-type explosion (*1). By some such process, a region about a parsec across may reach densities of 10^{10} to 10^{11} H_2 molecules per cubic metre. Its mass would thus be about 10^2 to $10^3 \mathcal{M}_\odot$. As previously discussed (§7A.3) molecular line emissions cool such clouds, and the efficiency of the cooling process increases rapidly with density. The temperature within such a cloudlet is therefore likely to fall to 50 to 100 K, and it will start to collapse under its own gravitational forces. The minimum mass for such a collapse is known as the Jeans mass, and is given by

$$\mathcal{M}_J = \frac{4\pi^{5/2}k^{3/2}T^{3/2}}{3(\mu m_H)^2 G^{3/2} N^{1/2}} \tag{8A.1.1}$$

$$\simeq 4 \times 10^5 (T^3/N)^{1/2} \quad (\mathcal{M}_\odot) \tag{8A.1.2}$$

where T is the kinetic temperature of the material and N is the number density.

During the collapse, the internal temperature remains low, and may fall to 10 K. Thus, a short while into the collapse, the Jeans mass may fall to a few solar masses. Shock waves and gravitational instabilities are then likely to lead to the fragmentation of the cloudlet into tens or hundreds of stellar mass units with dimensions of a few tenths of a parsec. Each of these, provided that it exceeds the appropriate Jeans mass, will be contracting individually, and may now be identified with a proto-star. Observationally, this stage may be represented by the Bok globules (figure 6A.1). These are compact, roughly spherical dark nebulae some 0.001 to 0.1 pc in size, best seen silhouetted against a bright nebula. They tend however to be associated with regions where the gas density is lower than expected for star forming regions, and so they may not be quite at this stage. They have also been suggested as candidates for identification with later stages of the proto-star's collapse.

Once such a proto-star has formed, we may consider its further progress more or less in isolation from the rest of the cloud; only something as radical as a nearby supernova or a collision with another proto-star would be likely to change its behaviour. The initial collapse of the proto-star takes its number density to the region of 10^{12} to 10^{13} m^{-3}, the temperature down to 10 K, and the size to 0.01 pc. The interstellar magnetic field is dragged with the collapsing material and its strength increases as the square root of the density increase. The time scale for this stage of the collapse is perhaps 10^6 to 10^7 years.

The second phase of the collapse occurs more rapidly, the material falling inwards at near the free-fall rate. In some 10^4 to 10^5 years, the size of the proto-star decreases to about 0.001 pc (200 AU), or comparable with the outermost reaches of the present solar system. The internal temperature remains close to 10 K throughout most of this collapse. Towards the end, however, the density rises towards 10^{18} m^{-3}, and the material becomes optically thick to the molecular emissions. The energy released by the collapse then goes into heating the interior of the proto-star, and the resulting gas pressure starts to slow the collapse. The collapse is probably non-homologous, with the outer regions collapsing much more slowly than the core. The dense centre is thus probably still surrounded by a lower density nebula stretching out to 0.1 pc, and still collapsing. Conservation of angular momentum may lead to instabilities within the collapsing proto-star, and its further fragmentation. The Jeans mass could fall as low as 0.001 $\mathcal{M}_{\odot}$, and so close binary or multiple star systems, and even planetary systems, could originate at this stage.

Once the core of the proto-star becomes optically thick to the molecular radio emission, it enters the final phase of the collapse. The temperature at the centre rises to 1500 to 2000 K as the released potential energy goes partially into the kinetic energy of the particles (§2P.5). By this time the central density approaches 10^{26} m^{-3}, and the size of the core of the proto-star is down to less than 1 AU. Eventually pressure equilibrium would be reached if this temperature rise were to continue unaltered, and the collapse would stop. At about 2000 K, however, the hydrogen molecules start to dissociate. This requires the input of some 2×10^8 J kg^{-1} for a pure hydrogen molecular gas. The temperature therefore remains constant at 2000 K, and the collapse proceeds isothermally so long as significant levels of hydrogen dissociation are occurring. For a solar-mass proto-star, about 4×10^{38} J are required to dissociate the hydrogen, and this would permit the collapse to continue for a further factor of two or so.

Eventually most of the hydrogen will have been dissociated, and the central temperature will start to rise again. As it approaches 5000 to 6000 K, however, the hydrogen starts to ionise. This requires the input of 1.5×10^9 J kg^{-1}, and so another isothermal contraction follows. Once the hydrogen has been ionised, single and double ionisations of helium continue to allow isothermal contractions at central temperatures around 10 000 and 20 000 K respectively. Although the helium is present in the material only to the extent of 8 to 10% by number, its complete ionisation requires six times the energy required for hydrogen, and so a similar total amount of energy will be absorbed. After the exhaustion of these energy sinks, the collapse again leads to an increasing central temperature. As this rises above about 10^6 K, nucleosynthetic reactions can begin (§2A.2), and the object becomes a star rather than a proto-star.

We have seen (§3A.7) that most stars rotate slowly; conservation of angular momentum during a collapse such as that just described would lead to rotational velocities averaging a hundred times the observed values. Thus, at some stage, stars must lose most of their angular momentum. In a few cases, it is possible that the collapsing cloudlet might form a rapidly rotating ring of material. If this then fragmented into several proto-stars, the angular momentum could be taken up by the stars' relative spatial velocities, rather than by their rotations. More normally we may expect that in the pre-nuclear burning phase the proto-star will be several times brighter than its eventual main sequence luminosity, while still being surrounded by large quantities of gas and dust 'left behind' during the collapse. An attractive and more generally applicable scenario for the angular momentum loss is then for its transfer from the proto-star to this surrounding nebula. Conservation of the magnetic field during the collapse of a proto-star could lead to field strengths of 0.01 T or more at this stage. Rotation will have flattened the proto-star and its surrounding nebula into a thin disc, with the magnetic field frozen into the plasma of the proto-star and the nebula. Since the former is rotating rapidly, while the latter is likely to be in orbital motion, the angular velocity of the proto-star will far exceed that of the nebula. The field lines connecting the two will therefore become wound up, exerting forces which try to force the nebula to corotate with the proto-star. Thus angular momentum will be transferred from the proto-star to the nebula. Far from this leading to corotation however, the loss of angular momentum from the proto-star will enable it to contract further, and actually increase its angular velocity, while the increased angular momentum of the nebula will cause it to move to a higher orbit, where it will actually have a reduced angular velocity. Thus, rather than leading to equalisation of the angular velocities, the magnetic field exacerbates their differences. The net effect is to transfer angular momentum outwards, allowing the proto-star to collapse (in the equatorial plane), and driving the nebula slowly away from it. During this stage, it may be that the material to form the planets condenses out around the nuclei provided by the interstellar dust grains. The differences between the jovian and terrestrial planets would arise naturally from the temperature structure of the nebula. Such an ejection process may also underlie the bi-polar outflows of material which have recently been observed to be associated with many proto-stars (§7A.3), though a definitive mechanism for their production remains to be found at the time of writing.

Eventually the proto-star becomes nuclear-burning. The increasing temperatures, radiation pressure, and the development of a stellar wind then drive off any remaining nebulosity, taking with it most of the magnetic field and angular momentum, and leaving a system comparable perhaps to the early solar system. The higher rotational velocities for the hotter stars (figure 3A.22) may arise through the initiation of hydrogen burning, before the angular momentum transfer has had a chance to operate to the same extent. Even then, however, the angular momentum must be reduced by a factor of ten or so in most cases, in order to match the observations. Turbulence, tides and viscosity could act in a similar way to transfer angular momentum if a magnetic field were absent.

On such a model, which it must be stressed remains largely qualitative and hypothetical, all single stars and well separated binaries cooler than F0 might be expected to have planetary systems. We might therefore expect planetary systems to occur around one star in three, some 10^{12} planets in the whole Galaxy. Very little direct evidence of such widespread occurrence of planetary systems is available. Some confirmation, however, comes from astrometric measurements which reveal sub-stellar companions for several nearby stars, and from the recent detection of gas and dust discs around some stars such as Vega and β Pictoris by the IRAS spacecraft (*I*nfra*R*ed *A*stronomy *S*atellite). Planetary-mass halos have been found by infrared speckle interferometry around the pre-main sequence stars HL Tau, and R Mon, probably due to material gravitationally bound to the stars. This technique has also revealed a companion smaller than Jupiter for the star van Biesbroeck-8, although that is a 'brown dwarf' with a surface temperature over 1300 K.

The observational evidence for the pre-nuclear burning phases is scanty. On the HR diagram (figure 8A.1), the expected track comes in from the bottom right-hand side, above the main sequence. At every early stages, the proto-star probably appears as a strong source of infrared and maser emission within a molecular cloud. It may be distinguished at this stage from the early development of nascent H II regions by the lack of the radio recombination lines in its spectrum. Possible candidates for this stage (position 1 on figure 8A.1) are the Becklin–Neugebauer object in the Orion nebula, which has an infrared luminosity of $10^3 L_{\odot}$, and W3-IRS5 in Cassiopeia some ten times brighter. Recently, however, faint radio emission from ionised gases has been detected from the former, and so it may be better interpreted as a new star, rather than as a proto-star. The nearby and associated Kleinmann–Low object may contain several possible proto-stars, but as yet this is unproven. At a slightly later stage, the optical luminosity may be sufficient to illuminate some of the surrounding gas, or to excite it into emission via shock waves, producing the faint nebulosities known as Herbig–Haro objects within the cloud. These latter may also arise, however, via the interaction of bi-polar outflows from the proto-star with the surrounding gas cloud (§7A.3). Such outflows are a relatively recent discovery but seem likely to prove to be a reliable sign of the presence of proto-stars of one solar mass and above.

The last stages of the evolution to the zero-age main sequence are phases of a continuing collapse. Theoretically, the first of these occurs with the star in convective equilibrium, and with a nearly constant surface temperature. During the second, the star is in radiative equilibrium, with more or less constant luminosity and an increasing surface temperature. The two phases are known as the Hayashi and Henyey tracks respectively (sections *A* and *B* on figure 8A.1). A very small number of objects have been tentatively suggested as candidates for the Hayashi phase. FU Orionis and the related

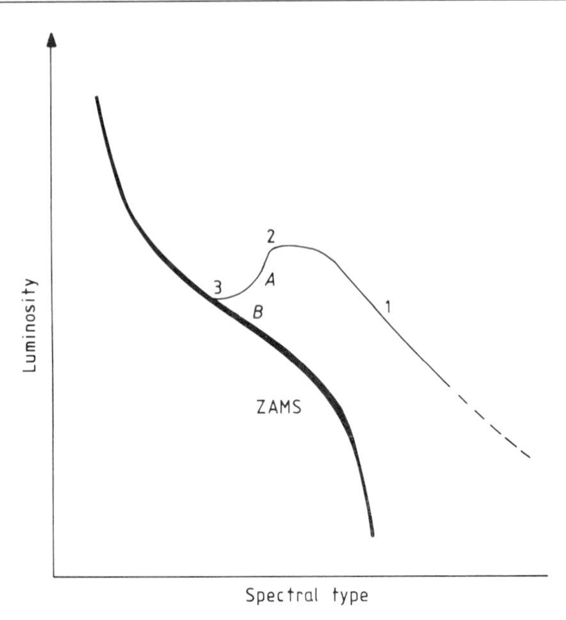

Figure 8A.1 Schematic track of the evolution of a proto-star on to the zero-age main sequence.

systems V1057 Cyg and V 1515 Cyg are strong infrared sources which have brightened rapidly. Their optical spectra are those of F- and G-type supergiants with evidence such as shell lines and P Cygni profiles for strong outflows of material from them. They may be at the top of the Hayashi track (point 2 on figure 8A.1), but this is by no means certain. The final approach to the main sequence is relatively slow, probably taking 10^6 to 10^7 years. There is therefore a good chance of observing stars in this phase. The T Tauri stars, and their possible hotter analogues, the Herbig Ae and Be stars, are fairly certainly identified with this stage (point 3 on figure 8A.1).

After the T Tauri phase, the star settles down to its main sequence existence. Since the fragmentation of the cloudlet is likely to have resulted in the near simultaneous formation of a number of stars, this stage is observed as a cluster of stars, many of them hot and luminous. H II regions will form around the hotter stars, and the cloud will eventually be disrupted and evaporated back to the intercloud medium. Galactic clusters, such as the Pleiades (figure 7A.2), and OB associations are examples of this phase. Eventually gravitational perturbations disrupt the cluster, and the stars merge with the other single and close binary stars which make up most of the Galaxy.

8A.2 POST-MAIN-SEQUENCE EVOLUTION

We have seen that stars spend the major portion of their lives on the main sequence (Chapters 1A, 3A). During this time they have a relatively simple internal structure, with the energy source being the conversion of hydrogen to helium at the star's centre. Eventually, even for low-mass stars with main sequence lives of 10^{12} years or more, this stage must come to an end as the hydrogen supply becomes exhausted. The behaviour of the star after this has happened is the concern of this section. It varies much more with mass than does the pre-main-sequence behaviour. Our understanding of many of the processes leaves a lot to be desired, since amongst the high-mass stars processes occur very rapidly and observational support is limited by the very small numbers of such stars, whilst at the low-mass end, the Galaxy, and possibly even the Universe, are not yet old enough for such stars to have evolved off the main sequence. Much of the following discussion is therefore speculative and without much observational basis. Nonetheless, a coherent and reasonable, if not completely established, picture of the likely behaviour patterns of stars after their main sequence lives has been achieved. It is convenient to divide the discussion here into the low-mass stars ($< 0.75\mathcal{M}_\odot$), the mid-range stars (0.5–$5\mathcal{M}_\odot$) and the high-mass stars ($> 3\mathcal{M}_\odot$).

Low-mass stars

The mass/luminosity relationship (equation 2A.4.6) for the middle and low end of the main sequence has the form

$$L \propto \mathcal{M}^{3.8}. \qquad (8A.2.1)$$

The main sequence lifetime of the Sun is thought to be around 10^{10} years (Chapter 4A). Thus, if all other things are equal, the main sequence lifetimes of stars, τ, would vary as

$$\tau = 10^{10}\mathcal{M}^{2.8} \text{ yr} \qquad (8A.2.2)$$

where $\mathcal{M}$ is the star's mass in solar mass units. Thus a star with a mass of $0.5\mathcal{M}_\odot$ may be expected to have a main sequence lifetime of about 7×10^{10} years. Now the best estimates for the age of the Galaxy place it at about 9 to 10×10^9 years. Even the Universe as a whole has an age of only 12 to 18×10^9 years if one of the 'Big Bang' cosmologies is correct. Thus even a star formed very early on in the life of the Universe or Galaxy will have completed only a quarter of its main sequence life, at most, if its initial mass was half that of the Sun or less. Ideas on the post-main-sequence evolution of such stars are therefore entirely theoretical.

Low-mass stars are expected to be converting hydrogen to helium via the proton–proton chain (§2A.2), though very small stars may not progress all the way along the basic set of reactions (equations 2A.2.1 to 2A.2.3). As the star ages, the proportion of helium in the central regions builds up, leading eventually to an almost pure helium core. The nuclear reactions then cease at the centre of the star and it is no longer a main sequence star. This occurs when a mass of hydrogen about 12% of the original mass (the Schönberg–Chandrasekhar limit) has been consumed. The reactions, however, still continue further out where hydrogen is available, in a shell around the core. The helium core will become isothermal and collapse slowly towards a partially electron-degenerate state (§2P.1). At the same time the outer parts of the star will expand slowly so that its position on the HR diagram will rise away from the main sequence. At the very lowest masses, the convective envelope may extend right to the centre of the star, leading to mixing, and a uniform composition for the whole star changing with time. The evolutionary pattern may thus be expected to differ markedly in some as yet undefined manner from the slightly more massive stars.

As the core contracts, its temperature rises, but not far enough to initiate nuclear reactions in the helium. Thus, after consuming as much of the hydrogen as possible, by which time it may have reached ten to a hundred times its main sequence luminosity, the nuclear reactions cease. The whole star then contracts, initially at constant luminosity as its surface temperature rises, later at constant temperature and decreasing luminosity. Eventually it becomes electron-degenerate and stabilises as a low-mass white dwarf. Thereafter it radiates away its stored energy and becomes fainter and cooler. In perhaps 10^{12} years it will end as a cold degenerate 'black dwarf'. The time scale for this last stage is uncertain, and it could be speeded up considerably by neutrino energy losses.

Medium-mass stars

These are stars whose behaviour is essentially similar to that expected for the Sun, and thus are of more than usual interest to human astronomers. After the collapse of the proto-star (§8A.1) and its arrival on the zero-age main sequence, the star is of uniform composition and is converting hydrogen to helium via the proton–proton and CNO reactions (§2A.2). These stars will have only a small convective core at most, and generally convection will only be important in their outer layers. The energy-generating region will therefore be unmixed. As the proportion of helium at the centre increases, the opacity of the material, which is largely due to electron scattering, decreases. This follows from the consumption of two electrons during the production of the two neutrons in each helium nucleus. The transfer of radiation from the centre to the surface of the star by radiation thus becomes easier with time. The star therefore slowly brightens during its main sequence life (section 1 to 2 of figure 8A.2). For the Sun this takes about 10^{10} years, and its surface temperature rises during that time from about 5600 to 6100 K, and its luminosity from about two-thirds to twice its present value. This may be expected to raise temperatures on the Earth by some 50 to 60 K (*2).

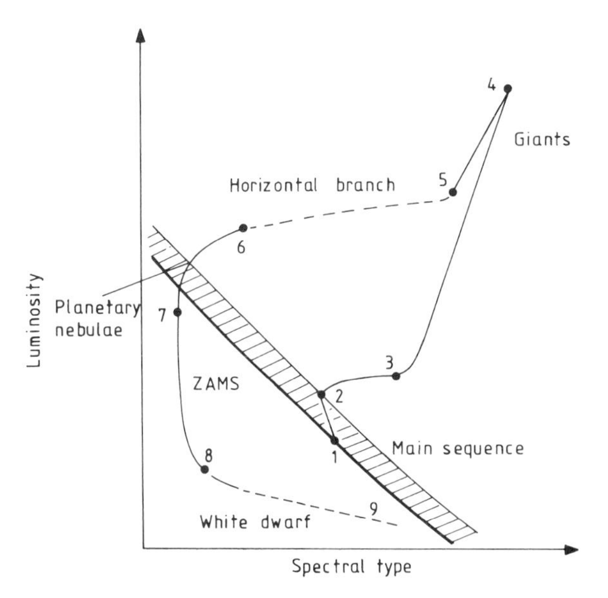

Figure 8A.2 Schematic path of a medium-mass star on the HR diagram during its post-main-sequence evolution.

The radiative interiors of these stars mean that their composition changes smoothly towards the centre, and that the centre becomes depleted of hydrogen before the outer layers of the core. As in the case of the low-mass stars, therefore, there is a smooth transition to a helium core surrounded by a hydrogen-burning shell. The helium core then slowly contracts, releasing gravitational energy. The energy goes partly into heating the core, and partly is radiated away as required by the virial theorem (§2P.5). The hydrogen-burning shell remains roughly in equilibrium with the core and so also increases its temperature and density. The sharp increase in the reaction rates engendered by these increases goes towards compensating for the decreasing availability of hydrogen, and so the luminosity of the star remains constant, or even rises for the higher mass stars.

As the core grows it approaches an electron-degenerate condition (§2P.1), and conduction becomes important in keeping the centre of the star in a near isothermal condition. The physical position of the shell energy source is more or less fixed, since if it were to move inwards, the reaction rates would increase rapidly with the higher temperatures and densities. The pressure would therefore rise and tend to push the shell back towards its initial position. Similarly an outward movement would lead to a reduction in pressure and a fall back downwards again. Now, for hydrostatic equilibrium, the density of the core must increase sharply towards its centre since degeneracy pressure depends only on $\rho^{4/3}$ (equation 2P.1.76). Given the fixed position for the edge of the core, and a core mass which changes only very slowly as more helium is added by the hydrogen-burning shell, the average core density must be constant. The increasing central density must therefore lead to a reduction in the density at the outer edge of the core. The continuity of temperature and pressure over the core/envelope boundary then ensures a reduced density for the base of the envelope. The mass of the envelope again is effectively constant, and so the outer radius of the star must expand in order for it to contain its material at the reduced mean density. The energy for this expansion is typically 5 to 10% of the star's luminosity. The lost energy is at least partially replaced by the release of potential energy from the increasing central condensation of the core, and so the observed stellar luminosity is hardly affected. With the constant luminosity of the star, the expansion leads to a fall in the surface temperature (figure 8A.2, section 2 to 3; higher mass stars may zig-zag over this section).

As the outer layers expand, an extensive convection zone develops within them. This improves the energy transport, and the star's luminosity starts to rise. On the HR diagram the star moves upwards (figure 8A.2, section 3 to 4). This takes the star towards the giant and red giant region where it may be radiating energy at a thousand times its main sequence rate. The hydrogen-burning shell is hot enough by now for the CNO cycle to predominate, and helium is added at an ever increasing rate to the core.

Eventually the rate of increase in the mass of the helium core and the consequent collapse outstrip the energy transport mechanisms. The core temperature then starts to rise above that of the hydrogen-burning shell. At a temperature of about 10^8 K, when the core mass is perhaps $0.5\ \mathcal{M}_\odot$, the triple-alpha reaction (equations 2A.2.15 and 2A.2.16) can start to convert the helium to carbon. The star thus develops two energy sources: hydrogen to helium in a shell around the core, and helium to carbon at the centre. This second energy source, however, is occurring inside electron-degenerate material, and so the pressure is almost independent of the temperature (equations 2P.1.74 and 2P.1.76). The temperature can therefore rise without changing the other conditions inside the core. Now, as we have seen (equation 2A.2.19), near 10^8 K, the energy generation rate of the triple-alpha reaction varies as about the fortieth power of the temperature. The reaction therefore starts to run away, with the energy being released explosively. Energy release rates of 10^{14} W kg^{-1} and total luminosities of 10^{39} W may be reached (compared with 10^{-3} W kg^{-1} and 4×10^{26} W for the present Sun). Eventually the electron energies rise high enough to release the degeneracy. The reversion to normal matter is accompanied by a rapid increase in the pressure in the core. The core expands and cools, slowing the helium runaway reactions and halting the hydrogen burning. The star is then left with just the helium burning at its centre as the only energy source. The process is known as the core helium flash, and its latter stages occur on time scales of seconds. Little evidence of all this hectic activity appears at the surface, however; probably the whole of the energy release is contained within the star. Changes to the outside appearance of the star do occur, though, but rather more slowly than the internal changes. The expansion of the core increases its share of the star's internal energy, allowing the outer layers to contract again. The convective outer zone diminishes, and the star's luminosity falls (figure 8A.2, section 4 to 5).

During the red giant phase, the star loses mass via a stellar wind (§5A.2), and this may be sufficient to affect the subsequent evolution. Numerical models become unreliable after the helium flash. There is some suggestion, however, of instability in the outer layers of the star, and this is confirmed by the observation of the horizontal-branch stars. In globular clusters the stars are expected to have formed together and thus all to be of about the same age. The HR diagram of a globular cluster therefore has the appearance of figure 8A.3, since the more massive stars will have proceeded further along their evolutionary tracks then the less massive ones. The horizontal-branch stars have absolute magnitudes of about + 0.5, and from their position must have proceeded past the helium flash. They occupy a part of the instability strip, a region of the HR diagram occupied by many varieties of pulsating stars. The RR Lyrae stars in particular may correspond to the medium-mass stars at this stage (figure 8A.2, section 5 to 6).

The final stages are rather uncertain. Probably solar-mass stars will re-establish a hydrogen burning shell and a helium-burning centre, the carbon perhaps combining with some of the remaining helium to produce oxygen (equation 2A.2.20). This situation is unstable; normally the hydrogen-burning shell will predominate, but 'mini helium flashes' will occur in the (non-degenerate) core, and/or similar runaway reactions may develop further out leading to shell flashes. These, or the increasing effect of radiation pressure on the outer layers, may lead to the expulsion of the outer layers of the star and to the formation of a planetary nebula (§6A.5; figure 8A.2, section 6 to 7). The more massive stars in this range may undergo further nuclear reactions (equations 2A.2.21 to 2A.2.23) before reaching this stage. Eventually, as with the lower-mass stars, the nuclear reactions die out. The star contracts (figure 8A.2, section 7 to 8) towards an electron-degenerate configuration. The star becomes a white dwarf and eventually perhaps a black dwarf (figure 8A.2, section 8 to 9 and further). Most stars in this mass range will have lost sufficient mass during their red giant, pulsating variable, and planetary nebula stages to be below the white dwarf mass limit. It is just possible, however, at the top end of the range for neutron stars to be the result in a very few cases.

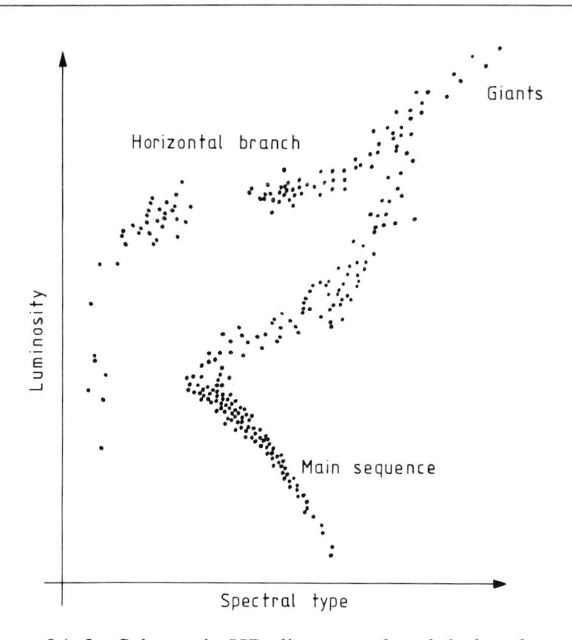

Figure 8A.3 Schematic HR diagram of a globular cluster.

Binary systems

The evolutionary patterns just outlined are applicable to single stars and to multiple systems with widely separated ($>$ 100 AU) components. Close binary systems often seem to have behaved contrary to these expectations. Thus in the Sirius system, for example, the primary is an Al V star with a mass of $2.14\mathcal{M}_{\odot}$, while the companion is a DA white dwarf with a mass of $0.94\mathcal{M}_{\odot}$. Now, by equation (8A.2.2), the age of such a white dwarf must be in excess of 10^{10} years, while the primary can at most be 10^{9} years old. While the formation of a close binary system from the close passage of two single stars is not impossible if a third body intervenes to carry away the excess energy, it is extremely unlikely. At most one or two such systems might be expected to be found in the entire Galaxy. Instead, systems like Sirius, or with other peculiarities, are found to be commonplace. We have only to cite novae (§3A.5), which are believed to be formed of an early giant star and a white dwarf with similar masses, to see that there must be at least tens of thousands of such systems in the Galaxy. An alternative to their formation from the coming together of two single stars must therefore be sought.

The answer lies in the binary nature of the stars and in the resulting distribution of potential around them. A small mass at some point within the system will experience three accelerations: gravitational attractions towards each of the stars, and a centrifugal acceleration due to any participation that it may have in the orbital motion of the system. Close to the stars, material may be expected to rotate at the same rate as the stars (corotation). In a frame of reference rotating with the stars (corotating frame of reference) a plot of the total potential arising from the action of all three forces reveals that each star's volume of influence is limited. The loci of equal potential are known as the Roche equipotential surfaces. The critical surface within which material is bound to one or other star defines the Roche lobes (figure 8A.4). Corotating material outside the Roche lobes remains bound to the system as a whole, until it gets beyond the surface through L3, when it may be lost entirely. The points marked L1 to L5 are known as the Lagrangian points. For our purposes, the inner Lagrangian point, L1, is of the most significance. Viewing the system in the space–time

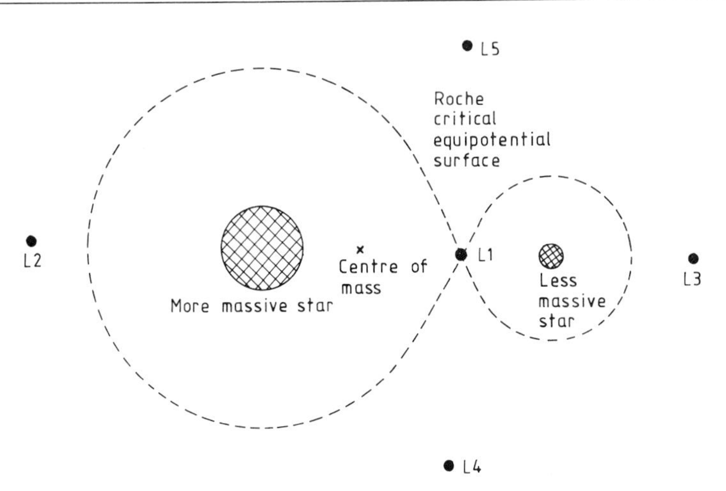

Figure 8A.4 Schematic Roche critical surface (cross section in the plane of the orbit) and Lagrangian points for a close binary system.

continuum (figure 8A.5), L1 can be seen to be the 'pass' between the two 'wells' of the stars, and to be entirely hemmed in by higher potentials. Thus if one of the stars expands physically, its outer layers will reach higher and higher potentials until they arrive at the critical surface. The material can then flow across the 'pass' at L1, and down into the potential well of the other star. In physical terms, if one star of a binary expands to fill its Roche lobe, then material will be lost from it at L1, and eventually (through turbulent loss of angular momentum), be accreted on to the other star. If the rate of expansion is sufficient to push material beyond L3, then it may be lost to the system entirely.

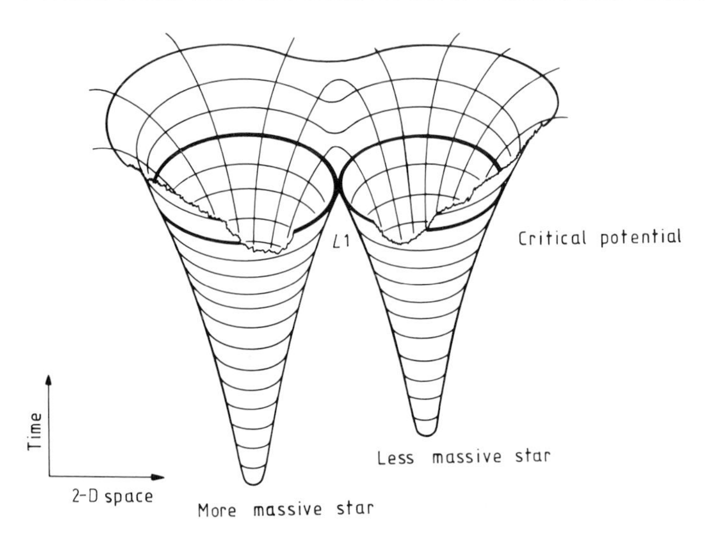

Figure 8A.5 Cutaway view of the space–time continuum near a binary system.

We may now follow through the development of stars in a binary system to see how their apparent conflict with single-star evolutionary behaviour arises. The vast majority of binary systems are thought to form through the late fragmentation or fission of the collapsing proto-star (§8A.1), though there is no good model for this. If the idea is correct, however, both stars will arrive on the main sequence nearly simultaneously. They will then generally be far smaller than their Roche lobes, and will evolve essentially as single stars. One possible exception to this may be the RS CVn stars which may arise through tidally enforced rapid rotation of the primary at this stage. In other systems, the initially more massive star (star 1) will progress through the main sequence more quickly (equation 8A.2.2) than the star with the lower initial mass (star 2). At the end of its main sequence phase, star 1 will start to expand towards the giant region as previously discussed. Fairly rapidly, however, its outer layers will reach the critical Roche surface, and mass exchange to star 2 will commence. This will continue for some time, since as mass is lost from star 1, its Roche lobe will shrink, enhancing the effect of the expansion. Star 1 will decrease in temperature and luminosity due to the mass loss, and evolve towards the subgiant region. Star 2 will increase in temperature and luminosity while remaining on or near the main sequence. Most known subgiants are observed to occur in binary systems, in partial confirmation of this model. At this stage, the system would resemble Algol (β Per), with a bright main sequence star and a fainter subgiant still filling its Roche lobe.

Continuing mass exchange can lead to the loss of the whole of the outer layers of star 1, leaving just its helium or carbon/oxygen core. Star 2 will by this time be the more massive star in the system. Nuclear reactions will cease in star 1, and it will settle down as a white dwarf. We will then have a Sirius-type situation, with a massive bright main sequence star, and a faint low-mass white dwarf. In higher mass binaries ($\mathcal{M}_1 + \mathcal{M}_2 > 20\mathcal{M}_\odot$) the system may appear as a Wolf–Rayet or Of star at this point, since the stripped core may remain hot enough for nuclear reactions to continue, making it the more luminous component.

Binary x-ray sources and x-ray bursters may be systems containing a main sequence star, perhaps filling its Roche lobe, and a neutron star. An even more extreme example may be Cyg X-1, where the evolved component may have progressed through to a black hole (§3A.9).

Star 2, being now much more massive than before, will start to evolve rapidly, and will soon move off the main sequence towards the giant region. As before, when the Roche lobe is filled, mass exchange will occur, this time from star 2 to star 1. The situation does not simply repeat, however, since this time the material is accreting on to a degenerate white dwarf. The system is thus likely to become a nova, dwarf nova, AM Her star, symbiotic star, etc, according to the details of the configuration of the system and its magnetic fields.

The final result is likely to be a system containing two collapsed objects. White dwarfs are perhaps the most likely end product, as in AM CVn, but neutron stars and black holes are also possible. Thus we have the binary pulsar (PSR 1913 + 16) which is a neutron star in orbit around another collapsed object, SS433 which is probably a white dwarf in orbit around a black hole, and some plausible models of type I supernovae (§3A.5). If one of the stars becomes a supernova, then it is quite possible that the binary will be disrupted. Some high-velocity single stars and pulsars may be the result of such a case.

Evolution in a binary system is a very complex process, and the above qualitative account, though attractive in the way in which it synthesises many different types of observation, will undoubtedly vary considerably from the truth in most actual cases. Real situations will depend critically upon the separations, magnetic fields, rotations, chemical compositions, etc, as well as upon the more basic parameters of the masses and age. Today's computer models cannot deal with such complexities even in outline, and so the reader should beware of accepting the account uncritically.

High-mass stars

Stars with masses over five or so times that of the Sun are very few and far between. They have attracted considerable interest, however, for their complex and spectacular reactions and phenomena. The later and more interesting stages are only qualitatively understood, so a note of caution about the accuracy of the discussion must again be sounded.

Massive stars go through their main sequence lives rapidly: typically in ten million years or less. Then, like the medium-mass stars, a helium core forms and they move off the main sequence, in this case towards the supergiant region. There is, however, one significant difference from the medium-mass case, and that is that the high-mass main sequence stars have a large central convective region. Thus the central region remains of uniform, though changing, composition as the hydrogen is converted to helium, and a very large helium core is formed immediately upon the exhaustion of the hydrogen. This collapses rapidly, and the outer regions expand equally rapidly, pushing the star into the supergiant region in only about 10^6 years, compared with the 10^8 years or so required for a solar-mass star to reach the giant region. Eventually helium burning starts and supplies the bulk of the star's energy. The pattern then continues to follow that for the medium-mass stars. A carbon core forms at the centre as the helium becomes exhausted, and two nuclear burning shells are left inside the star: hydrogen to helium, and helium to carbon. During this time, the star's position on the HR diagram has been oscillating slightly in temperature, with the luminosity gradually and more uniformly increasing.

This stage, possibly with additional conversion of the carbon to oxygen and neon, marks the end of nucleosynthesis in medium-mass stars. In these higher-mass stars, however, the carbon core collapses, and the central temperature rises to 6×10^8 K or more. The carbon nuclei can then combine directly to magnesium (equation 2A.2.30). The star may then have three energy sources: hydrogen to helium, helium to carbon, and carbon to magnesium. This process can continue in sequence until a very complex situation is reached. The reactions after the formation of carbon are not pure, and many different elements and isotopes will be formed. Some of the possible reactions were summarised in §3A.2. Towards the end, for stars of masses of 8 or $10\,\mathcal{M}_\odot$ or more, there are probably four major phases of nucleosynthesis: hydrogen to helium; helium to carbon, oxygen, neon;

helium, carbon, oxygen, etc to neon, sodium, magnesium, silicon, sulphur etc; and magnesium, silicon etc to nickel, iron, cobalt, etc.

The last phase probably occurs via the slow build-up of mass by the addition of neutrons in the *s* process (§2A.2) at temperatures in the region of 2×10^9 K, rather than by the direct interaction of heavy nuclei. The precise distribution of the reaction zones, and which of them predominates in terms of the energy generation at any given instant, remain to be understood at the time of writing.

All these phases occur much more rapidly than the star's main sequence phase; typical lifetimes are 1% or less of the main sequence lifetime. The later stages are further quickened through the abstraction of significant amounts of energy directly from the centre of the star by neutrinos. These mostly originate in the URCA process (§2A.2, note 6), and may decrease the lifetimes of the later stages by as much as a factor of a hundred.

Now, as we have seen (figure 2P.3), there are no exothermic nuclear reactions starting from the iron-group nuclei. Thus, as an iron core forms, it can only grow in mass with time, and no further nucleosynthetic phases will develop. As in previous cases, the core will collapse to an electron-degenerate state. However, in a massive star, the core may exceed the Chandrasekhar limit for the mass possible for degenerate material (§2A.1). This limit will in any case be lower than for an isolated mass of degenerate material, because of the pressure of the overlying layers of the star. When the limit is exceeded, the iron core will collapse further, the electrons and protons combining to give neutrons and neutrinos, as the material becomes baryon-degenerate (§2P.1). This collapse will be very rapid: roughly at free-fall rate, with an e-folding time of a hundredth of a second or less; and it is generally expected to trigger a type II supernova.

The observational details of a supernova have already been reviewed (§3A.5). The precise nature of its underlying processes is uncertain, but we may continue the above speculation to give a 'best guess' at what is happening. The collapsing core is eventually halted towards its centre as the density approaches 10^{18} kg m^{-3} by the developing baryon degeneracy pressure. The outer portions of the core are still collapsing rapidly, however, and vast amounts of energy are released and an extremely strong shock wave formed, as they impact on to the stationary central region of the core. The energy release heats the material to temperatures of 10^{10} K or more when the *r* and *p* processes (§2A.2) build up elements heavier than iron. Any remaining lighter elements will be consumed, adding to the energy release. The shock wave propagates ouwards from the central regions, raising the temperatures of the outer layers to 10^{10} K. Explosive nucleosynthesis results amongst the light elements still present in these regions, and the *r* and *p* processes again build up elements beyond iron. The energy release may be sufficient at this stage to eject the outer layers of the star and to provide the first external indication of the supernova. More probably, however, this requires the assistance of neutrino pressure. The density towards the centre will be so high that even neutrinos will undergo frequent scatterings, exerting an outward pressure on the descending material as they do so. The neutrino flux may briefly reach 10^{15} times the present solar total luminosity, through the myriad reactions occurring near the centre. The neutrino pressure may thus be the most significant influence on the subsequent motions of the outer regions of the star. The end result, of course, is that the outer layers of the star are flung off at some 10 000 km s^{-1}, and are the main observed result of the event, producing the supernova and the subsequent SNR. The collapsed core in some cases remains as a neutron star, and may become visible as a pulsar. In other cases, a white dwarf or black hole may be left.

As a final addendum to this story of the evolution of stars, there is the remarkable thought that we have all had personal experiences of many of these phenomena, including the interiors of supernovae! As we have seen, nucleosynthesis builds up the heavier elements from hydrogen inside all stars. It is only in supernovae, however, that the whole gamut of elements is produced *and* distributed out into space. The heavier elements comprising the Earth and much of our bodies must therefore have been generated in supernovae preceding the formation of the solar system. The resulting SNRs then gradually merged with the interstellar medium. Another supernova may then have initiated the collapse of the proto-Sun from an interstellar cloud containing these heavier elements. Thus we may justifiably join with Beatrice in thinking

'... a star danced, and under that was I born.'

(*Much Ado about Nothing*. Act II, Scene 1. W Shakespeare)

Appendix I Distances of the Stars

Distances away from the surface of the Earth are not amenable to determination by tape measure. Astronomers have therefore had to resort to other methods to find how far away astronomical objects may be. Outside the solar system, there are two main methods, and a number of less widely applicable techniques. The two main methods are triangulation and the use of the 'standard candle'.

AI.1 TRIANGULATION

This is the most fundamental and reliable of the various techniques. It is simply the extension of the methods of surveyors on Earth to the wider Universe. Triangulation requires a base line whose length is known accurately, and the measurement of one or more angles (figure AI.1). No base line on the Earth is long enough to permit the determination of the distances of even the nearest stars. The orbit of the Earth is therefore used to form the base line. As can be seen from figure AI.1, over the selected three-month interval, the Earth changes its position with respect to the nearby star by one orbital radius across the line of sight. The observed position of that star against the very much more distant background stars therefore also changes. The shift in the star's position is called the parallax angle, P, and its measurement gives the star's distance, provided that the Earth–Sun distance is known (§4A.1):

$$\text{Earth–star distance} = (\text{Earth–Sun distance})/(\tan P)\,. \quad \text{(AI.1.1)}$$

Now P is always very small, so that when measured in radians the angle and

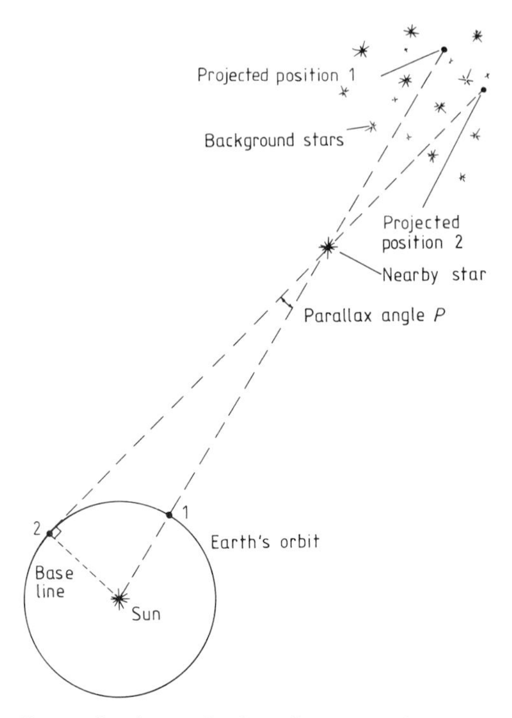

Figure AI.1 Determination of the distance of a nearby star by triangulation.

its tangent are almost equal. Furthermore, the astronomical unit has a value of 1.496×10^{11} m (§4A.1), and so we obtain

$$D = 1.496 \times 10^{11} P^{-1} . \tag{AI.1.2}$$

For a value of P of one arcsecond (4.8×10^{-6} radians), D has a value therefore of 3.08×10^{16} m. This distance is used by astronomers as the unit of measurement, and is known as the parsec (*par*allax—*sec*ond of arc; abbreviated pc). It is slightly less than the distance to the nearest star other than the Sun. Working in units of parsecs for D, and seconds of arc for P, equation (AI.1.2) becomes simply

$$D = 1/P \quad \text{(pc)}. \tag{AI.1.3}$$

A secondary, derived unit is often to be encountered, especially in more popular astronomical texts, and this is the light year (ly). It is a *distance*, not a *time* unit, and it is the distance *in vacuo* travelled by light in one year. Its value is 9.46×10^{15} m. Thus

$$1 \text{ ly} = 0.307 \text{ pc} \tag{AI.1.4}$$

and

$$1\ \text{pc} = 3.26\ \text{ly}. \tag{AI.1.5}$$

The shift in the position of a nearby star is obtained in most cases by measuring its position with respect to the background stars on two photographs taken a few months apart (in practice, dozens of photographs are likely to be used). The errors in this process are about 0.01″ for high-quality work, and 0.005″ at the very best. Thus a measured parallax angle of 0.01″ is actually 0.01″ ± 0.01″. That is to say its probable value lies in the range 0.02″ to 0.00″, and the star's distance therefore could be 50 pc or infinity! The useful range of triangulation for determining stellar distances is thus limited to about 50 pc ($P > 0.02''$), and even then large errors are possible for the outer half of the range. Thus given the star density near the Sun of about 0.5 stars per cubic parsec, only some 10 000 stellar distances can have reliable direct determinations. Beyond this limit, some of the methods discussed below may be useful. In general, however, their accuracy is lower than that of triangulation, and as a result uncertainties of 50% or more in the distances of stars, nebulae and galaxies are not uncommon.

AI.2 STANDARD CANDLE METHODS

There are a variety of objects which for one reason or another have absolute magnitudes (§3A.1) known through some property other than their distances. They can therefore be used as sources of known luminosity (standard candles). Equation (3A.1.2) can be inverted to give

$$D = 10^{0.2(m-M+5)}\ (\text{pc}) \tag{AI.2.1}$$

and so their distances can be determined. If interstellar absorption is significant, then equation (3A.1.2) becomes, using equation (7A.1.2),

$$M = m + 5 - \log_{10} D - AD. \tag{AI.2.2}$$

If the interstellar extinction is known, then the distance of the star can still be found by iteration using the formula

$$D_{n+1} = (1/A)(m - M + 5 - 5\log_{10} D_n) \tag{AI.2.3}$$

which converges reasonably rapidly.

Many objects can act as standard candles. Cepheids provide probably the best known example. Their period/luminosity relationship (equation 3A.5.2) enables their mean absolute magnitudes to be estimated from their periods. RR Lyrae stars have a nearly constant mean absolute magnitude of about +0.8. Mira-type variables exhibit a relationship between their absolute visual magnitudes and colour indices of the form

$$M_V \simeq (V-R) + 0.14(R-I)^3 - 3.8 \qquad \text{(AI.2.4)}$$

where the colour indices have been corrected for interstellar reddening, and a relationship between their maximum absolute magnitudes, $M_{V,\max}$ and their 'periods' P, of the form

$$M_{V,\max} \simeq -2.3 \log_{10}P + K_1S + K_2 \qquad \text{(AI.2.5)}$$

where S is the star's spectral subclass at maximum, and K_1 and K_2 are constants with the values

$$K_1 = 0.5, K_2 = 1.4 \text{ for } S \geqslant \text{M3}$$

$$K_1 = 0.2, K_2 = 2.4 \text{ for } S < \text{M3}.$$

The peak absolute magnitudes of novae are related to their initial rates of decline (equation 3A.5.7). Type I supernovae have relatively constant peak absolute magnitudes of about -19, enabling the distances of even quite remote galaxies to be estimated.

Slightly less reliable standard candles include the globular clusters around a galaxy. These often appear to have a limiting upper brightness. Usually the third or tenth brightest globular cluster of a galaxy is used rather than the brightest to reduce the possibility of exceptional individuals affecting the result. An accurate spectral and luminosity type (§3A.2) for a star allows its position on the HR diagram to be plotted, and so its absolute magnitude estimated. This last method provides what is called the spectroscopic parallax of the star, and is often used, since it is the only method widely applicable to many stars. By no means finally, though the last example to be mentioned here, there is the Wilson–Bappu method. This is based upon the empirical relationship observed between the widths of the emission cores to the Ca II H and K lines (and some other lines), in the spectra of cooler stars, and their absolute magnitudes. For example, the full width at half maximum of the Ca II K emission, W, measured in km s^{-1}, is related to the absolute magnitude by

$$M_V = 27.6 - 14.9 \log_{10} W \qquad \text{(AI.2.5)}$$

over a fifteen-stellar-magnitude range of luminosity. The basis of the relationship is not understood, though it obviously depends upon some change in the chromosphere with luminosity, but it nonetheless provides yet another way of estimating a star's actual brightness.

AI.3 OTHER METHODS

This section outlines some of the additional methods which may be applicable in a few cases. They form a very disparate group, and vary widely in their availability and reliability. This is not intended to be an exhaustive treatment of the subject, but just a summary to give the reader an introduction to the topic. Further details must be sought from the specialist literature and journals.

Nebulae: Very few nebulae are close enough for triangulation to determine their distances. Sometimes the distance to a star embedded in the nebula can provide a clue, otherwise one or more of the following methods may be useful. First, as we have seen (Chapter 6A), the sizes of H II regions are related to their electron number densities and the temperatures of their exciting stars (equation 6A.1.10). If these can be estimated, then the angular size of the H II region leads to its distance. Alternatively, within a galaxy the maximum size of an H II region is likely to be about 200 pc (§6A.4), so that its angular size will give both its distance and that of the galaxy. Secondly, in an expanding gas shell such as a planetary nebula, nova, or SNR, the line-of-sight expansion velocity may be determinable in absolute terms from the spectrum, while over a period of several years, the expansion of the nebula provides the angular velocity across the line of sight. The rate of expansion of the nebula's radius in seconds of arc per year, $\dot{\alpha}$, and the line-of-sight expansion velocity in $\mathrm{km\,s^{-1}}$, v, then give the distance as

$$D = v/5\dot{\alpha}\ (\mathrm{pc})\,. \qquad \text{(AI.3.1)}$$

One of the major problems with this method is that some or all of the angular expansion of the nebula may be due to an expanding ionisation front, rather than to physical motion of the material. The third method is applicable only to absorption nebulae. It is based upon the Wolf diagram (figure 7A.4). If sufficient stars in each magnitude range are available, then their average absolute magnitudes will be similar. The nebula reduces the brightnesses of the stars behind it. The degree of this reduction gives an estimate of the thickness of the nebula, and the mean apparent magnitude where it occurs gives the distance.

Pulsars: The dispersion measure (equation 7A.3.8) of a pulsar's radio signals, together with an assumption of the electron number density of the interstellar medium (typically $5 \times 10^4\ \mathrm{m^{-3}}$) leads to the pulsar's distance.

H I regions: The 21 cm and other radio and microwave line emissions from the molecular clouds, H II regions, etc will be Doppler shifted by the radial motion of the material with respect to the observer. If a model for the rotation of the Galaxy is assumed, and the intrinsic velocity of the object is small, then its distance may be estimated.

Star clusters: The HR diagram of a galactic or globular cluster may be plotted out in terms of apparent magnitude, since the stars within it are all at roughly equal distances from the Earth. The scaling required to fit this diagram to an absolute magnitude HR diagram will provide the distance modulus, and so the cluster's distance in an extension of the standard candle method.

Moving clusters: Apart from relatively small intrinsic motions, the stars in a cluster are comoving around the Galaxy. They therefore all have similar space velocities with respect to the Sun. Their proper motions through the effect of perspective appear to originate from a single point in the sky, known as the cluster's convergent point. If the angular distance of any star in the cluster from the convergent point is β, then simple geometry gives

$$D = (v_R/5\mu)\tan\beta \text{ (pc)} \tag{AI.3.2}$$

where v_R is the star's radial velocity in km s^{-1} (obtained spectroscopically) and μ is the star's proper motion in $''$/yr.

Centre of the Milky Way: Reliable estimates of the Sun's distance from the centre of the Galaxy, until recent infrared measurements, were based upon the position of the centre of the globular cluster distribution, H II regions etc, since it could not be observed directly.

Galaxies: The redshift in the spectra of galaxies beyond the local cluster enables their distances to be found from the Hubble law

$$D = (1/H)v_R \text{ (Mpc)} \tag{AI.3.3}$$

where H is the Hubble constant, currently variously estimated to lie in the region of 50 to 100 km s^{-1} Mpc^{-1}.

Statistical parallaxes: The motion of the Sun and stars through the Galaxy can lead to estimates of the mean distances of stars of particular types by an extension of the moving cluster method.

Colour excess: Interstellar reddening (§7A.2) of a stellar spectrum can be estimated via the colour excess (equations 7A.1.3, 7A.1.4) and colour factor (equation 7A.1.6). An assumption of the rate of interstellar extinction (typically 0.002 magnitudes per parsec in the plane of the Galaxy) then provides the distance to the star.

Appendix II Notes

PREFACE

1 Astronomers primarily use the term 'evolution', in this context, to mean the various stages within the life cycle of a single star. It is *not* used in the biological sense of the changes in the overall characteristics of a large population of broadly similar individual objects.

CHAPTER 1A

1 Einstein has of course shown that the other laws and even Newton's law of gravitation are only approximations as well. However, for the purpose of this chapter, the additional corrections are quite negligible, and we will remain with the classical results.

The gravitational redshift can also provide an estimate of the ratio between the mass and radius of an object. This effect has been detected in the spectra of white dwarfs, but so far with insufficient accuracy to provide any useful estimates of their masses.

2 The term *double star* may also be encountered. This includes the binary stars, but also in addition those pairs of stars which are close together in the sky as seen from the Earth, but which are actually at very different distances. The latter group of stars have no significant mutual gravitational interaction, and do not provide any useful information to astronomers (except to act as tests of the resolution of telescopes when their angular separations are very small).

3 In the Galactic plane about 45% of the stars are single, and 55% double or triple. In the halo, by contrast, the majority of stars occur singly. The reason for this difference is not known.

4 By convention, radial velocities are positive when the object is receding from the Earth, i.e. when the lines are shifted to wavelengths longer than their laboratory values. Velocities are negative for approaching objects, when the wavelengths are decreased.

5 Ideally, total and annular eclipses lead to flat-bottomed minima, partial eclipses to round-bottomed minima, and no eclipse to no minimum. Limb darkening, reflection, and non-spherical shapes for the stars (particularly in very close binaries), etc, confuse the situation very considerably, however. If they cannot be explicitly determined in the solution of the light curve, then these details add very greatly to the uncertainties in the final mass estimates for the system.

6 This is a 'bootstrap' process which would not be possible unless some stellar masses were already known unequivocally.

7 Stars are divided into two main groups, the Baade classes, on the basis of their metal abundances, and labelled Population I and Population II. Population I stars are young, have a high proportion of the heavier elements, and are found primarily in the spiral arms of galaxies (hence their occasional alternative name of Arm Population). Population II stars (also known as Halo Population) are older, with lower heavy element abundances, and occur mostly in the nuclei of galaxies and in globular clusters.

CHAPTER 2P

1 amu: atomic mass unit $= 1.66 \times 10^{-27}$ kg

$$= 0.9922 m_{\mathrm{H}}.$$

2 A mole is a quantity of a substance whose mass in *grams* is numerically equal to the molecular weight of the particles comprising the substance. The number of particles in a mole is fixed, whatever the substance might be, and that number is called Avogadro's number,

$$N_{\mathrm{A}} = 6.022 \times 10^{23}.$$

3 h is Planck's constant, and has a value of

$$h = 6.626 \times 10^{-34} \mathrm{~J~s}.$$

$h/2\pi$ is commonly given the symbol, $\hbar$, and has a value of

$$\hbar = 1.056 \times 10^{-34} \mathrm{~J~s}.$$

4 The Pauli exclusion principle states that no two fermions may have quantum numbers which are the same as each other, within limits imposed by the Heisenberg uncertainty principle. Therefore, if p and q are two conjugate quantum numbers, then their values for two

separate fermions must differ by at least δp and δq, where

$$\delta p \delta q \geqslant h \, .$$

The principle determines the electron structure of atoms, and so underlies all the gross properties of matter which permit our existence and govern our lives. Additionally, however, it also applies to individual fermions, whether they form a part of an atom or not. In the latter case, which is the one of importance in this section, the quantum numbers are somewhat different in form from those describing electrons within atoms, but they still exist (see main text).

5 The polytropic index, n, is related to the density exponent, γ (equation 2P.1.43) by

$$n = 1/(\gamma - 1) \, .$$

6 Thermodynamic equilibrium is the state in which each and every physical process which may be occurring is exactly balanced by its inverse.

7 Radiation density, u, is related to the flux by

$$u = (4/c)\mathscr{F}$$

so that

$$u = aT^4$$

where

$$a = 8\pi^5 k^4 \mu^2 / 15 h^3 c^3$$
$$= 7.5646 \times 10^{-16} \, \mathrm{J \, m^{-3} \, K^{-4}} \, .$$

CHAPTER 2A

1 The definition of a star becomes problematical at very low luminosities. Thus the obvious definition:

> A body which radiates more energy than it receives

would result in Jupiter being defined as a star, since it radiates away about 2.5 times the energy that it receives from the Sun, probably due to continuing gravitational contraction, while the definition

> A body which radiates energy that has been generated by fusion reactions in its interior

would exclude white dwarfs and neutron stars. A more cumbersome, but less unsatisfactory definition might be

> A body which radiates energy that has been generated by nuclear fusion reactions in its interior; or the high-temperature, dense central remnant of such a body.

2 Allowances for differential rotation, magnetic fields, etc increase the Chandrasekhar limit in theory, but do not seem to be important in practice.

3 'Minor' changes in the Sun are likely to render the Earth uninhabitable in under a thousand million years (see §8A.2).

4 Nuclei can have differing quantum levels for their constituent particles, somewhat on the lines of the various energy levels occupied by electrons in atoms. Nuclei in an excited state are denoted here by a superscript asterisk: ${}^{12}_{6}C^*$ and they revert very rapidly to their lowest (ground) state by the emission of a gamma ray, or undergo radioactive decay to other nuclei.

5 The ${}^{15}_{9}F$ nucleus decays by electron capture (EC) of an inner orbital electron in about 110 minutes.

6 A particularly important case of the *breakdown* of the *e* process occurs at very high temperatures. During the last moments leading up to a supernova explosion (§§3A.5, 8A.2) neutrino production is not balanced by neutrino destruction, leading to very high energy loss rates, and possibly to the neutrino pressure being the driving force of the supernova explosion itself. The main neutrino-producing reactions in this situation are thought to be

$$e^- + e^+ \rightarrow \nu_e + \tilde{\nu}_e$$

and the URCA processes. These latter produce neutrinos by a variety of reactions. The 'ordinary' URCA process is

$${}^{\alpha}_{\beta}X + e^- \rightarrow {}_{\beta-1}{}^{\alpha}X'$$

$${}_{\beta-1}{}^{\alpha}X' \rightarrow {}^{\alpha}_{\beta}X + e^- + \nu_e + \tilde{\nu}_e$$

where X and X′ are nuclei of odd mass number. The process is named for the URCA casino in Rio de Janeiro which reputedly carried money away in as untraceable a fashion as these neutrinos carry energy away from a star!

7 This very roughly calculated rate is in reasonable agreement with the postulated type II supernova rate of about one to three per century per galaxy, and the final stages of nucleosynthesis are thought to underlie the mechanism for these supernovae (§8A.2). Type I supernovae, which are slightly more frequent, are probably due to a different process.

8 Conduction of heat by electrons is a very efficient and rapid process in degenerate matter. White dwarfs therefore have almost isothermal interiors with the bulk of the energy transfer being by electron conduction.

9 Neutrinos, except possibly momentarily during the collapse phase of type II supernovae, are exceptions to this. They escape directly into

space, so that their production cannot be balanced by their destruction. Strict conditions of TE cannot therefore be said to exist inside any star. The energy of the neutrinos, however, is so small that the deviation is negligible, except possibly again during supernovae, and the assumption of TE for the interiors of stars is sufficiently precise for almost all purposes.

CHAPTER 3P

1 The energies of electrons in atoms etc are quantised; only certain values are allowed. An individual allowed set of quantum numbers is called a *state*. Often several states may have identical energy values, only becoming distinguishable in the presence of a magnetic field. These degenerate states combine to form a *level*. In some cases two or more related levels form a *term*.

Most optical spectrum lines are due to transitions between levels in atoms, ions and molecules. At longer wavelengths, lines can arise through changes in the quantised vibrational or rotational energies of molecules.

A detailed account of quantum theory, level notation, transition rules etc is beyond the scope of this book. The reader is referred to the bibliography for further reading if interested.

2 There are two main systems of notation for the level of ionisation of an element. Here the chemical symbol followed by a roman numeral is used. The roman numeral has a value of I for the neutral atom, II for the singly ionised atom, III for the doubly ionised atom etc. Negative ions are symbolised by the addition of a negative sign as a following superscript to the chemical symbol. Thus we have

H I H II He II Fe V etc

and

H^-.

CHAPTER 3A

1 The rapidity of the variation in brightness of an object places an upper limit on its size. To see how this occurs, imagine an optically thin object one light year across. Even if it were to double in its brightness instantaneously, it would still appear to take a year to change as seen by an external observer. This is simply because the radiation from the far side would take a year longer to reach the observer than that from

the nearest point. Although some geometries and situations may be imagined to circumvent this (for example, an optically thick plane surface perpendicular to the line of sight) their improbability, and the impossibility anyway of an instantaneous change over the whole object, mean that for all practical purposes we may write

$$\text{size of the emitting region} \leqslant c\,\Delta t$$

where Δt is the time required for a significant change in the observed brightness of the object and c is the velocity of light (see also note 6P.1).

2 Infall on to a neutron star releases about 15% of the rest-mass energy of the accreting material. As an energy production process it can be outstripped in efficiency only by accretion on to a black hole, when 50% of the rest-mass energy may be released in some circumstances, or by the combination of matter and antimatter which provides 100% of the rest-mass energy. The normal nuclear fusion reactions which power most stars are far less efficient; the total energy released during the nuclear fusion of hydrogen into iron releases only about 1% of the rest-mass energy of the material.

3 Be stars are B type stars whose spectra contain emission lines, not, as is often mistakenly assumed, beryllium stars. Their properties have been extensively reviewed by the author in *Early Emission Line Stars* (Adam Hilger, 1982).

4 A local inertial frame of reference is one in which Newton's first law of motion holds. It may be physically realised, approximately, by the interior of a small spacecraft falling freely under gravity, and not acted on by any external force such as rockets, solar wind, atmospheric drag, etc. With such a frame of reference, gravity is seen to be a pseudo-force, like centrifugal force, arising from the choice of a non-inertial frame of reference. In a non-uniform gravitational field, numerous local inertial frames of reference are required to cover the space–time region.

5 Virtual particles are a consequence of the Heisenberg uncertainty principle (note 2P.4 above). In one form this states that the energy of a system cannot be determined accurately over a short interval of time. Thus a particle/antiparticle pair, including photon pairs, can spring into existence out of nothing, without violating the principle of conservation of mass–energy, providing that they then annihilate each other within the time interval required by the uncertainty principle. For a proton/antiproton pair, this time would be about 10^{-24} s.

A vacuum may thus be conceived of as being filled with multitudinous varieties of pairs of particles. Each pair exists for only the briefest of instants before mutual annihilation occurs. Unlike normal particle/antiparticle annihilations, however, there is no release of energy; it goes to 'pay back' the energy 'debt' caused by the formation of the particle pair.

CHAPTER 4A

1 A spectrohelioscope or spectroheliograph is a specialised monochromator which enables an image of the Sun to be built up in the light of a single spectrum line. It comprises an entrance slit to a spectroscope which scans slowly or rapidly (for photography or vision respectively) over the image of the Sun produced by a telescope. The resulting spectrum moves concurrently with the motion of the entrance slit. A specific wavelength is then selected by a second slit placed in the image plane of the spectroscope and moving in phase with the first slit in order to remain centred on the selected spectrum line (see Chapter 5 of *Astrophysical Techniques*).

2 A coronagraph is a specialised telescope which enables the solar corona and other phenomena to be observed at times other than during solar eclipses. It does this by producing an artificial eclipse. In the original version, an occulting disc is placed in the image plane of the telescope to obscure the bright solar photosphere, and the light passing this then re-imaged. More recently, the occultation has been by baffles placed in front of the telescope. In either case extreme care has to be taken to eliminate scattered and diffracted light from the photosphere, which would otherwise swamp the observations. The best results are now obtained from spacecraft-borne instruments, but there are still a few coronagraphs working from high quality, high altitude sites on Earth (see Chapter 5 of *Astrophysical Techniques*).

3 The zodiacal light is a faint band of light in the sky, which is centred on the ecliptic, and which hence follows the zodiacal constellations. It may best be seen extending up from the eastern horizon soon after sunset, under good observing conditions. Its intensity falls smoothly with increasing angular distance from the Sun, until near the anti-solar point when it increases by a small amount. The latter feature is sometimes called the *Gegenschein* (counter-glow), and is a very difficult naked-eye observation.

The zodiacal light is due to scattered light from the Sun. The dust particles causing the scattering are concentrated into the plane of the solar system, so leading to the association with the zodiac. The particles are typically 10 microns in size, and occur at a rate of about one particle in 5 000 000 cubic metres. The *Gegenschein* is due to the greater efficiency of direct back-scattering in comparison with high-angle scattering, and originates from particles on the far side of the Earth from the Sun.

4 The magnetosphere is the region around the Earth wherein the terrestrial magnetic field is effective. Its boundary is the magnetopause, which is the surface over which the solar wind pressure equates to the terrestrial magnetic pressure. The Earth's magnetic field effectively stops at the magnetopause. Within the magnetopause, the motions of charged particles are governed by the configurations of the terrestrial magnetic field.

5 In a true whip, the thickness of the lash decreases from the handle to the tip. The transverse wave energy imparted to the handle by the wielder is thus shared out amongst smaller and smaller quantities of material as the wave travels down the whip. Since the wave energy is essentially constant, the portions of the whip must move transversely at greater and greater speeds in order to contain the energy. Eventually the transverse motions may exceed the (atmospheric) speed of sound. The whip-crack is then just the sonic boom as the tip of the whip passes through the sound barrier.

CHAPTER 5A

1 A stream line is the locus of all points in the material which originated from the same part of the Sun (or, in general, any source of a fluid flow).

2 There is considerable opportunity for confusion over terminology here. As discussed in §3A.2, the stars at the hot end of the spectral sequence are often called the early stars. This does not (now) carry any significance with respect to the star's age. The terminology arose at the start of attempts at spectral classification, when stars which we now know to be hot were indeed thought to be young. To avoid confusion, the term 'early star' is avoided in this section.

3 Particles experience two other significant forces which serve to complicate this analysis. The first is the directed pressure of the solar wind, the second is the Poynting–Robertson effect. This latter effect causes particles to fall towards the Sun (or star), and arises from the momentum of the photons which are impacting the particle having a component, due to aberration, in a direction opposite to the direction of motion of the particle. The particle thus loses angular momentum, and spirals in towards the Sun. A one micron particle, near the Earth, would be lost in this way in about ten thousand years. The continuing existence of submicron particles in the solar system as evidenced by the zodiacal light (note 4A.3 above) and by meteors must therefore be due to their continuing production. Probable sources of such particles include comets, asteroid/meteoroid collisions, and meteorite impacts on airless planets and satellites.

4 This is very similar to the situation encountered in jet engine and rocket design; using too high a burn rate can lead to increased turbulence and lower the exhaust velocity rather than increase it.

CHAPTER 6P

1 A similar effect underlies the apparent faster-than-light (superluminal) expansion of some extragalactic radio sources. The observed interval between two events is compressed with respect to the rest frame interval. If the two events are, say, the separations of a jet and a galactic nucleus at two different times, then they may appear to be moving apart faster than light, though the true situation is that the jet is moving almost towards us at perhaps $0.5c$ or more.

2 Observations of polarised astronomical sources are usually by means of a photometer which measures the source's brightness as seen through a linear polariser set at various different position angles. If $I(\max)$ and $I(\min)$ are the maximum and minimum observed intensities of the source as the polariser rotates, then the degree of linear polarisation is given by

$$\pi_L = \frac{I(\max) - I(\min)}{I(\max) + I(\min)}.$$

(See Chapter 5 of *Astrophysical Techniques.*)

CHAPTER 6A

1 The ionisation coefficient for hydrogen in the ground state varies approximately as the cube of the wavelength for Lyman continuum photons. Thus any gas cloud must become optically thin to photons of sufficiently short wavelengths. However, the number of photons at these short wavelengths varies approximately as $e^{-1/\lambda}$. Thus for most H II regions the number of photons whose wavelengths are sufficiently short to enable them to escape from the gas cloud will be very small indeed. Little error is therefore introduced into the analysis by the assumption that all photons with wavelengths shorter than 91.2 nm are absorbed within the H II region.

CHAPTER 7A

1 Although the absolute magnitudes of individual stars vary enormously, if sufficient numbers are used in each apparent magnitude range,

then the average absolute magnitude will be constant. The apparent magnitude will then be a function of distance alone. This was the principle upon which W Herschel's method of mapping the extent of the Galaxy was based, known as star gauging. In his case, interstellar extinction in general meant that his map covered only the local part of the spiral arm near the Sun.

CHAPTER 8A

1 Seyfert galaxies form about 10% of the observed spiral galaxies. They may be ten to a hundred times more luminous than the average spiral galaxy, with most of this excess originating in a small intensely bright source at the centre of the galaxy. They resemble miniature quasars in many ways. The source of the excess energy appears to be an explosion, but the mechanism for such an event is unknown. Possibly all spiral galaxies go through a Seyfert phase one or more times in their lifetimes.

The infrared and radio observations of the centre of the Milky Way Galaxy are suggestive of such a phenomenon having just finished within our own Galaxy. The explosion may be expected to produce expanding pressure/density waves within the Galaxy, and there is some evidence for the existence of at least one such at about 3 to 4 kpc from the Galactic centre.

2 This is a simple-minded calculation based upon black body effective temperatures. It ignores changes in the Earth's albedo which may result from its increased temperature, and the probable initiation of a 'runaway greenhouse effect' which will raise the surface temperature towards that of Venus (700 K). According to some expectations the latter effect might start to operate following a general increase in the Earth's temperature by 5 to 10 K, in which case life as we know it has 'only' some five hundred million years left on Earth, irrespective of its own efforts at destruction.

Appendix III Bibliography (since 1975)

Reference	*Chapters to which it has most relevance*
Allen C W (1973) *Astrophysical Quantities* London: Athlone Press (3rd edn)	All
Allen D A (1975) *Infrared: the New Astronomy* Reid	3A, 6A, 7A
Aller L H (1984) *Physics of Thermal Gaseous Nebulae* Dordrecht: Reidel	6A, 7P, 7A
Avrett E H (ed.) (1976) *Frontiers of Astronomy* Cambridge, Mass.: Harvard University Press	2P, 2A, 3A 6A, 7P, 7A 8A
Balian R, Encrenaz P and Lequeux J (eds) (1975) *Atomic and Molecular Physics and the Interstellar Matter* Amsterdam: North Holland	6A, 7P, 7A
Bartholomew-Briggs M C (1982) *Essentials of Numerical Competition* Chartwell-Bratt	2A, 3A, 4A 8A
Beckmann J E and Phillips J P (eds) (1982) *Submillimetre Wave Astronomy* Cambridge: Cambridge University Press	7P, 7A

Bowers R L and Deeming T (1984) — 1A, 2P, 2A, 6P, 6A, 7A, 8A
Astrophysics I: Stars
Astrophysics II: Interstellar matter and galaxies
Jones and Bartlett

Bumba V and Kleczec J (eds) (1976) — 4A
Basic Mechanisms of Solar Activity
(*IAU Symposium 71*)
Dordrecht: Reidel

Clark D H and Stephenson F R (1977) — 3A, 6A, 8A
The Historical Supernovae
Oxford: Pergamon

Collins G W II (1978) — 2P, 8A
The Virial Theorem in Stellar Astrophysics
Pachart Publishing

Conti P S and de Loore C W H (eds) (1979) — 1A, 2A, 3A, 5A, 6A, 8A
Mass Loss and Evolution of O-Type Stars
(*IAU Symposium 83*)
Dordrecht: Reidel

Cox J P (1980) — 3A
Theory of Stellar Pulsation
Princeton University Press

Duley W W and Williams D A (1984) — 6A, 7P, 7A
Interstellar Chemistry
New York: Academic Press, 1984

Dyson J E and Williams D A (1980) — 5A, 6A, 7P, 7A
Physics of the Interstellar Medium
Manchester: Manchester University Press

Eddy J A (ed.) (1978) — 4A
The New Solar Physics
(*AAAS Selected Symposium 17*)
AAAS

Eddy J A (1979) — 4A, 5A
A New Sun: Solar Results from Skylab
Washington DC: National Aeronautics and Space Administration

Eggleton P, Mitton S and Whelan J (eds) (1976) — 1A, 2A, 3A, 6A, 8A
Structure and Evolution of Close Binary Systems
(*IAU Symposium 73*)
Dordrecht: Reidel

Fuhr J R, Miller B J and Martin G A (1978) — 1A, 3P, 3A, 7P
Bibliography on Atomic Transition Probabilities (1914 through 1977)
(NBS special publication 505)
National Bureau of Standards [Washington DC: US Department of Commerce]

Gartenhaus S (1975) — 2P, 3P, 6P
Physics: Basic Principles, Vols. I and II
New York: Holt Reinhart and Winston

Gurzadyan G A (1980) — 3A, 4A, 7A
Flare Stars
Oxford: Pergamon

Hillier R (1984) — 3A, 4A, 6A
Gamma Ray Astronomy
Oxford: Clarendon Press

Hudson A and Nelson R (1982) — 2P, 3P, 6P
University Physics
New York: Harcourt Brace Jovanovich Inc.

Irvine J M (1978) — 2P, 2A, 3A, 8A
Neutron Stars
Oxford: Clarendon Press

Jager C de (1980) — 1A, 2A, 3A, 5A, 6A
The Brightest Stars
Dordrecht: Reidel

Jaschek M and Groth H-G (eds) (1981) — 3A
Be Stars
(IAU Symposium 81)
Dordrecht: Reidel

Jong T de and Maeder A (eds) (1977) — 6A, 7A, 8A
Star Formation
(IAU Symposium 75)
Dordrecht: Reidel

Kane S R (1978) — 4A
Solar Gamma-, X-, and EUV Radiation
(IAU Symposium 68)
Dordrecht: Reidel

Kitchin C R (1982) — 3A, 5A, 6A
Early Emission Line Stars
Bristol: Adam Hilger

Kitchin C R (1984) *Astrophysical Techniques* Bristol: Adam Hilger	1A, 2A, 3A 4A, 5A, 6A 7A
Kruger A (1979) *Introduction to Solar Radio Astronomy and Radio Physics* Dordrecht: Reidel	4A
Lang K R (1978) *Astrophysical Formulae* Berlin: Springer	All
Longair M S (1981) *High Energy Astrophysics* Cambridge: Cambridge University Press	7P, 7A
Loore C W H de and Willis A J (eds) (1982) *Wolf-Rayet Stars; Observations, Physics, Evolution* (*IAU Symposium 89*) Dordrecht: Reidel	1A, 3A, 5A
Manchester R N and Taylor J H (1977) *Pulsars* San Francisco: Freeman	2P, 2A, 3A 8A
McDonnell J A M (ed.) (1978) *Cosmic Dust* New York: Wiley	7A
Meadows A J (1978) *Stellar Evolution* Oxford: Pergamon (2nd edn)	2A, 3A, 8A
Mihalas D (1978) *Stellar Atmospheres* San Francisco: Freeman (2nd edn)	3P, 3A, 5A
Mitton S (ed.) (1977) *Cambridge Encyclopaedia of Astronomy* London: Jonathan Cape	1A, 2A, 3A 4A, 5A, 6A 7A, 8A
Mitton S (1981) *Daytime Star* London: Faber and Faber	4A
Narlikar J (1977) *Structure of the Universe* London: Oxford University Press	2A, 3A, 8A

Nicolson I K M (1982) — 4A
The Sun
London: Mitchell Beazley

Noyes R W (1982) — 4A
The Sun, Our Star
Cambridge, Mass.: Harvard University Press

Open University (1985) — 6A, 7A
Matter in the Universe, (S256)
Milton Keynes: The Open University

Rowan-Robinson M (ed.) (1976) — 3A, 6A, 7A
Far Infrared Astronomy
Oxford: Pergamon

Sahade J and Wood F B (1978) — 3A, 8A
Interacting Binary Stars
Oxford: Pergamon

Sanford P W, Laskarides P and Salton J (eds) (1984) — 3A, 5A
Galactic X-ray Sources
Chichester: Wiley

Schramm D N (1977) — 3A, 6A
Supernovae
Dordrecht: Reidel

Sexl R and Sexl H (1979) — 1A, 2P, 2A, 3A, 8A
White Dwarfs–Black Holes; an Introduction to Relativistic Astrophysics
New York: Academic Press

Shapiro S L and Teukolsky S A (1983) — 2P, 2A, 3A, 8A
Black Holes, White Dwarfs and Neutron Stars
New York: Wiley-Interscience

Shklovski I S (1978) — 1A, 2A, 3A, 6A, 8A
Stars: their Birth, Life and Death
San Francisco: Freeman

Smith F G (1977) — 2P, 2A, 3A, 8A
Pulsars
Cambridge: Cambridge University Press

Solomon P M and Edmunds M G (1980) — 6A, 7P, 7A
Giant Molecular Clouds in the Galaxy
Oxford: Pergamon

Spitzer L Jr (1978) — 7P, 7A
Physical Processes in the Interstellar Medium
New York: Wiley

Svestka Z (1976) — 4A
Solar Flares
Dordrecht: Reidel

Tassoul J-L (1978) — 3A
Theory of Rotating Stars
Princeton University Press

Tucker W H (1975) — 2P, 2A, 3P, 7P
Radiation Processes in Astrophysics
Cambridge, Mass.: MIT Press

Underhill A and Doazan V (1982) — 3A, 5A, 6A
B Stars with and without Emission Lines
(*NASA SP 456*)
Washington DC: National Aeronautics and Space Administration

Vandergraft J S (1978) — 2A, 3A, 4A, 8A
Introduction to Numerical Computations
New York: Academic Press

Wilson J G (1976) — 7A
Cosmic Rays
London: Wykeham

Wilson T L and Downes D (1975) — 6A
H II Regions and Related Topics
Berlin: Springer

Woerden H van (ed.) (1977) — 6A, 7P, 7A
Topics in Interstellar Matter
Dordrecht: Reidel

Major popular and specialist journals

Annual Reviews of Astronomy and Astrophysics
Astronomical Journal
Astronomy and Astrophysics
Astrophysical Journal
Monthly Notices of the Royal Astronomical Society
Nature
New Horizons in Astronomy
New Scientist

Publications of the Astronomical Society of the Pacific
Science
Scientific American
Sky and Telescope

Appendix IV Symbols and Constants

Some of the widely used symbols and constants are listed here for convenience. They are also defined within the text when first encountered, as are those which are less universally defined.

amu — Atomic mass unit: one twelfth of the mass of the $^{12}_{6}C$ atom $= 1.660\,53 \times 10^{-27}$ kg

AU — Astronomical unit; mean Earth–Sun distance $= 1.495\,978\,923 \times 10^{11} \pm 1.5 \times 10^{3}$ m

B — B magnitude in the *UBV* system; magnetic field strength

c — Velocity of light *in vacuo* $= 2.997\,925\,0 \times 10^{8}\ \mathrm{m\,s^{-1}}$

$C_P\ C_V$ — Specific heats of a gas at constant pressure and constant volume

$e^-\ e^+$ — Symbols for electrons and positrons

eV — Electron volt $= 1.602\,192 \times 10^{-19}$ J

G — Gravitational constant $= 6.670 \times 10^{-11}\ \mathrm{N\,m^2\,kg^{-2}}$

h — Planck's constant $= 6.626\,20 \times 10^{-34}$ J s

$\hbar$ — Atomic unit of angular momentum $= h/2\pi = 1.054\,59 \times 10^{-34}$ J s

k — Boltzmann's constant $= 1.380\,62 \times 10^{-23}\ \mathrm{J\,K^{-1}}$

$L_\odot$ — Solar luminosity $= 3.8478 \times 10^{26}$ W

ly — Light year; the distance travelled by light in a vacuum in one year $= 9.460\,530 \times 10^{15}$ m

M — Absolute magnitude

m — Apparent magnitude

$\mathcal{M}$ — Macroscopic mass (usually stars)

$\mathcal{M}_\odot$ — Solar mass $= 1.990 \times 10^{30}$ kg

m_e — Mass of the electron $= 9.109\,56 \times 10^{-31}$ kg

m_p — Mass of the proton $= 1.673\,52 \times 10^{-27}$ kg

P — Pressure

p	Momentum
p^+	Symbol for a proton
pc	Parsec; distance at which the parallax angle of a star is $1'' = 3.085\,678 \times 10^{16}$ m
$R_\odot$	Solar radius $= 6.961 \times 10^8$ m
T	Temperature
U	U magnitude of the UBV system
V	V magnitude of the UBV system
X	Mass fraction of hydrogen
Y	Mass fraction of helium
Z	Mass fraction of all elements other than hydrogen and helium
α	Right ascension
γ	Ratio of the specific heats of a gas
δ	Declination
ε	Mass emission coefficient
ε_0	Permittivity of a vacuum $= 8.85 \times 10^{-12}$ F m^{-1}
$\varkappa$	Mass absorption coefficient
λ	Wavelength
μ	Molecular weight; refractive index; cos θ
ν	Frequency
ρ	Density
σ	Stefan's constant (Stefan–Boltzmann constant) $= 5.6696 \times 10^{-8}$ W m^{-2} K^{-4}
τ	Optical depth

Appendix V Answers to the Problems

1A.1 2.7×10^{30} kg ($1.35\ \mathcal{M}_\odot$)
1.4×10^{30} kg ($0.7\ \mathcal{M}_\odot$)

1A.2 1.7×10^{30} kg ($0.85\ \mathcal{M}_\odot$)
1.2×10^{30} kg ($0.6\ \mathcal{M}_\odot$)

1A.4 $a = 1.5 \times 10^{10}$ m (0.1 AU)
$e = 0.3$
3×10^{30} kg ($1.5\ \mathcal{M}_\odot$)
1×10^{30} kg ($0.5\ \mathcal{M}_\odot$)
The system is viewed at 45° to its major axis.

2A.1 $\rho_c = 5.6 \times 10^{3}\ \mathrm{kg\,m^{-3}}$
$P_c = 4.5 \times 10^{14}\ \mathrm{N\,m^{-2}}$

2A.2 1.79×10^{6} K
1.72×10^{6} K

2A.3 5.8×10^{30} kg ($2.9\ \mathcal{M}_\odot$)

2A.4 0.043″
8.1×10^{11} m ($1200\ R_\odot$)

3A.2 See figure AV.1

3A.3 17.4 pc
180 000 pc

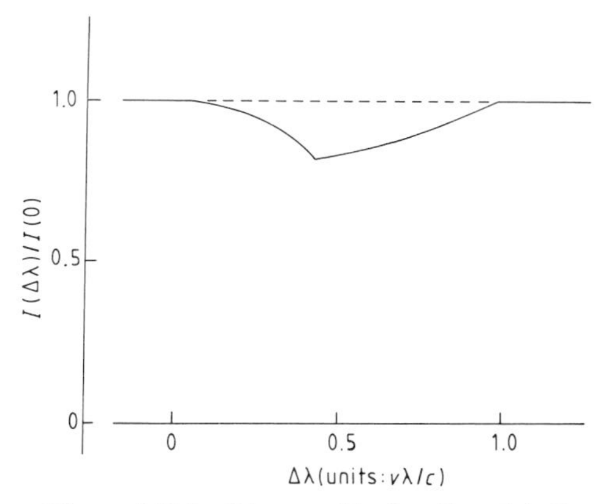

Figure AV.1 Line profile for $R_2 = 1.1\ R_1$.

3A.4 $2.3\ R_\odot$
$0.038\ R_\odot$
$600\ R_\odot$
$0.000\,74\ R_\odot$

3A.5 1.5×10^{30} kg ($0.7\ \mathcal{M}_\odot$).
This is likely to be an underestimate because the moment of inertia will be smaller in reality, due to some central condensation.

4A.1 2% of the central pressure

4A.2 1.4960×10^{11} m

5A.1 10^{14} yr

6A.1 320

6A.2 Interstellar medium around the B0 star has only 6% of the density of that around the O5 star.

6A.3 0.77 s approximately

7A.1 Grains destroyed in 1.4×10^{23} kg of the interstellar medium per second ($2.2\ \mathcal{M}_\odot$ per year).
3×10^{10} yr

Index

Page numbers shown in bold type denote the start of an extended section on the topic.